# BUILDING ENERGY MANAGEMENT SYSTEMS AND TECHNIQUES

Principles, Methods, and Modelling

# BUILDING ENERGY MANAGEMENT SYSTEMS AND TECHNIQUES

## Principles, Methods, and Modelling

**FENGJI LUO**
School of Civil Engineering, The University of Sydney, Sydney, NSW, Australia

**GIANLUCA RANZI**
School of Civil Engineering, The University of Sydney, Sydney, NSW, Australia

**ZHAO YANG DONG**
School of Electrical and Electronics Engineering, Nanyang Technological University, Singapore

ELSEVIER

Elsevier
Radarweg 29, PO Box 211, 1000 AE Amsterdam, Netherlands
125 London Wall, London EC2Y 5AS, United Kingdom
50 Hampshire Street, 5th Floor, Cambridge, MA 02139, United States

**Notices**

Knowledge and best practice in this field are constantly changing. As new research and experience broaden our understanding, changes in research methods, professional practices, or medical treatment may become necessary.

Practitioners and researchers must always rely on their own experience and knowledge in evaluating and using any information, methods, compounds, or experiments described herein. In using such information or methods they should be mindful of their own safety and the safety of others, including parties for whom they have a professional responsibility.

To the fullest extent of the law, neither the Publisher nor the authors, contributors, or editors, assume any liability for any injury and/or damage to persons or property as a matter of products liability, negligence or otherwise, or from any use or operation of any methods, products, instructions, or ideas contained in the material herein.

ISBN: 978-0-323-96107-3

For information on all Elsevier publications
visit our website at https://www.elsevier.com/books-and-journals

*Publisher:* Megan Ball
*Acquisitions Editor:* Peter Adamson
*Editorial Project Manager:* Debarati Roy
*Production Project Manager:* Maria Bernard
*Cover Designer:* Greg Harris

Typeset by STRAIVE, India

Working together
to grow libraries in
developing countries

www.elsevier.com • www.bookaid.org

# Contents

# Preface

This book provides an introduction to building energy management techniques and the fundamental design schemes of building energy management systems. In a field where continuous research advancement and commercial system development occur, the book covers the latest directions and methodologies that are studied in both academia and industry in building energy management and energy demand side management. Fundamental concepts related to building energy management systems are introduced with reference to and in the context of smart grids, demand response and demand side management, and distributed renewable energy. This will enable the reader to gain a clear understanding of building-side energy systems and the related technologies that drive the emergence and development of building energy management systems.

Advanced topics are introduced to enable the reader to become familiar with the different energy management scenarios and procedures for modern buildings in an automatic and highly renewable-penetrated building environment. For example, different categories of energy management techniques are presented for different building-side energy resources (including battery energy storage systems, plug-in appliances, and HVAC systems) while providing considerations for accounting for the satisfaction and indoor comfort of the building occupants. The basic principles of evolutionary computation are covered and applied to building energy management problems. Other advanced topics deal with the exploitation of building energy management in the context of building-to-grid integration, peer-to-peer energy trading, and microgrid. The book also introduces concepts related to occupant-to-grid integration and its implementation through personalized recommendation technology to guide the occupants' choices on energy-related products and their energy usage behaviors that can contribute to enhancing the energy efficiency of buildings.

The application examples presented throughout the book aim at introducing the key aspects of the algorithms that support different designs of building energy management systems and their possible implementation. The complexity of the examples is kept to a minimum to enable detailed descriptions of the problem variables and the solution process. The proposed algorithms can be further extended to account for more sophisticated model representations of renewable energy sources, energy storage systems, building models, weather conditions, and indoor comfort, as well as to account for a larger number of problem settings, for example, involving a larger number of appliances, building occupants, and buildings. Pseudocodes of the algorithms used in the examples are provided with explanatory notes.

It is expected that, in the coming decades, a growing number of buildings will be equipped with building energy management systems to enhance their operational energy efficiency and that an effective engagement of the renewable energy sources deployed in buildings will significantly contribute to satisfying the energy demand of buildings.

The authors would like to acknowledge the support of their respective institutions, the University of Sydney and Nanyang Technological University, and of the publishing team from Elsevier.

**Fengji Luo**
**Gianluca Ranzi**
**Zhao Yang Dong**

# CHAPTER 1

# Introduction

## Contents

## 1.1 Introduction to building energy management systems

### 1.1.1 Background

In the last decades, the building sector has been one of the main consumers of energy worldwide. In 2020, the energy consumption of this sector accounted for 36% of the global energy consumption worldwide and contributed to 37% of the global energy-related carbon dioxide ($CO_2$) emissions [1]. Fig. 1.1 provides an overview of the $CO_2$ emissions produced by the building sector between 1990 and 2019 [2]. The direct and indirect emissions from electricity and commercial heat used in buildings have increased to 10 GtCO2 in 2019, the highest level ever recorded.

The large energy consumption of the building sector requires a careful evaluation on how energy is produced and consumed in this sector. In recent years, extensive efforts have been placed in this direction worldwide in establishing dedicated strategies, for example, through initiatives such as the net-zero carbon buildings' commitment [3] or by the World Green Building Council [4]. Governmental agencies have significantly contributed to implementing strategies to reduce energy consumptions in buildings, such as the European guidelines on Nearly Zero-Energy Buildings (NZEBs) and zero-emission buildings [5,6]. The implementations of these strategies require a collaborative approach among different sectors. In the following and throughout this textbook, we consider how building energy management systems can contribute to minimize the energy consumption of buildings, to support the operations of smart grids, and to enhance the wider acceptable of energy renewables while accounting for the comfort and satisfaction of building occupants.

*Building Energy Management Systems and Techniques*
https://doi.org/10.1016/B978-0-323-96107-3.00007-2

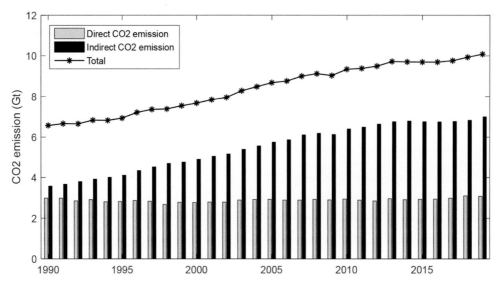

**Fig. 1.1** $CO_2$ emissions produced by the building sector worldwide between 1990 and 2019 [2].

In the first part of the chapter, we provide an overview of the structure and operations of building energy management systems (BEMSs). Particular attention is devoted to highlight some of the key developments that have supported the increasing popularity of BEMSs, such as Building Internet-of-Things (IoT) technology and smart grids. Recent trends of BEMSs' designs dealing with the energy management of groups of buildings at a precinct and/or community level are also presented to demonstrate their growing applicability and impact in supporting energy-efficient practices. An overview of the benefits of BEMSs is then outlined to recognize their positive impact on buildings' operations. The chapter terminates with a brief description of the layout of the chapters included in the book.

## 1.1.2 Building energy management systems

*Building energy management systems*, also denoted as BEMSs, are centralized computer systems that monitor and control energy-related resources and devices in building systems. The development of BEMSs has been supported by advances taking place in the fields of microelectronics and information and communication technologies (ICTs).

The early BEMSs can be traced back to the 1970s, e.g., [7,8], when they consisted of a computer-based central station and a group of outstations that consisted of boxes and/or cabinets for relays and connections to sensors and actuators. The central station had the ability to perform simple calculations and perform control decisions. These BEMSs were costly and were mostly used for large installations [8]. In the 1980s, microcomputers have been used to support the deployment of BEMSs because of their increasing computational power. In these earlier installations, BEMSs were typically designed for controlling the major energy components in a building, mostly heating, ventilation, and air conditioning (HVAC) systems, lighting systems, and lifts. In recent years, with the wider deployment of IoT technology [9] in buildings, usually denoted as a BIoT, the BIoT infrastructure enabled controllability of a variety of plug-in electrical appliances and equipment that were not typically managed with traditional BEMSs.

Buildings have been traditionally regarded as pure energy consumers. From this viewpoint, a traditional BEMS would manage a building's energy consumption while aiming to maintain comfortable indoor conditions for the building occupants. The growing acceptance of distributed renewable energy sources (typically wind turbines and solar panels) and energy storage systems installed on building sites has enable buildings to be regarded as prosumers, i.e., both consumers and producers of energy. Such a new role (prosumer) has been supported in the strategies and implementations of BEMSs, therefore moving to manage not only a building's energy consumption but also a building's energy production and storage.

With the penetration of renewable energy sources, buildings can now have a significant impact on the operations of external electric power grid. Now they can also generate and feed energy back to the grid. To satisfactorily operate in this two-way arrangement, a building BEMS is now required to interact with the grid and to support the grid's operations while aiming to optimize the building's energy performance.

The above technical developments need intelligent and sophisticated energy management strategies and algorithms to be implemented in BEMSs to manage the various controllable resources and devices efficiently.

### 1.1.3 Energy management strategies and algorithms in BEMSs

When referring to a BEMS, it is common to denote the set of computer and automation systems deployed in a building as well as the energy management algorithms, the computing hardware, and the sensing, actuation, and communication facilities.

For the purpose of this textbook, and consistently with available literature (e.g., [10–16]), with the term BEMS, we intend to relate to its energy management strategies and algorithms. Driven by the enhanced controllability in the building's operational environment, extensive effort has been placed in recent years in developing sophisticated energy management strategies for enabling adaptive operational optimizations in the management of building energy resources and devices, e.g., HVAC systems, energy storage systems, electric vehicles, renewable energy sources, and plug-in appliances. Many of these strategies rely on machine learning (e.g., [11–13]) and optimization (e.g., [14–16]) techniques to find optimal operation solutions for energy resources and devices installed in buildings while accounting for specific objective requirements. These sophisticated energy management strategies enable buildings to better adapt to the dynamics of their operational environment (such as the intermittency of the on-site renewable energy and the time-varying energy prices). BEMSs have also found wide applicability in smart buildings [17,18].

Current trends in building energy management strategies and algorithms are to consider the management of group of buildings at a precinct and/or community level, e.g., [19–24]. These energy management techniques consider the coupling of the operations between multiple buildings as well as the impact of their operations on the external environment, e.g., the grid. Underpinning this broader implementation, BEMSs' strategies have been proposed in recent years to enable energy sharing mechanisms among buildings (e.g., [19–22]) and to aggregately control appliances in multiple buildings part of the same community (e.g., [23,24]). These recent advancements enhance the responsibility and role played by BEMSs in supporting energy-efficient implementations at a large scale in the built environment.

## 1.2 BEMSs in smart grids

### 1.2.1 Smart grids

The current energy shortage and climate crisis impose urgent actions for transforming power and energy systems in the way these generate, transform, and distribute energy, as well as for minimizing the use of energy. In this context, *smart grids* have been proposed at the beginning of the century [25,26] and have found a wide acceptance worldwide [27,28] because of their ability in supporting the implementation of sustainable practices. Traditional power grids heavily rely on fossil fuels to serve the energy demand, while smart grids can rely on a widespread integration of clean and renewable energy sources that can also be distributed geographically, e.g., wind energy, solar energy, and wave energy.

In traditional power systems, the power grid aims at generating the power required to serve the energy loads of buildings according to their energy demand. As a result, considerable costs are required to build the power infrastructure for catering for peak demands of the energy loads. Smart grids follow a different approach and support the interaction between energy loads required by buildings (and other entities) and the grid to establish a more efficient energy system that incorporates both the grid and buildings.

### 1.2.2 Grid-interactive building energy management

An important feature of smart buildings is their capability of communicating with smart grids and of consuming and producing energy while accounting for the requirements of the grid. This approach is usually referred to as the *demand response* (see Chapter 4 for more details).

BEMSs provide an effective platform for facilitating the interaction between buildings and the grid. In this context, a BEMS acts on behalf of building occupants to perform automatic control actions on the energy resources and devices installed in the building while considering both the operational requirements of the smart grid, and the needs and preferences of the occupants. Supported by the underlying information and communication infrastructure (briefly introduced in Chapters 3 and 4), a BEMS can perform two-way communications with the grid: (i) a BEMS can receive information from the grid and be aware of the grid's operational requirement; and (ii) a BEMS can report real-time building energy performance to the grid (through sensors and devices deployed in the building), therefore enabling the grid to be aware of the building's energy expected consumption and production. In such a framework, it is possible to implement grid-aware energy management strategies in the design of a BEMS and, therefore, contribute toward an enhanced energy-efficient smart grid.

## 1.3  Benefits of BEMSs

BEMSs bring multiple benefits to their stakeholders (e.g. building occupants, building owners and managers, and the smart grid) that can be summarized as follows:
- *minimization of operational energy costs*—with an intelligent energy management, it is possible to reduce the overall energy consumption of buildings and, when available, to enable them to exploit dynamic energy tariffs;
- *improved comfort and productivity of building occupants*—the operations of a BEMS can increase the comfort level for building occupants by monitoring and controlling the indoor conditions such as temperature, air quality, and lighting. A BEMS can automatically operate plug-in appliances and equipment following the occupants' requirements and inferred preferences. An enhanced comfort level is expected to lead to increased productivity;
- *staff savings and reduced maintenance costs*—manual operations on energy resources and devices can be replaced by the automatic communications and controls performed by a BEMS. For example, a BEMS could support a real-time monitoring of the operations of the energy resources and devices and promptly identify faults and potential risks of the energy resources and devices. These methodologies are expected to lead to reduced human intervention and reduced maintenance costs; and
- *supporting sustainable energy practices*—through an optimized management of energy generation, production, and storage, a BEMS can contribute toward a better utilization of on-site renewable energy sources to locally serve the energy demand of a building. With such an approach, it is possible to reduce the dependency of a building on energy generated from fossil fueled plants, to avoid the energy losses due to long-distance energy transmissions from the grid to buildings, and thereby contributing toward sustainable and energy-efficient practices for buildings and grids.

## 1.4 Layout of the book

The first part of the book provides an overview of the fundamental concepts related to BEMSs that include the energy resources (Chapter 2), information infrastructure (Chapter 3), and power demand response and demand side management (Chapter 4). The second part of the book (Chapters 5–10) is dedicated to the introduction of building power load forecasting, to BEMSs, and to energy management strategies for major energy devices in buildings. Simple application examples are provided to highlight the implementation of different building power load forecasting methods and different BEMSs' designs. Pseudocodes are presented to clarify the algorithms underpinning these procedures. The third part of the book (Chapters 11–13) covers the interaction among buildings and between buildings and the grid. In the final chapter (Chapter 14), latest trends in BEMSs' strategies are presented with a specific focus on engaging building occupants into the building energy management loop. An attempt is made throughout the book to keep the complexity of the application examples to a minimum to enable the discussion of how specific variables vary within a small problem domain. The techniques and topics covered in this book are expected to provide a reference to the research and development of BEMSs for energy-efficient buildings.

## References

[1] Global Status Report for Buildings and Construction: Towards a Zero-Emission, Efficient and Resilient Buildings and Construction Sector, United Nationals Environment Programme, Nairobi, Kenya, 2021. (Online). Available from: https://globalabc.org/sites/default/files/2021-10/GABC_Buildings-GSR-2021_BOOK.pdf. (Accessed 10 June 2023).

[2] Buildings Sector Energy-Related CO2 Emissions in the Sustainable Development Scenario, 2000–2030, International Energy Agency, Paris, France, 2020. (Online). Available from: https://www.iea.org/data-and-statistics/charts/buildings-sector-energy-related-co2-emissions-in-the-sustainable-development-scenario-2000-2030. (Accessed 10 June 2023).

[3] Net Zero Carbon Buildings Commitment, World Green Building Council, London, UK, 2021. (Online). Available from: https://www.worldgbc.org/thecommitment. (Accessed 10 June 2023).

[4] World Green Building Council, WorldGBC.org. Available from: https://www.worldgbc.org/contact-us (Accessed 7 March 2022).

[5] European Commission, Recommendations on guidelines for the promotion of nearly zero-energy buildings and best practices to ensure that, by 2020, all new buildings are nearly zero-energy buildings, Off. J. Eur. Union L208 (2016) 46–57.

[6] Proposal for a Directive of the European Parliament and of the Council on the Energy Performance of Buildings, European Commission, Brussels, Belgium, 2021, December. Rep. COM (2021) 802 final.

[7] J.P. Bentley, Principles of Measurement Systems, fourth ed., Longman Scientific and Technical, London, UK, 1988.

[8] G.J. Levermore, Building Energy Management Systems: Applications to Low-Energy HVAC and Natural Ventilation Control, first ed., CRC Press, Boca Raton, Florida, USA, 2013.

[9] S. Greegard, The Internet of Things, MIT Press, Cambridge, MA, USA, 2015.

[10] P. Zhao, S. Suryanarayanan, M.G. Simoes, An energy management system for building structures using a multi-agent decision-making control methodology, IEEE Trans. Ind. Appl. 49 (1) (2013) 322–330.

[11] L. Yu, S. Qin, M. Zhang, C. Shen, T. Jiang, X. Guan, A review of deep reinforcement learning for smart building energy management, IEEE Internet Things J. 8 (15) (2021) 12046–12063.

[12] M.R. Sunny, M.A. Kabir, I.T. Naheen, M.T. Ahad, Residential energy management: a machine learning perspective, in: Proc. 2020 IEEE Green Technologies Conference (GreenTech), Oklahoma City, USA, 2020.

[13] F. Magoules, H.X. Zhao, Data Mining and Machine Learning in Building Energy Analysis, John Wiley and Sons, Hoboken, New Jersey, USA, 2016.

[14] R. Yang, L. Wang, Multi-zone building energy management using intelligent control and optimization, Sustain. Cities Soc. 6 (2013) 16–21.

[15] J.K. Gruber, F. Huerta, P. Matatagui, M. Prodanovic, Advanced building energy management based on a two-stage receding horizon optimization, Appl. Energy 160 (2015) 194–205.

[16] F. Wang, L. Zhou, H. Ren, X. Liu, S. Talari, M. Shafie-khah, J.P.S. Catalao, Multi-objective optimization model of source-load-storage synergetic dispatch for a building energy management system based on TOU price demand response, IEEE Trans. Ind. Appl. 54 (2) (2018) 1017–1028.

[17] J. Sinopoli, Smart Building Systems for Architects, Owners and Builders, Elsevier, Amsterdam, Netherlands, 2009.

[18] R. Bakker, Smart Buildings: Technology and the Design of the Built Environment, RIBA Publishing, London, UK, 2020.

[19] Z. Zhao, F. Luo, C. Zhang, G. Ranzi, A social relationship preference aware peer-to-peer energy market for urban energy prosumers and consumers, IET Renew. Energy Gener. (2021), https://doi.org/10.1049/rpg2.12349. early access.

[20] S. Cui, Y.W. Wang, J.W. Xiao, Peer-to-peer energy sharing among smart energy buildings by distributed transaction, IEEE Trans. Smart Grid 10 (6) (2019) 6491–6501.

[21] B.R. Park, M.H. Chung, J.W. Moon, Becoming a building suitable for participation in peer-to-peer energy trading, Sustain. Cities Soc. 76 (2022) 1–16.

[22] S. Cui, J.W. Xiao, Game-based peer-to-peer energy sharing management for a community of energy buildings, Int. J. Electr. Power Energy Syst. 123 (2020) 1–9.

[23] A. Barbato, A. Capone, G. Carello, M. Delfanti, M. Merlo, A. Zaminga, House energy demand optimization in single and multi-user scenarios, in: Proc. 2011 IEEE International Conference on Smart Grid Communications (SmartGridComm), Brussels, Belgium, 2011, October.

[24] S. Kakran, S. Chanana, Energy scheduling of residential community equipped with smart appliances and rooftop solar, in: Proc. 7th International Conference on Power Systems (ICPS), Pune, India, 2017, December.

[25] Energy Independence and Security Act 2007, U.S. Government Printing Office, Washington DC, USA, 2007. (Online). Available from: https://www.govinfo.gov/content/pkg/BILLS-110hr6enr/pdf/BILLS-110hr6enr.pdf. (Accessed 10 June 2023).

[26] H. Farhangi, The path of the smart grid, IEEE Power Energy Mag. 8 (1) (2010) 18–28.

[27] M. Hashmi, S. Hanninen, K. Maki, Survey of smart grid concepts, architectures, and technological demonstrations worldwide, in: Proc. of 2011 IEEE PES Conference on Innovative Smart Grid Technologies Latin America (ISGT LA), Medellin, Colombia, 2011, October.

[28] M.L. Tuballa, M.L. Abundo, A review of the development of smart grid technologies, Renew. Sust. Energ. Rev. 59 (2016) 710–725.

# CHAPTER 2

# Energy sources in building systems

## Contents

## 2.1 Introduction

Energy consumption in buildings has gained significant attention in recent years due to its growing trend produced, among the others, by the effects of climate change and population growth, and by the increasing installations and use of heating, ventilation, and air conditioning (HVAC) systems. On a positive side, this trend has also encouraged the development of new technologies capable of reducing the dependency of energy generation on fossil fuels by promoting, for example, the deployment and high penetration of renewable energy technologies.

In the last couple of decades, *distributed renewable energy sources* have been gaining popularity in building applications. Distributed renewable energy sources mainly include small wind turbines and photovoltaic (PV) solar panels that capture energy from wind and solar radiation and convert it into electricity. The term *distributed* intends to reflect the condition that these energy sources are geographically distributed, for example, in different buildings, and that they are not installed at the macro energy supplier side.

Natural sources, such as solar and wind energy, are intermittent and stochastic in nature. For example, the wind speed at a specific site can frequently vary over time, or the solar radiation can be highly intermittent due to the movements of the sun and clouds. As a result, the power output from a distributed renewable energy resource is highly variable and does not necessarily match the power demand of an energy consumer, such as a building or household. At times, the renewable power output is not sufficient for covering the power demand, while, at other times, the renewable power output is larger. The latter scenario provides opportunities for storing the surplus renewable energy for later use or for sharing it. Energy storage is usually implemented by relying on energy storage systems (ESSs). ESSs can provide energy backup support to buildings in power outage events. With the growing integration of distributed renewable energy sources and ESSs, buildings can now perform self-energy supply operations to serve their own energy demand. In this process, they can reduce their dependency on external energy systems, e.g., electric power grids and gas networks, as it is the case for *zero-emission buildings* [1,2].

This chapter intends to provide a brief overview of the energy sources that are typically deployed in buildings to provide the basis of some of the technologies that will be considered in later chapters when dealing with the design of building energy management systems. In the first part of the chapter, we present some key features related to wind power and outline the common typologies of wind turbines together with considerations related to their design in urban environments. We then present a mathematical model that can be used to simulate the wind power output when evaluating different building energy management strategies. This is followed by an introduction to solar energy and how this can be harvested with different technologies available for buildings. A simple mathematical model for the PV solar power output is also introduced. In the final part of the chapter, we present key features of ESSs that have been gaining popularity in recent years with the wider penetration of renewable energy sources.

## 2.2 Wind power

### 2.2.1 Introduction to wind turbines

Wind energy has been widely used to date for a wide range of applications, such as sailing, windmills, and windpumps [3]. In recent years, wind farms have been gaining popularity, and many large-capacity installations have been built worldwide and integrated in electric power grids. According to the Global Wind Report 2021 [4], the worldwide wind power capacity has reached 743 gigawatt (GW) that is equivalent to $CO_2$ production savings in the order of 1.1 billion tons.

Wind power is generated by means of wind turbines. These use blades to capture the mechanical power of the wind and to convert it into electrical power through generators. Wind turbines are often classified into two types according to the axis around which the turbine blade rotates, i.e., horizontal axis wind turbines (HAWTs) and vertical axis wind turbines (VAWTs) as depicted in Fig. 2.1. In HAWTs, the turbine blades are located far away from the ground. This arrangement enables a HAWT to capture the wind at a height with high speed and, hence, possessing large energy. Due to this advantage, most

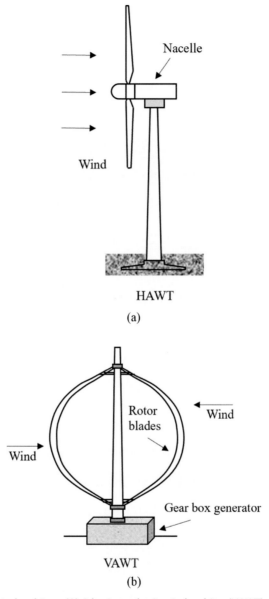

**Fig. 2.1** Examples of wind turbines. (A) A horizontal axis wind turbine (HAWT). (B) A vertical axis wind turbine (VAWT).

large wind turbines, such as those used in large-capacity wind farms, are HAWTs. VAWTs are more suitable for small-capacity wind turbines (e.g., several kilowatts), and the main market of VAWTs has been focused at local electricity productions, including installations on buildings.

## 2.2.2 Wind power integration in buildings

Wind turbines can also be classified according to their level of energy generation. Those installed for large-scale energy generation are often denoted as macro wind turbines, while those used for local energy production are referred to as micro wind turbines. It is common for macro wind turbines to be located in suburbs and rural areas (Fig. 2.2A), and for micro wind turbines to be installed on buildings in urban areas (Fig. 2.2B).

(a)

(b)

**Fig. 2.2** Examples of wind turbines. (A) A macro wind turbine [5]. (B) A micro wind turbine [6].

Examples of factors that need to be accounted for when considering the installation of a micro wind turbines on buildings include:

- *energy generation efficiency at low wind speed*—the use of a wind resource in urban areas is still limited when compared to suburbs and rural areas due to the relatively lower wind speeds occurring in urban areas. There is an exponential increase in the energy that can be generated by wind turbines when moving from low wind speeds to high wind speeds [7];
- *noise*—building-integrated wind turbines need to operate as quiet as possible to minimize the disturbance to the building occupants and to the neighborhood; and
- *aesthetics*—when integrating micro wind turbines in urban areas, the visual appearance of the wind turbine system becomes an important design consideration as it needs to integrate within the urban landscape.

### 2.2.3 Wind power output model

The amount of kinetic energy (KE) that can be captured by a wind turbine depends on the amount of air mass that can flow through its blade's area. A simple model that can be used for energy management simulations is presented in the following to quantify the power output obtained from a wind turbine. In this model representation, the KE of the wind turbine is expressed as [3]:

$$KE = \frac{1}{2} \times m \times v^2 \tag{2.1}$$

with the rate of the air mass flowing through the blade ($m$) being calculated as:

$$\dot{m} = \frac{m}{\Delta t} = \rho \times A^{bld} \times v \tag{2.2}$$

where $KE$ is the kinetic energy contained in the wind flowing through the cross-sectional area swept by the wind turbine's blades (in J); $m$ is the air mass flowing through the cross-sectional area swept by the blades (in kg); $v$ is the wind speed (in m/s); $\rho$ is the air density (in kg/m$^3$); $A^{bld}$ is the cross-sectional area swept by the wind turbine's blades (in m$^2$).

From Eqs. (2.1 and 2.2), we can derive the power extracted by a wind turbine's blade (denoted by $P^{wind}$, in watt) as follows:

$$P^{wind} = \frac{KE}{\Delta t} = \frac{1}{2} \times \dot{m} \times v^2 = \frac{1}{2} \times \rho \times A^{bld} \times v^3 \tag{2.3}$$

that highlights how the power generated by a wind turbine is dependent on three main factors that consist of the air density, the blade's surface area, and the wind speed (with the power output of the turbine increasing with the cube of the wind speed). For example, when the wind speed reaching a turbine doubles, the consequent produced power increases by a factor of eight. Wind turbines are usually built away from ground level to harvest the higher wind speeds available at greater heights. A wind turbine's power output is also proportional to the blade's area. For a conventional wind turbine, the area swept by the wind turbine can be calculated as $A^{bld} = \pi \times D^2/4$, where $D$ is the diameter of this area (in m).

A wind turbine is usually operational only for a certain range of wind speeds. There are three critical wind speed parameters that are associated with a wind turbine's power output, and these consist of the cut-in speed, the rated speed, and the cut-out speed. In particular, the cut-in speed represents the lower wind speed that can drive the wind turbine for the power generation. When the wind speed is lower than the cut-in speed, the wind turbine does not produce any power. When the wind speed is larger or equal to the cut-in speed and lower than the rated wind speed, the wind turbine's power output increases as the wind speed becomes larger as described in Eq. (2.3). When the wind speed falls between the rated speed and the cut-out speed, the wind turbine output equals its rated power. When the wind speed is larger than the cut-out speed, the protection device of the wind turbine will stop it from operating to prevent the wind turbine from being damaged by the possible over-fast spinning of the blades. Based on this representation, the power output model of a wind turbine can be described as:

$$P^{wind} = \begin{cases} 0 & \text{if } v < v^{in} \\ \dfrac{1}{2} \times \rho \times A^{bld} \times v^3 & \text{if } v^{in} \leq v \leq v^{rate} \\ \dfrac{1}{2} \times \rho \times A^{bld} \times (v^{rate})^3 & \text{if } v^{rate} < v \leq v^{out} \\ 0 & \text{if } v > v^{out} \end{cases} \tag{2.4}$$

where $v^{in}$, $v^{rate}$, and $v^{out}$ are the cut-in, rated, and cut-out wind speeds of the wind turbine (in m/s), respectively.

### 2.2.4 Worked example

Consider a small-capacity wind turbine with following manufacturing parameters: $D=4$ m; $v^{in}=2$ m/s; $v^{rate}=10$ m/s; and $v^{out}=15$ m/s. The air density at the wind turbine's site is $1.3 \text{ kg/m}^3$. Determine the power output of the wind turbine at the following three wind speed values: (1) 4.8 m/s, (2) 12.0 m/s, and (3) 18.4 m/s.

**Solution**

The area swept by the wind turbine's blades is:

$$A^{bld} = \pi \times \frac{D^2}{4} = 3.14 \times \frac{4^2}{4} = 12.56 \text{ m}^2$$

**(1)** When the wind speed is 4.8 m/s, the wind power output is:

$$P^{wind} = \frac{1}{2} \times \rho \times A^{bld} \times v^3 = \frac{1}{2} \times 1.3 \times 12.56 \times 4.8^3 = 902.87 \text{ W}$$

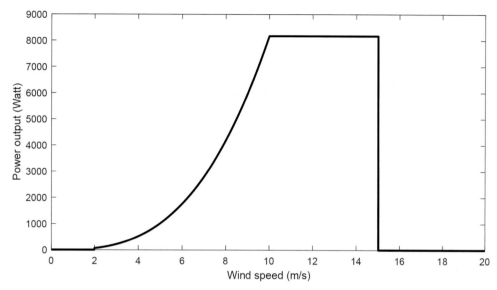

**Fig. 2.3** Power output of the wind turbine in the worked example for different values of wind speed.

**(2)** The wind speed of 12.0 m/s is larger than the wind turbine's rated wind speed and smaller than its cut-out wind speed and, therefore, the wind turbine produces power at its rated wind speed, i.e., 10 m/s:

$$P^{wind} = \frac{1}{2} \times \rho \times A^{bld} \times v^3 = \frac{1}{2} \times 1.3 \times 12.56 \times 10^3 = 8{,}164.0 \text{ W}$$

**(3)** The wind speed of 18.4 m/s is larger than the wind turbine's cut-out wind speed and, therefore, its power output is zero.

Fig. 2.3 shows the power output curve of the wind turbine in this worked example for different values of wind speed.

## 2.3 Solar energy

### 2.3.1 Overview

The sun consists of a valuable energy source that reaches our planet in the form of solar radiation, and a large number of technologies have been developed to date for its harvesting [8]. In the following, we provide a brief introduction to solar thermal energy solutions that harness solar energy to generate heat. This is followed by an overview on PV solar power that absorbs solar radiation to generate electricity.

### 2.3.2 Solar thermal energy and its applications in buildings

Solar thermal energy [9,10] refers to thermal energy (heat) obtained from solar radiation that can either be directly used for specific applications or be further transformed into electricity.

**Fig. 2.4** A typical flat plate solar collector [11].

*Solar thermal collectors* (also known as solar collectors) are widely used to absorb solar radiation. These are usually available in the form of flat plate collectors or evacuated tube collectors. Fig. 2.4 shows a typical solar thermal collector formed by a flat plate painted black to maximize the absorption of solar radiation. In the built environment, the heat generated by solar thermal collectors is commonly used for water heating. Fig. 2.5 illustrates an operating diagram of a solar thermal water heating system that is typically installed on roofs.

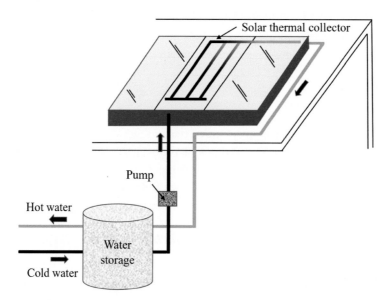

**Fig. 2.5** Diagram of a solar thermal water heating system typically installed on roofs.

The heat produced from solar thermal collectors can also be used to generate electricity. The technology performing the energy conversion is referred to as the *Concentrating Solar Power* (CSP) technology [12]. In CSP systems, mirrors or lenses are typically used as the aperture to concentrate a large area of solar radiation onto a targeted location. The heat is then absorbed by a working fluid to create steam to power a turbine and to produce electricity. The electricity generated by CSP power plants is delivered by power transmission networks to serve the energy demand of buildings and other energy loads.

### 2.3.3 Photovoltaic solar power and its applications in buildings

The PV solar power technology relies on PV materials to capture the solar radiation and to generate electricity based on the photovoltaic effect. A PV solar panel is made up by many photoelectric cells (Fig. 2.6). When the solar radiation hits one of the faces of a photoelectric cell, it produces an electric voltage difference between its faces. The voltage difference drives the electrons in the PV material to flow from one face to the other, therefore generating an electric current.

There are mainly two technologies for integrating PV solar power sources into buildings: building-applied photovoltaics (BAPV) and building-integrated photovoltaics (BIPV). Both technologies aim at exploiting building envelope components (e.g., roofs, facades, and sunscreens) to actively capture sunlight and to generate electricity. The main difference between them relates to the way they are designed and integrated into the building. With the use of BAPV, PV solar modules are installed on surfaces of existing buildings, while with BIPV traditional construction element are replaced with modules

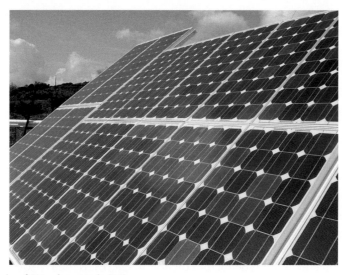

**Fig. 2.6** Example of PV solar panels [13].

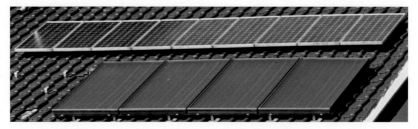

**Fig. 2.7** Example of BAPV: PV solar panels mounted on rooftop [14].

**Fig. 2.8** Example of BIPV: A solar photovoltaics façade [15].

incorporating PV. A common BAPV system installed on a rooftop is shown in Fig. 2.7, and Fig. 2.8 gives an example of a BIPV.

### 2.3.4 PV solar power model

The energy conversion that occurs in PV solar panels relies on a complex chemical-electrical process, e.g., [16,17]. To account for the influence of the energy conversion performed by PV panels in building energy management simulations, we introduce a simple empirical model that relates the power output of a PV solar panel, the solar radiation, and the surface area of the solar panel as follows:

$$P^{pv} = I \times A \times \eta \tag{2.5}$$

where the adopted notation is defined in the following, $P^{pv}$ is the PV solar power output (measured in W), $A$ is the surface area of the PV solar panel (measured in m$^2$), $I$ is the solar radiation density on the plane of the PV solar panel (expressed in $W/m^2$), and $\eta$ is the energy conversion factor of the solar panel that provides an estimate on its efficiency.

The solar radiation density $I$ depends on several factors, such as the location of the building, the position of the solar panel, and the amount of light reflected by the solar panel. Based on these, the components of the solar radiation density received by a solar panel (i.e., direct beam radiation, diffuse radiation, and reflected radiation) can be estimated [3]. The energy conversion factor $\eta$ is a function of the temperature of the PV solar panel. Higher temperatures of the panel reduce the energy conversion efficiency. For illustrative purposes, we assume a constant value for the energy conversion factor $\eta$ to be used in the analysis presented in the following. For accurate predictions of the solar panel performance, it is important to keep track of the temperature variations taking place in the PV solar panels.

### 2.3.5 Worked example

Consider two PV solar panels A and B. The areas of the two panels are 60 and $45\,m^2$, respectively. The energy conversion factors of the two panels are 20% and 28%, respectively. At a particular date and time, the solar radiation received by the two panels is the same and equal to $880\,W/m^2$. Determine which solar panel can generate more power.

**Solution**

Based on the given parameters, the power output of the two PV solar panels (denoted by $P_A^{pv}$ and $P_B^{pv}$, respectively) can be calculated as:

$$P_A^{pv} = I \times A \times \eta = 880 \times 60 \times 0.2 = 10,560 \text{ W}$$

$$P_B^{pv} = I \times A \times \eta = 880 \times 45 \times 0.28 = 11,088 \text{ W}$$

Since $P_B^{pv} > P_A^{pv}$, the PV solar panel B can generate more power than the panel A.

## 2.4 Energy storage systems
### 2.4.1 Overview

A distinguishing feature of most of the renewable energy sources, such as wind power and PV solar power, consists in the fact that they are highly stochastic and intermittent. For example, the PV solar panel only generates power during daytime when there is sufficient solar radiation. The wind speed frequently varies at different points in time. This variability and intermittency are not desirable for the following reasons:

- *low utilization of renewable energy* due to the mismatch between building's power demand and the renewable energy output—this is particularly significant for residential buildings that exhibit temporal patterns of peak and valley energy consumption such as those illustrated in Fig. 2.9 (data source: the Australian "Smart Grid, Smart City" customer trial dataset [18]). In such cases, the output of the PV solar source reaches its peak

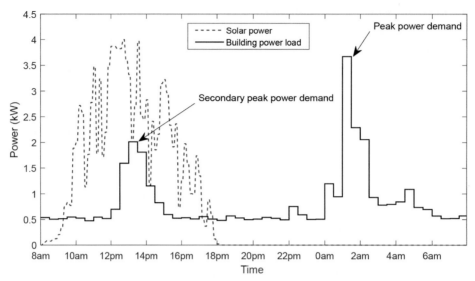

**Fig. 2.9** Example of a mismatch between the profiles of the PV solar power output and the peak power load of a building.

at noon, and it can serve only the secondary peak consumption of the building, while the building's peak power consumption occurs in the evening when little or no solar power is available; and

- *reverse power flow*—the surplus power generated by a renewable energy source that cannot be absorbed by a building can be directed to the external power grid. This process is denoted as reverse power flow. In power systems, the power supply and demand need to be balanced in real time. Unexpected and intermittent reverse power flows would affect this power supply-demand balance and, consequentially, affect the stability of the power grid. This aspect poses challenges in maintaining the stable and secure operation of the grid.

The energy storage technology has been recognized as an effective approach for accommodating renewable energy sources and for reducing the impact of their intermittence on energy entities (e.g., buildings and grids). An ESS can store the surplus energy produced by renewable energy sources when the produced renewable energy is greater than the energy demand. The ESS can then release the energy to contribute to the load demand when no sufficient renewable power is available. Apart from accommodating renewable energy, ESSs can also provide emergent power backup services to energy customers when not sufficient energy is delivered from the grid, e.g., during power outage events.

There are different types of ESSs, and representative ones can be grouped as follows:

- *Battery energy storage systems (BESSs)*—they rely on electrochemical reactions to store energy;
- *hydroelectric reservoirs*—they store energy as water's gravitational potential energy and release water to drive a generator to produce electricity;

- *supercapacitor ESSs*—they use a high-capacity capacitor to store energy;
- *flywheel ESSs*—they accelerate a rotor to a very high speed, and the energy is stored in the form of rotational energy; and
- *compressed air ESSs*—the energy is used to compress air that is stored in suitable storage vessels. The energy stored in the compressed air can then be released to drive an expander, which in turn drives a generator to produce electricity.

### 2.4.2 BESSs and their application in buildings

Among the different kinds of energy storage technologies, BESSs have the advantages that can have relatively small dimensions and that they are mobile, i.e., they are not tight to a specific geographical location. These features underpin the growing popularity in the installation of BESSs for building applications.

BESSs can be classified into different types based on the chemical material used, e.g., [19], and selected typologies include:

- *lead-acid batteries*—these have been used to support uninterruptible power supplies, i.e., to provide power backup functions in the occurrence of power outages. When compared to other types of batteries, the cost of lead-acid battery is relatively low;
- *lithium-ion batteries*—these have been commonly adopted in small electronic devices, such as smart phones and tablets. Lithium-ion batteries have high energy density;
- *sodium nickel batteries*—these batteries can be operated in high-temperature environments because they use molten salt as the electrolyte; and
- *liquid metal batteries*—these batteries use liquid metal as the electrodes and electrolyte. Their main advantage relies on the fact that the electrodes do not degrade over time, while the current high cost of liquid metal affects the cost performance of this type of battery.

### References

[1] C. Riedy, A. Lederwasch, N. Ison, Defining Zero Emission Buildings, Sydney, Australia, Australian Sustainable Built Environment Council, 2011. (Online). Available from: https://www.asbec.asn.au/files/ASBEC_Zero_Carbon_Definitions_Final_Report_Release_Version_15112011_0.pdf. (Accessed 10 July 2023).

[2] B. Dean, J. Dulac, K. Petrichenko, P. Graham, Towards Zero-Emission Efficient and Resilient Buildings, Global Alliance for Buildings and Construction, 2016. (Online). Available from: https://wedocs.unep.org/bitstream/handle/20.500.11822/10618/GABC-Report_Updated.pdf?sequence=1&amp%3BisAllowed=. (Accessed 10 July 2023).

[3] M. Gilbert, Renewable and Efficient Electric Power Systems, John Wiley & Sons, Hoboken, New Jersey, USA, 2013.

[4] J. Lee, F. Zhao, Global Wind Report 2021, Belgium, Global Wind Energy Council, Brussels, 2021. (Online). Available from: https://gwec.net/wp-content/uploads/2021/03/GWEC-Global-Wind-Report-2021.pdf. (Accessed 10 July 2023).

[5] This file is made available and licensed under the Creative Commons Attribution-Share Alike 4.0 International license. Attribution: Spike. https://commons.wikimedia.org/wiki/File:Wind_Turbine_Enercon_E-82_at_Wind_Park_Mausdorf_02.jpg (Accessed 21 July 2023).

[6] This file is made available and licensed under the Creative Commons Attribution–Share Alike 4.0 International license. Attribution: Rehman Abubakr. https://commons.wikimedia.org/wiki/File:VAWT-SanFrancisco-October2011.JPG (Accessed 10 July 2023).

[7] S.N. Akour, M. Al-Heymari, T. Ahmed, K.A. Khalil, Experimental and theoretical investigation of micro wind turbine for low wind speed regions, Renew. Energy 116 (2018) 215–223.

[8] C.J. Chen, Physics of Solar Energy, John Wiley & Sons, Hoboken, New Jersey, USA, 2011.

[9] L. Chandra, A. Dixit (Eds.), Concentrated Solar Thermal Energy Technologies: Recent Trends and Applications, Springer Singapore, Singapore, 2017.

[10] M. Asif, Fundamentals and application of solar thermal technologies, in: Encyclopedia of Sustainable Technologies, Elsevier, 2017, pp. 27–36.

[11] This file is made available in Wikimedia Commons. This file is licensed under the Creative Commons Attribution-Share Alike 3.0 Unported license. https://commons.wikimedia.org/wiki/File:Solar_panels,_Santorini2.jpg (Accessed 10 July 2023).

[12] K. Lovegrove, W. Stein (Eds.), Concentrating Solar Power Technology: Principles, Developments, and Applications, Elsevier, Amsterdam, Netherland, 2020.

[13] This file is made available in Wikimedia Commons. This work has been released into the public domain by its author, el. This applies worldwide. https://commons.wikimedia.org/wiki/File:Solar_tracker_in_Lixouri.jpg (Accessed 10 July 2023).

[14] This file is made available in Wikimedia Commons. This file is licensed under the Creative Commons Attribution-Share Alike 3.0 Unported license. Attribution: Tiia Monto. https://commons.wikimedia.org/wiki/File:Solar_panels_in_Mittenwald.jpg (Accessed 10 July 2023).

[15] This file is made available in Wikimedia Commons. This file is licensed under the Creative Commons Attribution-Share Alike 3.0 Unported license. Attribution: Tove Heggo. https://commons.wikimedia.org/wiki/File:Oseana_kunst-_og_kultursenter.jpg (Accessed 10 July 2023).

[16] Y.A. Jieb, E. Hossain, Photovoltaic Systems: Fundamental and Applications, Springer, Berlin, Germany, 2022.

[17] S. Deambi, Photovoltaic System Design: Procedures, Tools and Applications, CRC Press, 2020.

[18] "Smart grid, smart city customer trial dataset." September 09, 2015 (Online). Distributed by Australian Government Department of the Environment and Energy. Available from: https://data.gov.au/dataset/ds-dga-4e21dea3-9b87-4610-94c7-15a8a77907ef/details (Accessed 10 July 2023).

[19] "Battery energy storage systems." Powercontinuity.co.uk (Online). Available from: https://powercontinuity.co.uk/knowledge-base/battery-energy-storage-systems-bess (Accessed 15 February 2022).

# CHAPTER 3

# Information infrastructure for building energy management

## Contents

## 3.1 Introduction

Recent technological advancements have enabled the integration of Information and Communication Technologies (ICTs) in buildings that has led to improvements in the building operational performances and to a wider acceptance of the smart building paradigm [1,2]. From an energy viewpoint, the advanced information infrastructure deployed in buildings has contributed to a wider acceptance of Building Energy Management Systems (BEMSs). A BEMS acts as the intelligence of a building in dealing with energy-related aspects and is capable of controlling energy resources deployed in a building, so as to enhance the building's energy efficiency and to contribute at improving the efficiency of the smart grid.

This chapter presents the information infrastructure that supports the operations of BEMSs in buildings. In the first part of the chapter, building automation systems (BASs) are introduced, and their components are briefly presented to outline how these can support the operations and strategies of BEMS. This is followed by the introduction of the concept of Building Internet-of-Things (BIoT) within the Internet-of-Things (IoT) paradigm and how this can assist the implementation of effective BEMSs' strategies. In the second part of the chapter, cloud and edge computing are described to highlight how they can optimize the computations required by a BEMS. The final part of the chapter is dedicated to the information infrastructure that combines the paradigms of BAS/BIoT, cloud computing, and edge computing.

*Building Energy Management Systems and Techniques*
https://doi.org/10.1016/B978-0-323-96107-3.00016-3

## 3.2 Building automation systems

Automation in buildings has gained popularity in the past decades because it is supported by the technological advancements occurred in the domain of BASs, e.g., [3,4]. A BAS is a platform that connects various devices in a building's structure, including the heating, ventilation, and air conditioning (HVAC) systems, the lighting systems, the video surveillance system, and others. In this manner, a BAS is able to enhance the performance of buildings under a number of aspects that involve the security and energy efficiency of the building as well as the indoor comfort of the building occupants.

A BAS typically consists of the components listed in the following:

- *Building Management Systems* (*BMSs*)—they consist of centralized software systems that represent the intelligence of a BAS. A BMS makes control decision on the building operations based on information collected from sensors and on preferences specified for the building operations;
- *sensors*—they monitor the real-time operations and conditions of a building (e.g., temperature, relative humidity, noise, indoor air quality) and transmit the data to the BMS;
- *actuators*—an actuator is a component that performs operations. As an example, these can activate a simple light when a certain scenario occurs or can move sunshades under particular sunlight conditions;
- *communication protocols*—these are the framework in which a BMS can communicate and interact with different devices; and
- *user interfaces*—these are digital portals that enable the different stakeholders, such as the building owners, managers, and occupants, to interact with the BAS. Through user interfaces, it is possible to monitor the building's conditions, to submit control commands, and to change settings.

BASs provide the necessary underlying infrastructure for implementing management and control strategies of the devices and actuators deployed in a building system. A BEMS can be regarded as the part of the BMS that focuses on managing the energy consumption and production in a building or multiple buildings. By taking advantage of the underlying infrastructure provided by a BAS (i.e., sensors, actuators, communication protocols, and user interfaces), a BEMS can monitor the energy consumption and production of the building energy resources and devices and implement strategies for their control.

The recent developments of IoT technology support the deployment of cost-effective solutions for building automation and provide new opportunities to integrate the building devices into a digital platform. These advancements have also contributed to the further developments of BEMSs' designs and applications.

## 3.3 Building Internet-of-Things

The IoT technology [5,6] enables to connect diverse *things* (e.g., household appliances, vehicles, and wearable devices) in a network and to the Internet that consists of a global system of interconnected computer networks that use the Transmission Control Protocol/Internet Protocol (TCP/IP) suite to communicate between networks [7]. The Internet can be regarded as a human-centric technology as it enables people around the world to communicate with each other and to exchange information through connected computers. IoT embraces this vision by providing a wide range of *things* (or objects) with functionalities that enable communications with humans and other objects.

The devices in an IoT system are equipped with processors, sensors, and communication hardware. They can collect, send, and act on data they acquire from the environment. By relying on the embedded processors, the devices in an IoT system can exchange messages with each other as well as act on the data they receive. The devices usually work without human intervention. People can interact with the devices, for example, by manually changing the settings of the devices and by sending specific control instructions.

The BIoT [8,9] is an application of the IoT paradigm that supports the operations of smart buildings. The BIoT connects a wide range of objects in buildings with each other and with the building occupants to enhance the operational efficiency of the buildings.

Although the BIoT technology can significantly enhance the capability of BEMSs in managing buildings' energy, it also imposes challenges to BEMSs in storing and processing the data sent from the BIoT system. Since the number of the diverse energy resources and devices that are managed by a BEMS can be large, the resulted energy management logics and algorithms can become computationally intensive and, therefore, require significant computing power of the IT equipment that hosts the BEMS. Cloud computing and edge computing provide a support to enable BEMSs to perform computationally- and data-intensive energy management tasks.

## 3.4 Cloud computing

### 3.4.1 Introduction to cloud computing

Cloud computing [10,11] refers to a computing paradigm in which computing resources are delivered through Internet as services on a pay-as-you-go basis. The computing resources can include processing power, data storages, and software applications. The computing resources can be accessed through web portals. In cloud computing, a Cloud Service Provider (CSP) typically invests in powerful IT equipment and aggregately manages it in datacenters. The virtual computing resources are produced from and associated with real computing resources through virtualization technologies [12], and these can typically deliver multiple levels of computing services that can be grouped as follows [13]:

- *Infrastructure-as-a-Service (IaaS)*—IaaS delivers virtualized computing infrastructure (e.g., servers and data storages) to the user in an on-demand manner;
- *Platform-as-a-Service (PaaS)*—PaaS offers a programmable environment in which the user can develop applications directly on the cloud side (through web portals); and

– *Software-as-a-Service (SaaS)*—SaaS consists of software that is delivered to users through web portals. SaaS shifts users from using locally installed computer programs to online software services that offer equivalent functionalities.

### 3.4.2 Introduction to edge computing

Backboned by datacenters, cloud computing provides powerful platforms for storing data transmitted from different sources and for performing computations on the data. Cloud computing that handles a large number of distributed devices needs to be carefully designed to avoid or limit possible issues related to: (1) bandwidth bottleneck due to the large amount of data collected by IoT devices; (2) high electricity consumption due to frequent communications between datacenters and the devices; (3) privacy concerns on storing the devices' data in datacenters managed by third parties; and (4) slow response time that could occur due to sending/retrieving data to/from the cloud. In the context of these possible issues, edge computing has been gaining popularity for its ability to perform computations in the close proximity of devices on the edge of the IT network [14]. In this manner, computational tasks can be offloaded from the cloud to the edge side, therefore significantly reducing the load of the cloud's network and reducing the consequent response time. The reduced communication frequency can also contribute to decrease the energy consumed by the edge devices.

### 3.4.3 Cloud- and edge-computing-supporting BEMSs

The operations of BEMSs can benefit from both cloud and edge computing. BEMSs need to perform computationally- and data-intensive operations to monitor and control energy management tasks. The large data handled in these computations is generated by the pervasive BIoT devices and sensors, and these can generate data at medium-to-high frequencies. Such data collection poses challenges on the BEMSs for storing and analyzing the data. In addition, energy management algorithms and strategies implemented in BEMSs are usually computationally demanding (e.g., relying on sophisticated machine learning and optimization techniques) and, therefore, require high-performing hardware.

Cloud computing can provide powerful support to enhance the computing and data storage capabilities of BEMSs. For example, BEMSs can benefit from the IaaS-level service delivery by renting virtual computing resources to store data and perform computationally- and data-intensive energy management operations. BEMSs can also benefit from the PaaS- and SaaS-level services of cloud computing. For example, a BEMS could rely on the cloud-side analytics tools, available through the IaaS, to analyze the data collected for a building and could also use the programmable environment of the PaaS to build up customized BEM logics.

BEMSs can also benefit from edge computing. A BEMS can optimize its computational performance by sharing operations and data storage on the edge side (i.e., stored locally in the building or its vicinity) or in the cloud. For example, a BEMS can have the individual energy devices in the building preprocessing the data (e.g., noise removal and re-sampling) once the data is sampled by the sensors (by means of embedded processors present in the devices). The BEMS can also process part of the data and perform simple BEM computations on the edge side (i.e., on local computer systems), while the computationally- and data-intensive BEM tasks can be carried out on the cloud.

## 3.5 An integrated information infrastructure for BEMSs

Fig. 3.1 depicts a schematic of the information infrastructure in building systems that combine the paradigms of BAS/IoT, cloud computing, and edge computing. The facilities of these ICT paradigms interact with each other to collaboratively provide underlying information services to buildings. The fine-grained data collected by the BAS and/or BIoT sensors can be stored and processed by the energy devices and BEMSs locally, or they can be stored remotely in the cloud. Computationally-intensive and complex analytics can be performed on the cloud by utilizing virtual computing resources. The integrated information infrastructure representation of Fig. 3.1 also provides the

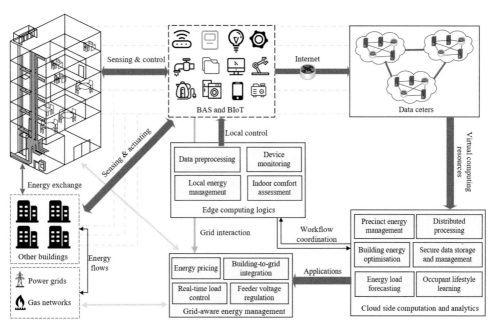

**Fig. 3.1** Illustrative schematic of an integrated information infrastructure for building energy management.

required flexibility to enable building-side edge computing units to collaborate with the cloud-side analytics programs and to decompose a complex workflow into interrelated subtasks that can be processed efficiently.

## References

[1] J. Sinopoli, Smart Building Systems for Architects, Owners and Builders, Elsevier, Amsterdam, Netherland, 2009.
[2] R. Bakker, Smart Buildings: Technology and the Design of the Built Environment, RIBA Publishing, London, UK, 2020.
[3] H. Merz, T. Hansemann, C. Hubner, Building Automation: Communication Systems with EIB/KNX, LON and BACnet, Springer International Publishing, NY, USA, 2018.
[4] S. Wang, Intelligent Buildings and Building Automation, CRC Press, Boca Raton, FL, USA, 2009.
[5] M. Kranz, Building the Internet of Things: Implement New Business Models, Disrupt Competitors, Transform Your Industry, John Wiley & Sons, Hoboken, NJ, USA, 2016.
[6] S. Greegard, The Internet of Things, MIT Press, Cambridge, MA, USA, 2015.
[7] D.E. Comer, The Internet Book: Everything You Need to Know About Computer Networking and How the Internet Works, CRC Press, 2018.
[8] M. Bajer, IoT for smart buildings – Long awaited revolution or lean evolution, in: Proc. 2018 IEEE 6th International Conference on Future Internet of Things and Cloud, Barcelona, Spain, 2018.
[9] C.S. Suriyarachchi, K.G.A.S. Waidyasekara, N. Madhusanka, Integrating internet-of-things and facilities manager in smart buildings: A conceptual framework, in: Proc. 7th World Construction Symposium 2018, Colombo, Sri Lanka, 2018, July.
[10] R. Puttini, Z. Mahmood, T. Erl, Cloud Computing: Concepts, Technology & Architecture, Prentice Hall, Hoboken, NJ, USA, 2013.
[11] R. Buyya, J. Broberg, A.M. Goscinski, Cloud Computing: Principles and Paradigms, John Wiley & Sons, Hoboken, NJ, USA, 2010.
[12] M. Portnoy, Virtualization Essentials, John Wiley & Sons, Hoboken, NJ, USA, 2016.
[13] M.J. Kavis, Architecting the Cloud: Design Decisions for Cloud Computing Service Models (SaaS, PaaS, and IaaS), John Wiley & Sons, Hoboken, NJ, USA, 2014.
[14] K.A. Kumari, G.S. Sadasivam, M. Niranjanamurthy, D. Dharani, Edge Computing: Fundamentals, Advances and Applications, CRC Press, Boca Raton, FL, USA, 2021.

# CHAPTER 4

# Power demand response and demand side management

## Contents

## 4.1 Introduction

Buildings have been traditionally considered as energy consumers in the broader energy systems, e.g. power grids and gas networks. In this framework, buildings consume the energy that they need and rely on external energy sources, e.g. power and gas plants, to deliver it through the available energy transmission and delivery systems. In recent years, the following technical developments have supported the engagement of buildings in contributing to the operations of energy systems:

- *wider deployment of building-side renewable energy sources* (see Chapter 2)—this enables buildings to transition from being energy consumers to become energy *prosumers* (i.e. to become both producers and consumers) [1]. Buildings are now able to generate energy and, when possible, feed surplus energy back into the external energy systems;
- *building electrification*—it refers to the process of replacing fossil fuel-powered appliances, such as space heating, water heating, cooking, and boilers, with electricity and other fossil fuel-free zero-carbon alternatives;
- *digitalization*—it refers to the fusion of advanced technologies and the integration of physical and digital systems within the building domain. The digitalization of a building is supported by the advances of Internet-of-Things (IoT) and ubiquitous computing technologies that enable the acquisition of fine-grained data from various devices and the real-time monitoring of the building environments; and
- *automation*—it refers to the automatic control of the devices and equipment installed in a building. A building automation system usually includes wireless communication systems, processors that are embedded in the devices, and actuators that perform control actions on the devices.

The technological advances previously introduced underpin the possibility of a wide range of interactions between buildings and the external power grid. In this context, this chapter focuses on the interaction strategies that can occur when considering power demand response (DR) and demand side management (DSM) techniques.

In the first part of the chapter, we introduce basic concepts related to electricity, electrical power, and energy, as well as to the basic structure of power systems. We then present the concept of smart grids and how these have been gaining popularity in recent years thanks to the advantages that they provide. The latter include the possibility of exploiting the high penetration of renewable energy, the larger number of players being able to contribute in the grid's operations within its highly deregulated structure, the self-healing and fault-tolerance characteristics, and the ability to establish a two-way communication between end energy consumers and the grid. The remaining part of the chapter is dedicated to highlight key strategies that can be implemented in a grid's setting and that consist of DR and DSM. These approaches are briefly introduced together with the concepts related to electricity tariffs. An application example is provided to gain insight into some of the characteristics and advantages of the strategies introduced in the chapter.

## 4.2  Basic concepts

### 4.2.1  Electricity

*Electricity* refers to the physical phenomena associated with the presence and motion of matter that has a property of electric charge.[a] Simply speaking, electricity can be defined as the flow of electric charge. There are two forms of electricity: *static electricity* and *current electricity*.

Static electricity exists when opposite electric charges are built up on objects separated by an insulator (air, glass, rubber, etc.). When the electric charges find a path to balance the system (referred to as "equalizing"), static electricity occurs. For example, when electrons travel through an air gap and collide with electrons in the air, energy would be realized in the form of visible light, and this is the lightning we are all familiar with.

Current electricity (also known as dynamic electricity) refers to an electrical charge in motion. Current electricity exists when negatively charged electrons flow from atom to atom through a conductor in an electrical circuit.

The comparison of static and current electricity is illustrated in Fig. 4.1. We use current electricity every day to power our electrical devices, and this is the electricity generated, transmitted, and distributed by power systems. Unless otherwise stated, the electricity mentioned in the rest of this book refers to as the current electricity.

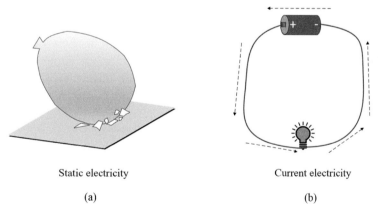

Static electricity                                      Current electricity

(a)                                                          (b)

**Fig. 4.1** Static electricity (A) and current electricity (B).

---

[a] Electric charge is the property of subatomic particles that causes a particle to experience a force when placed in an electric and magnetic field. There are two opposite types of electric charges: positive and negative, commonly carried by protons and electrons, respectively.

## 4.2.2 Electric current and voltage

There are two fundamental concepts associated with electricity: *electric current* and *electric voltage* (usually simply referred to as *current* and *voltage*). Current is the rate at which electric charge flows past a point of the electric circuit. If one coulomb charge (i.e. SI unit of electric charge) passes through a point in a circuit in 1 second, the amount of current is known as 1 ampere (1 A). 1 A of current consists of $6.24 \times 10^{18}$ electric charge carriers (electrons).

Current is produced by voltage. Voltage refers to as the amount of potential difference force between two points in a circuit. It is the force that pushes charged electrons through a conducting loop in a circuit, enabling them to do work (e.g. illuminating a light), i.e. voltage gives the force to electrons the flow through the circuit. The higher the voltage, the greater the force, and the more electrons flow through the circuit. 1 volt (1 V) of voltage is defined as the electric potential difference between two points in a circuit that releases 1 J of energy per coulomb of charge that passes through the circuit.

The relationship between voltage and current is described by Ohm's law:

$$V = I \times R \qquad (4.1)$$

where $V$ is the voltage (in volts); $I$ is the current in the circuit (in amperes); $R$ is the resistance in the circuit, which is a measure of the opposition to the flow of current (in ohms, denoted by "$\Omega$").

To gain a conceptual understanding of how the concepts of electric charge, current, and voltage relate to each other, it is common to draw their analogy to equivalent concepts describing the behavior of a hydraulic water system as shown in Fig. 4.2. The hydraulic water circuit of Fig. 4.2A is driven by a mechanical pump, which produces the pressure to push the water through the pipe. The kinetic energy carried by the

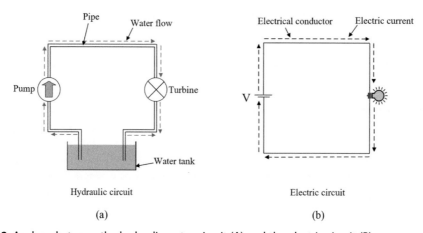

**Fig. 4.2** Analogy between the hydraulic water circuit (A) and the electric circuit (B).

flowing water is then able to be used to do work, such as driving the turbine shown in the figure. Similarly, the electric circuit is driven by a power source (e.g. a battery), which produces the voltage that induces the current to flow through the circuit, and the current electricity can be used to do work, such as lightning the lamp in the figure.

In this analogy, the voltage is equivalent to the hydraulic water pressure; the current is equivalent to the hydraulic water flow rate; the electric charge is equivalent to the amount of water; and the electrical conductor of the electric circuit is equivalent to the water pipe in the hydraulic water circuit.

### 4.2.3 Electrical power and energy

*Electrical energy* is generated by the movement of electrons from one point to another. When charged electrons flow in an electric circuit, they generate electric energy. *Electrical power* (or simply power) is the rate at which the electric energy is transferred through a circuit. The power can be calculated from the voltage and current as:

$$P = V \times I = I^2 \times R = \frac{V^2}{R} \qquad (4.2)$$

where $P$ denotes the electrical power (in watts). Power can be considered as the electrical energy generated by the circuit in a unit of time. Giving a constant value of power, the amount of electrical energy generated over a time period can be calculated as:

$$E = P \times \Delta t \qquad (4.3)$$

where $E$ denotes the electrical energy (in watt-hours) and $\Delta t$ is the duration of the time period (h).

### 4.2.4 Worked example

Consider a circuit with a resistance of 80 $\Omega$. The current flowing through the circuit is 5 A. Determine the electrical energy generated by the circuit over 140 min.
**Solution**
The power generated by the circuit is:

$$P = I^2 \times R = 5^2 \times 80 = 2000 \text{ W}$$

from which we can determine the energy generated by the circuit as:

$$E = P \times \Delta t = 2000 \times \frac{140}{60} = 4666.67 \text{ W}$$

### 4.3 Power systems

The delivery of electricity occurs through a power system that consists of a network of electrical components deployed to supply, transfer, and use electrical power. In the last decades, power systems have developed to be complex interconnected physical systems

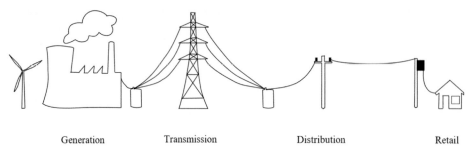

Generation          Transmission          Distribution          Retail

**Fig. 4.3** Structure of a conventional power system.

that serve hundreds of millions of people throughout the world. Although power systems vary in scale and complexity of their design, a typical power system can be considered to be formed by the following three-segment structures: power generation, transmission, and distribution [2]. It is common for power systems to include one more "retail" segment as depicted in Fig. 4.3 by the four-segment structure. The retail segment refers to the electricity retailing mechanism. In the power distribution side, the retailer purchases energy from the power generation companies at a wholesale price and delivers electricity to end customers. The end customers are charged by retail prices that are higher than the wholesale price.

The power generation side includes different types of power generation plants that produce electricity. Conventionally, power plants consist of coal-fueled plants, nuclear plants, and hydropower plants. The power transmission segment is composed of devices that include transformers, transmission lines, and power towers. The transformers raise the voltage of the electricity generated from power plants, and the high-voltage electricity is transmitted by the transmission lines over long distance to the power distribution side, usually referred to as the load areas. In the load areas, the power distribution network lowers the voltage of the power transmitted by the transmission lines through transformers placed in the substations and delivers electricity to end consumers (e.g. commercial/residential buildings, industrial electric loads, and public electric facilities) through overhead or underground feeders.

The three- or four-segment power system structure is considered to be a vertical structure because the power flows following a top-down pattern, i.e. top-down structure from the source to the demand side. Conventional power systems heavily rely on fossil fuels (e.g. coals, gas, and uranium minerals) for which we have limited resources, and their use leads to greenhouse emissions. The communication flow between the end energy customer and the grid occurs only in one direction where the utility monitors the energy demand and generates the amount of electricity required. The electricity generation is completely driven by the demand, and the energy customers simply consume electricity that they need without any interaction with the grid.

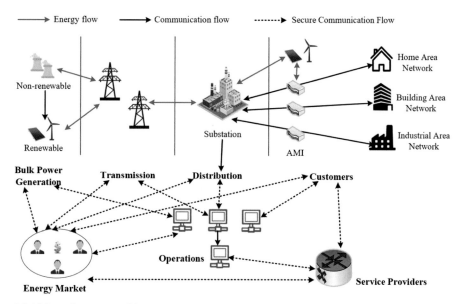

**Fig. 4.4** Vision of a smart grid.

## 4.4 Smart grids

### 4.4.1 Overview

The concept of *smart grid* was initially proposed at the beginning of this century with the vision to redefine the traditional power system (Fig. 4.4). Different definitions are available for smart grids, e.g. [3,4], and these recognize that a smart grid is an electricity network that integrates digital communication technology to enable widespread renewable energy integration and a two-way flow of electricity and data, and that is capable of detecting, reacting, and proacting to changes in a number of factors, e.g. electricity usage.

### 4.4.2 High penetration of renewable energy

In smart grids, renewable energy sources are widely deployed to reduce the penetration level of (or to fully replace) the fossil fueled power sources. Renewable energy sources can be generally integrated into smart grids in one of the following two arrangements:
- in the form of renewable power plants (usually wind farms and solar farms) that group a number of renewable power generators (e.g. wind turbines and photovoltaic (PV) solar panels) in the same location and connect them to the grid through a common coupling point or substation. The size of renewable power plants varies depending on the number of renewable power generators involved and on the geographical area covered; and
- in the form of distributed renewable energy sources in the power distribution system side, such as on-road wind turbines and rooftop solar panels. These distributed

renewable energy sources have usually small capacity and are mainly used for supporting the local energy production.

The integration of distributed renewable energy converts the traditional vertical structure of the power system into a distributed structure where the power generation does not only rely on bulk power plants as it makes use of dispersed renewable energy sources located at the edge of the grid, i.e. located in distribution networks and at the end energy customer side.

### 4.4.3  Highly deregulated structure

In traditional power systems, power generation, transmission, and distribution segments are vertically managed by a monopolized organization, e.g. a national or regional power companies. Such an organizational arrangement can have adverse effects on the drivers for improving operational efficiency of the grid and for optimizing the energy resource distribution. The lack of competition in the different segments of the power system might also have a negative effect on the incentives and drivers to promote technological innovations.

In a smart grid scenario, energy assets do no longer need to be centralized under one organization, and they can be owned by different organizations across the different segments of the power utility service, such as power generation, transmission, and distribution. In such a deregulated structure, different stakeholders operate their energy businesses in the grid environment. The main stakeholders include power generation companies, distribution companies, retailers, and auxiliary service providers. Power companies own power generation resources and sell electricity to power distributions or retailers. Power distributors own power delivery assets, and they deliver power to end customers or further sell energy to power retailers. End energy customers sign contracts with distribution companies or retailers, and they are charged at retailing price based on their electricity use. Power transmission infrastructures are owned by power transmission companies that charge the power generation companies, distributors, and retailers for the network use cost. A schematic of this deregulated structure is illustrated in Fig. 4.5.

Marketplaces, also known as *power markets* or *electricity markets* [5], are established to facilitate energy trading under the supervision of a market operator. In power markets, electricity is traded as a commodity. The participants of the power market reach electricity purchase/selling agreements under the regulation of the market operator.

In recent years, with the increasing penetration of distributed renewable energy sources, electricity trading at the edge of the grid, i.e. trading among energy customers, has been gaining popularity in various parts of the world. With such an arrangement, for example, a residential customer can sell the surplus electricity generated from the customer's rooftop PV solar panels to neighbors or to the grid. Such emerging distributed energy trading activities further deregulate smart grids.

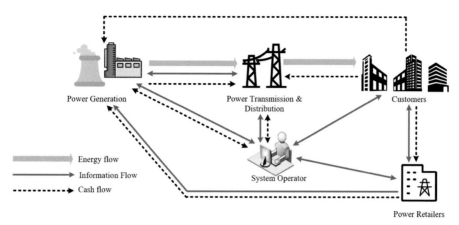

**Fig. 4.5** Schematic of the deregulated structure of a smart grid.

## 4.4.4 Self-healing and fault tolerance

Smart grids are deeply integrated with advanced computing, communication, and control technologies. The widespread deployment of sensors and other intelligent devices enable a smart grid to be capable of perceiving its operation conditions in a real-time manner and of performing self-assessment on its status. Using this information, abnormalities and faults can be detected at an early stage. The advanced computing and control technologies can enable the grid to automatically isolate a fault and protect the power infrastructure. With such arrangements in place, the grid can operate more securely and reliably than traditional energy systems.

The integration of advanced information and communication technologies (ICTs) has also changed the way in which utilities can respond to outage events. Through a fast localization and isolation of the fault (by relying on the real-time information collected from field devices), the grid can minimize the time required to restore power and, consequently, the number of energy customers affected by such an event.

## 4.4.5 Two-way communication between end electricity consumers and the grid

In traditional power systems, there is a one-way communication between the end electricity consumers and the grid. In this arrangement, the power generation is fully driven by the demand.

In smart grids, the communication relies on a two-way communication channel because the utility collects power consumption data from the customer and the customer can receive information from the utility. The latter information can include pricing rates, incentive signals, and power control instructions. The customer can then respond to the information received, for example, by actively adjusting and re-shaping its power consumption. This approach is usually referred to as DR. The implementation of such two-way communication is based on the deployment of the advanced metering infrastructure (AMI) [6].

## 4.5 Demand response and demand side management

### 4.5.1 Basics of demand response and demand side management

In traditional power systems, the amount of power generation is completely driven and decided by the demand. Such an approach can be costly and inefficient. For example, some power generators may need a long warm-up time to operate with full power and to serve the required energy demand. Some generators are expensive to operate and, sometimes, the energy demand can even be larger than the available power.

The concept of DR [7,8] has been proposed to operate power systems in a more affordable and flexible way. DR refers to the ability of changing the power consumption profile of energy customers to match the demand of power with the available supply. The word *response* refers to the fact that energy customers can actively re-shape their power consumption profiles to respond to certain requirements or signals from the external power grid (i.e. DR aims to support the power supply–demand balance by adjusting the demand instead of adjusting the power generators). Fig. 4.6 illustrates two power consumption profiles over time to highlight the advantages provided by DR in re-shaping the power consumption of an energy customer. In this re-shaping process, the peak power of the typical consumption profile is reduced (e.g. peak clipping), and

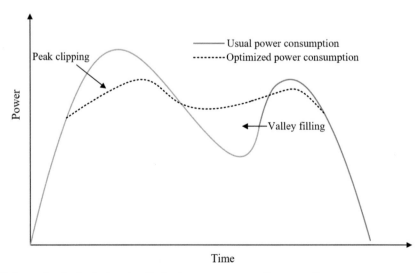

**Fig. 4.6** Example of a typical and optimized power consumption over time.

the reduced peak power is shifted to "fill" the power consumption in the valley hours. As a result, the whole power consumption profile becomes smoother and can thus alleviate the load at the generation side during peak hours.

Based on the concept of DR, the DSM [7] refers to the initiatives, techniques, and actions designed to manage and optimize a site's energy consumption. DSM can be classified as static or dynamic. In static DSM, the physical condition of the energy customer is permanently changed so that the customer's energy efficiency is improved. For example, a building could adopt more efficient heating, ventilation, and air conditioning (HVAC) designs or more efficient building components that enhance it is energy efficiency. A dynamic DSM is able to dynamically reduce or shift the energy consumption of a customer to achieve certain desired objectives. Most of the energy management techniques introduced in this book can be considered to fall within the dynamic DSM.

## 4.5.2 Advanced metering infrastructure

The widespread deployment of AMI in the years has been supporting the wider use of smart grids. AMI [6] is an architecture for an automated two-way communication between energy customers and utility companies. The goal of an AMI is to provide utility companies with real-time power consumption data of energy customers and to allow customers to make informed choices in the energy domain. Using digital technologies, AMI provides electricity customers with several automatic metering functions that can be performed manually in traditional energy systems, such as electricity meter readings, electricity theft detection, and fault and power outage identification.

The schematic of an example of AMI is presented in Fig. 4.7. A typical AMI deployment includes a number of components that consist of smart meters, communication networks, and information management systems, and these are now briefly introduced:

- *smart meters*—these are the core components of an AMI and can be regarded as one of the key enabling technologies for smart grids. Equipped with embedded microchips, a smart meter measures the customer's power consumption at a certain time interval (every 60-, 30-, 15-, or 5-min or at shorter intervals). To support automatic meter readings, smart meters also provide other functions including tamper detection, outage monitoring, and voltage monitoring. Smart meters have communication functions that support information exchanges between the customer and the utility;
- *communication networks*—these are needed to transmit the large amount of data generated by smart meters to other information management systems for further analysis and processing. Most utility companies install new or upgrade existing communication networks to support the deployment of smart meters. Based on the utility company's business objectives and practical considerations, wired or wireless communication systems (Table 4.1), or the combination of the two, can be deployed; and

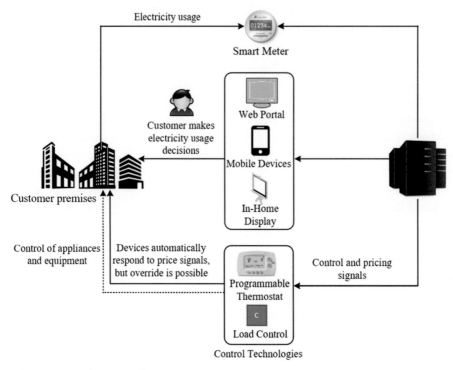

**Fig. 4.7** Schematic of an example of an AMI.

**Table 4.1** Examples of AMI communication technologies [6].

| Wired | Wireless |
|---|---|
| Fiber-optic cable | Radio frequency—mesh network |
| Power line communications | Radio frequency—point to multipoint |
| Telephone dial-up modem | Radio frequency—cellular |
| Digital subscriber line | |

- *information management systems*—these are integrated into the AMI to perform higher-level processing and analysis tasks on the smart meter data. There could be different information management systems, such as billing systems (to process interval load data to automate bill generation), customer information systems (to process and store customer data on customer locations, demographics, contact information, and billing histories), and distribution management systems (to process outage data and customer voltage-level data, and perform optimization procedures). The integration of information management systems opens opportunities to implement a wide range of new functions to enhance the efficiency of grid operations.

## 4.6 Demand response for grid peak power reduction

### 4.6.1 Introduction

The basic application of DR is to support the power utility to reduce the peak load of the grid. A daily regional power demand profile often shows a pattern where the demand is extremely high at some time intervals (known as the "peak" periods) and low at other time intervals (known as the "valley" period). A high peak demand tends to push a power network infrastructure to its capacity limit and this imposes security hazards to the power system's operations. The physical power system infrastructure is built for serving the peak demand, and a market with a higher peak demand would require larger investments to set up the power system infrastructure (e.g. more transmission lines and larger-capacity transformers) and higher costs for the power utility.

DR provides an affordable mean to reduce the power network's peak power through having the energy customers actively reduce their power consumption or shift their power demand from peak hours to off-peak hours. Peak power load reduction can be achieved through both static DSM and dynamic DSM. In the following, we provide two case studies that demonstrate how these two types of DSM can contribute to reducing the grid's peak power demand.

### 4.6.2 Case study of static DSM: The Binda-Bigga fuel substitution project

The case study of the Binda–Bigga fuel substitution project in Australia ([9]) provides an example of an effective implementation of a static DSM. Binda and Bigga are two rural towns located in New South Wales (Australia). There are around 250 electricity customers in the Binda and Bigga areas, most of them being residential. In 2004, at the time of the Binda-Bigga fuel substitution project, the electricity supply of the area was managed by the Country Energy Company. In the 1980s, the company installed an electricity line connecting Binda and Bigga, and other areas. In the winter of 2004, the peak electricity consumed in the area pushed the electricity line to reach its maximum capacity. Two peaks occurred in the electricity consumption profile of the area, i.e. one around midnight and one in the evening. The midnight peak occurred due to the hot water controlled loads, while the evening peak occurred probably due to the heating load activated to deal with the drop of outdoor temperature below $-9°C$. Once these two events occurred, the Country Energy recognized the need to find a solution to satisfy the future (growing) peak demands of the residents.

The typical and most straightforward approach to address growing electricity peak demands is to upgrade the existing power infrastructure, e.g. build more power transmission and distribution lines. This approach is expensive and can be uneconomical. In 2004, the Country Energy launched a fuel substitution project, which used DSM strategies to defer the possible infrastructure upgrade in the area. The main objectives of the project were:

**(1)** to reduce the electricity consumption of the area during winter evening peaks (approximately from 6 pm to 10 pm);

**(2)** to deliver real benefits to rural residents through reducing their energy consumption and improving the energy supply; and.

**(3)** to reduce greenhouse gas emissions through fuel substitution of electric appliances to gas appliances.

In the project, the Country Energy established a new energy solution, referred by them as an Energy Saver Package, that encouraged residents to support the possible transition from electric appliances to gas appliances. The package targeted appliances that would reduce the electricity demand of residents during the periods usually dedicated to the cooking of evening meals and heating of homes. The package offered residents several services to enable them to affordably replace electric appliances with gas appliances, such as those shown in Table 4.2. In addition, the package also offered the residents [5]: (1) free installation of gas appliances and gas bottles, and removal of electrical appliances for metal recycling; and (2) gas credits of AU$ 170 per appliance.

Over 70 customers joined the program and purchased an Energy Saver Package between July and October 2004, with a total of 106 appliances being purchased. As a result of this initiative, the peak electricity demand of the area significantly reduced. The total cost to Country Energy for executing this DSM project was AU$108,000, which was smaller than the estimated infrastructure upgrade cost of AU$412,500.

## 4.6.3 Application example of a dynamic DSM for grid peak power reduction

Peak power load reduction can also be obtained by dynamically shifting the power consumption of energy customers. In this section, we use a simple application example to

**Table 4.2** Cost of gas appliances offered in the Energy Saver Package in the Binda-Bigga fuel substitution project.

| Appliance | Recommended retail price | Price in the Energy Saver Package | Saving to the customer |
|---|---|---|---|
| Rinnai Granada (unflued heater) | AU$899 | AU$250 | AU$649 |
| Rinnai EnergySaver (flued heater) | AU$1699 | AU$1000 | AU$699 |
| Chef Stove | AU$699 | AU$250 | AU$449 |
| Rinnai Granada + Chef Stove | AU$1598 | AU$470 | AU$1128 |
| Rinnai EnergySaver + Chef Stove | AU$2398 | AU$1200 | AU$1198 |

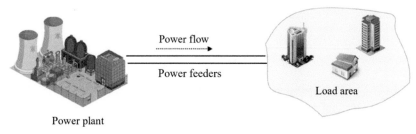

Power flow

Power feeders

Load area

Power plant

**Fig. 4.8** Schematic of the application example with a simplified power system consisting of a power plant and a load area.

illustrate how a dynamic DSM can assist the power utility to reduce the peak power demand of the grid.

Let us consider a highly simplified power system arrangement that consists of a coal-fired power plant and a load area formed by three buildings. The power utility company is planning to build power feeders between the power plant and the load area, so as to transmit electricity to feed the three buildings. The schematic of this simplified power system is shown in Fig. 4.8.

Let us assume that the investment cost of one feeder is \$500,000, and the maximum power transmission capacity of each line is 8 kW. The daily power consumption profile of the three buildings is shown as the blue lines in Fig. 4.9, and the power consumption profile of the whole load area, i.e. the sum of the power consumption of the three buildings, is depicted by the blue line in Fig. 4.10. The data adopted in this application example is kept as simple as possible for illustrative purposes as, in a real application scenario, the costs and complexity involved in the project would be higher.

Fig. 4.9 shows the peak load power consumption of the load area to be 8.4 kW. Since one feeder can transmit 8 kW at most, two feeders are needed and these require an investment cost of \$500,000×2=\$1,000,000.

Different daily power consumption profiles are shown in the red dotted lines in Fig. 4.9, and these provide a description of how a DR implementation can shift the energy consumption curves. The corresponding power consumption profile of the load area after the implementation of a DR strategy is shown in the red dotted line in Fig. 4.10. With the implementation of a DR strategy, the peak load power consumption of the load area reduces from 8.4 to 7.4 kW, based on which one feeder is sufficient for transmitting the power to feed the load area. As a result, the total investment cost of the feeder reduces as well as other associated costs, such as maintenance and protection costs.

This application example presents a simple scenario to highlight how a power grid can benefit from the implementation of DSM strategies, for example, by assisting in the dynamic reduction or shifting of the electricity load at the customer side.

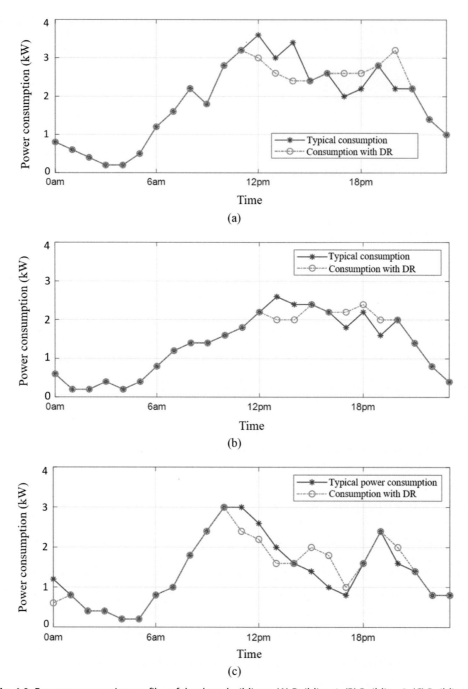

**Fig. 4.9** Power consumption profiles of the three buildings. (A) Building 1. (B) Building 2. (C) Building 3.

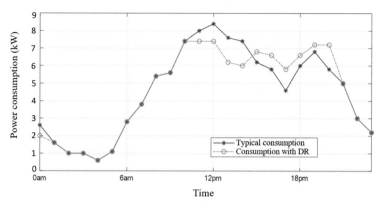

**Fig. 4.10** Aggregated power consumption profile of the three buildings.

In the following, we consider two types of DSM strategies that can be implemented to encourage power consumption shifting on the customers' side, and these are denoted as incentive-based DSM and price-based DSM.

## 4.7 Incentive-based demand side management

*Incentive-based DSM*, also known as *direct load control* (DLC), consists of a management strategy in which a utility company can remotely control the on/off switching of selected customer-side electric appliances (agreed with the customer) to assist the power utility to reduce power consumption during certain periods.

When implementing a DLC approach, the customer needs to sign a contract with the utility company in which both parties agree that the utility can remotely control specific appliances during certain periods of times, e.g. during hot summer mid-days when the grid is heavily loaded or during grid emergency events. For example, an agreement could specify that, between 1 and 3 pm in hot summary days, the utility company is allowed to shut down the customer's air conditioner occasionally with a cumulative shut down duration of 20 min to contribute to the reduction of the grid's peak load. For such an arrangement, the utility usually provides financial incentives, such as discounted electricity tariff or cash reward. DLC can be implemented on a variety of appliances, typically the appliances that consume significant power in a building.

## 4.8 Price-based demand side management
### 4.8.1 Introduction

Price-based DSM strategies, also known as indirect load control (IDLC) strategies, use time-varying electricity tariff signals to stimulate users to shift the usage time of their appliances. Price-based strategies set up time-varying electricity tariff structures that apply different rates during different time periods. In time-varying electricity tariffs, the electricity rate is higher in peak load and secondary peak load hours, and it is lower in valley load hours.

By using time varying electricity prices as signals, energy customers are encouraged to shift the use of some appliances from peak hours to other time periods to reduce their energy cost. For example, a user that usually likes to use the washing machine at noon (12–1 pm) might accept to run the machine after 10 pm if understanding and aware of the potential advantages offered by a time-varying price that has significantly lower rates in the late evening hours when compared to mid-day. By doing this, the user can reduce the energy expenses, and the grid can reduce the network load at noon when the network load is expected to be high.

In a price-based DSM, there is no DLC of appliances, and their usage shifting depends completely on the user's subjective willingness to take advantage of the reduced electricity rates. This is the reason as to why price-based DSM are denoted as IDLC.

## 4.8.2 Electricity tariffs for energy customers

An electricity tariff refers to the rate at which an energy customer is charged for electricity usage. It is common that a customer's electricity bill is comprised of two parts, i.e. a fixed charge and a variable charge. These are briefly introduced in the following:

- *fixed charge*—it is also known as "daily supply" or "service to property" charge. Once the customer is connected to a power distribution network, the customer pays a fixed price ($/day) regardless of the customer's electricity usage;
- *variable charge*—it is also known as "usage charge", and it requires a customer to pay for each unit of electricity consumed, usually listed as dollars/cents per kilowatt hour ($/kWh).

Examples of commonly adopted electricity tariffs can be grouped as shown below:

- *single rate*—the same rate for the electricity usage is applied at all times. This is also referred to as a static rate or a flat rate;
- Time-of-use (TOU)—different electricity usage rates apply to different periods in a day. Higher rates apply in high energy demand hours (e.g. noon and evening) and lower rates apply in low energy demand hours (e.g. morning and midnight);
- *block rate*—the electricity rate increases with the accumulated daily electricity consumption; and
- Real-time pricing (RTP)—the electricity rates vary from hour to hour or at even shorter intervals. When compared to other tariffs, RTP can better reflect the real-time electricity supply-demand balance that occurs in the power market, in which power generation companies and power distribution companies (or power retailers) trade electricity on an hourly or minute basis. Since RTP is a real-time tariff, it requires the installation of smart meters to enable the energy customer to receive the electricity

rate hour-by-hour. At present, the practice of RTP is still in the early stage. A representative case study of a RTP implementation consists of the residential RTP programs in Illinois (U.S.), reported in [10].

Apart from the above electricity tariffs that charge energy customers based on their energy consumption, there is another type of electricity tariff denoted as *feed-in tariff*. A feed-in tariff is a credit that an energy customer can receive for unused electricity that is reversely fed back to the grid (i.e. the power utility pays the customer for the electricity generated from the customer's site according to the feed-in tariff rate, e.g. in $/kWh). Feed-in tariffs are set for encouraging end energy customers to invest in and install renewable energy sources (e.g. PV solar panels and wind turbines) on their sites, so as to eventually improve the local energy supply in the customer side. For example, the feed-in tariff rate practiced by the EnergyAustralia (a power utility company) in New South Wales, Australia, is 7.6 Australian cents/kWh [11]. This means that every time a customer generates 1 kWh energy from his/her renewable energy source and sends it to the grid, the customer will be paid 7.6 cents by the utility company.

## 4.9 Application example

We now consider an application example to highlight how time-varying electricity tariffs can contribute to peak load reduction. Let us consider that there are three energy customers charged at the same TOU electricity tariff. The TOU tariff has the three-segment structure shown in Table 4.3, where it applies three different electricity prices to the peak-, shoulder-, and off-peak hours. The peak price is approximately five times larger than the off-peak price.

The typical power consumption profiles of the three customers are shown in Table 4.4. Being charged with the TOU tariff, the customers could choose to shift, for their financial benefit, a portion of their energy usage from peak and secondary peak hours to off-peak hours.

Let us assume that the changes in the power consumption of the customers in the selected hours are indicated by the values shown in brackets in Table 4.4. It can be seen that, for each customer, the total energy used over the day remains the same, and that the change only applies to the time at which the energy is consumed.

**Table 4.3** Overview of the three-segment TOU tariff structure.

| Time period | Electricity price ($/kWh) |
|---|---|
| Peak (6–8 pm) | 0.7 |
| Shoulder (1–3 pm, 8-10 pm) | 0.4 |
| Off-peak (other hours) | 0.15 |

**Table 4.4** Power consumption profiles of the three customers in the application example (values included in brackets account for the possible power consumption shifting).

| Time | Customer 1 (kW) | Customer 2 (kW) | Customer 3 (kW) |
|------|------|------|------|
| 0 am | 0.2 (2.1) | 0.4 (1.8) | 0.6 |
| 1 am | 0.2 (0.9) | 0.5 (1.6) | 0 |
| 2 am | 0.1 | 0.1 (1.1) | 0 |
| 3 am | 0.1 | 0 | 0 |
| 4 am | 0.1 | 0 | 0 |
| 5 am | 0.2 | 0 | 0 |
| 6 am | 0.2 | 0.3 | 0 |
| 7 am | 0.4 (1.2) | 0.6 | 0 |
| 8 am | 0.9 (1.3) | 1.0 | 1.5 |
| 9 am | 1.2 | 1.7 | 0.8 |
| 10 am | 1.4 | 1.2 | 2.3 |
| 11 am | 1.5 | 0.8 (1.1) | 2.5 |
| 12 pm | 1.8 | 0.5 (1.2) | 1.1 (2.2) |
| 1 pm | 2.5 (1.3) | 0.4 | 3.2 (2.1) |
| 2 pm | 1.6 (1.1) | 2.4d | 2.5 (2.3) |
| 3 pm | 1.3 | 3.3 (2.3) | 2.7 |
| 4 pm | 3.2 | 3.6 | 2.1 (2.3) |
| 5 pm | 0.7 (1.2) | 2.3 (2.7) | 1.7 (2.3) |
| 6 pm | 5.3 (4.1) | 1.7 (2.7) | 3.8 (3.2) |
| 7 pm | 4.8 (3.9) | 6.2 (4.1) | 4.1 (3.3) |
| 8 pm | 5.8 (5.1) | 6.0 (4.6) | 4.7 (3.9) |
| 9 pm | 6.7 (5.3) | 6.4 (4.4) | 3.2 (3.4) |
| 10 pm | 4.2 (4.5) | 5.3 (4.8) | 2.5 (3.3) |
| 11 pm | 1.3 (2.6) | 1.7 (2.8) | 0.7 (1.3) |
| Total energy | 45.7 kWh | 46.4 kWh | 40.0 kWh |

In this example, we can calculate the one-day electricity cost of each customer by performing the following summation of the energy cost in each hour:

$$C_i = \sum_{t=1}^{T}(E_{i,t} \cdot r_t) \tag{4.4}$$

$$E_{i,t} = P_{i,t} \cdot \Delta t \tag{4.5}$$

where the notations are defined as follows: $C_i$ is the energy cost of the $i$th customer (in $\$$) and, in the application example, $i \in \{1,2,3\}$. $T$ is the total number of time intervals in 1 day and, in the application example, $T = 24$. $\Delta t$ is the duration of each time interval (in hours) and, in the application example, $\Delta t = 1$ h, i.e. each time interval has a duration of 1 h. $r_t$ is the TOU rate in the $t$th time interval (in $\$$/kWh) and, in the application example, is provided in Table 4.3. $E_{i,t}$ is the energy consumed of the $i$th customer during the period of the $t$th time interval (in kWh), and it is calculated by multiplying the power

**Table 4.5** Summary of one-day electricity cost of the three customers by considering the base scenario (where no power consumption shifting is performed) and the revised scenario (that accounts for the possible power consumption shifting).

|  | Cost without power shifting | Cost with power shifting |
|---|---|---|
| Customer 1 | $38.0 | $17.4 |
| Customer 2 | $37.6 | $16.8 |
| Customer 3 | $32.5 | $15.2 |
| Total | $108.1 | $49.4 |

**Table 4.6** Overview of the peak load evaluation for the three customers.

|  | Peak load without power shifting | Peak load with power shifting |
|---|---|---|
| Customer 1 | 6.7 kW | 5.3 kW |
| Customer 2 | 6.4 kW | 4.8 kW |
| Customer 3 | 4.7 kW | 3.9 kW |
| 3-Customer system | 16.5 kW | 13.6 kW |

consumption with the time duration using Eq. 4.2. $P_{i,t}$ is the energy consumed by the $i$th customer in $t$th time interval (in kW) and, in the application example, its value is determined based on Table 4.4.

The one-day electricity cost of the three customers can then be calculated by considering the base scenario (where no power consumption shifting is performed) and the revised scenario (that accounts for power consumption shifting) by applying Eqs. (4.4) and (4.5) to the data of Tables 4.3 and 4.4. A summary of these results is provided in Table 4.5.

The cost of each customer is significantly reduced when shifting the electricity usage based on the TOU tariffs and the consequent reductions in total power consumptions are summarized in Table 4.6. Fig. 4.11 illustrates the power consumption of this three-customer system by considering the base scenario (where no power consumption shifting is performed) and the revised scenario (that accounts for possible power consumption shifting). In particular, the total peak power consumption of the three-customer system significantly reduces from 16.5 to 13.6 kW, and this reduction is expected to be beneficial to the external power grid as well.

## 4.10 Comparison between incentive- and price-based DSM strategies

Price-based DSM has found wide applicability in several countries to support the implementation of dynamic electricity tariffs (such as TOU). Incentive-based DSM (or DLC) strategies have found less popularity in the applied energy system domain. Despite this,

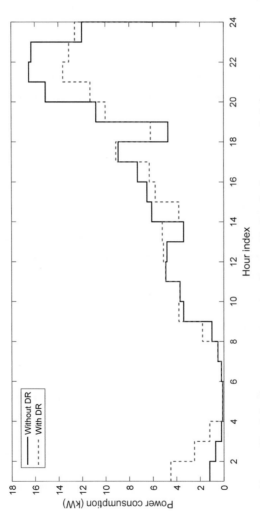

**Fig. 4.11** Power consumption profile of the three-customer system by considering the base scenario (where no power consumption shifting is performed) and the revised scenario (that accounts for possible power consumption shifting).

the main advantage of DLC is that it can produce reliable load reduction effect in a defined time period since the customer's appliances are directly controlled by the utility company. With this approach, the utility can perform control actions based on the operational requirements and considerations to ensure the energy demand is effectively reduced or shifted. The major limitation of DLC is that it is intrusive and could be disturbing to the customer. This happened, for example, in the PG&E's SmartAC project where many customers complained that their air conditioners were "rudely" shut down by the company and that this significantly affected their comfort [12]. In addition, the implementations of remote control in DLC programs are technically challenging.

Compared to DLC, price-based DSM is not intrusive for the customer, and it can be easily implemented. The utility company sets up dynamic electricity prices and uses them to encourage the customers to actively re-shape their power consumption profile. The limitation of price-based DSM is that its load re-shaping effect depends on the customer's subjective willingness to engage with it and can lead to uncertain outcomes.

## References

[1] Y. Hertig, S. Teufel, Prosumer involvement in smart grids: The relevance of energy prosumer behavior, in: Proceedings of 35th International Conference on Organizational Science Development, Portoroz, Slovenia, 2016, March.

[2] D.P. Kothari, I.J. Nagrath, Power System Engineering, Tata McGraw-Hill, NY, USA, 1994.

[3] Energy Independence and Security Act, U.S. Government Printing Office, Washington DC, USA, 2007, December 19.

[4] European Smart Grid Task Force Report, EU Commission, Brussels, Belgium, 2010, June.

[5] C. Harris, Electricity Markets: Pricing, Structures, and Economics, John Wiley & Sons, Hoboken, NJ, USA, 2006.

[6] Advanced Metering Infrastructure and Customer Systems, U.S. Department of Energy, Washington DC, USA, 2016, September. (Online). Available from: https://www.energy.gov/sites/prod/files/2016/12/f34/AMI%20Summary%20Report_09-26-16.pdf. (accessed 10 July 2023).

[7] P. Palensky, D. Dietrich, Demand side management: demand response, intelligent energy systems, and smart loads, IEEE Trans. Ind. Inform. 7 (3) (2011) 381–388.

[8] H. Zhong, P. Du, N. Lu, Demand Response in Smart Grids, Springer International Publishing, NY, USA, 2019.

[9] D. Crossley, International Experience in Using DSM to Support Electricity Grids (PowerPoint Slides), 2008, Available from: https://userstcp.org/wp-content/uploads/2019/12/1.Crossley_Workshop_Task15.pdf. (accessed 10 July 2023).

[10] "Residential real-time pricing program." Pluginillinois.org. https://www.pluginillinois.org/realtime.aspx (accessed 08 March 2022).

[11] Feed-in tariffs from EnergyAustralia (Online). Available from: https://www.energyaustralia.com.au/home/solar/feed-in-tariffs. (accessed 12 January 2023).

[12] D.R. Baker, PG&E's SmartAC Program Malfunctions. Sfgate.com. https://www.sfgate.com/business/article/PG-amp-E-s-SmartAC-program-malfunctions-4673820.php. (accessed 08 March 2022).

# CHAPTER 5

# Building energy management systems

## Contents

## 5.1 Introduction

Recent advances in building technologies have increased the level of sophistication that is required for their effective management and that relies on the support of dedicated control and monitoring solutions. The wider deployment of on-site renewable energy sources has contributed to this increased complexity in effectively managing the building's energy production and consumption. Over the last decade, we have observed an increased interest of building owners, building managers, and occupants in establishing energy-efficient strategies to reduce the overall energy consumption and cost.

In this chapter, we introduce the key features and the structure of a building energy management system (BEMS) and discuss how it is usually integrated within the building operations. The first part of the chapter gives an overview of BEMS. This is followed by a part in which attention is provided at describing how a BEMS interacts with energy resources deployed in a building (usually referred to as *building-side*) and with renewable energy resources, such as wind turbines or photovoltaic (PV) solar panels. The most common categories of building-site energy resources that can be managed by BEMSs are

*Building Energy Management Systems and Techniques*
https://doi.org/10.1016/B978-0-323-96107-3.00005-9

outlined to highlight their key characteristics. In the last part of the chapter, a taxonomy of the main functional features of BEMSs is provided, and most of these functional features are discussed in detail in later chapters.

## 5.2 Overview of a building energy management system and its operations

### 5.2.1 Overview

A BEMS represents the overarching framework that can be deployed at building level to monitor and control the energy needs of a building. For clarity, it is worth distinguishing this definition from the concept of a building management system (BMS) [1]. A BMS is a central control system that is responsible for the automatic regulation and control of equipment in a building. It links the functionality of individual pieces of equipment in the building so that they operate as one complete integrated system.

A widely accepted definition of a BEMS has been provided by the International Energy Agency (IEA) in 1997 [2], and it regards a BEMS as "an electrical control and monitoring system that has the ability to communicate data between control nodes (monitoring points) and an operator terminal. The system can have attributes from all facets of building control and management functions such as Heating, Ventilation and Air Conditioning (HVAC), lighting, fire, security, maintenance management and energy management." This definition is still valid even if the information and communication technologies supporting a BEMS have significantly enhanced over the last decades. Also, the number of sensors, equipment, and energy resources monitored and controlled by a typical BEMS in a building has increased. The stakeholders that deal with a BEMS include building occupants, building owners, building managers, and energy utility companies. From the viewpoint of a building occupant, the three main functions that are performed by a BEMS are to provide a comfortable environment and indoor climate, to ensure safety and security, and to support energy-efficient operations of building activities. Such a wide range of tasks can be achieved because a BEMS is able: (i) to monitor the status of the building operations and environmental conditions; (ii) to perform optimization operations for the scheduling of building and plant operations; (iii) to access data on energy management information; and (iv) to monitor and control energy resources and appliances, for example, by defining and deployment control schedules for the automatic switching on and off of appliances.

There has been a growing acceptability that a building managed by a BEMS can be referred to as a *smart building* because of the intelligence and smart features that it enables to support and perform. In reality, there are different levels of intelligent sophistication employed in a BEMS, and this depends on the building typology and requirements defined by the relevant stakeholders. For example, it is possible for a BEMS to be deployed to perform simple control operations on the building energy resources to ensure that, in parts of the buildings where no occupants are present, the lights and the HVAC systems are switched off to reduce their associated energy consumption. In such a simple logic implementation, the BEMS would not exploit possible advantages in the preheating or precooling of building spaces that could reduce later energy consumption needs near peak hours. Such a simple BEMS installation can certainly lead to significant operational energy savings but would not exploit the potentials of having a BEMS installed in a building with monitoring and control capabilities on a wide range of energy resources. In this context, it is desirable to further exploit the use of BEMSs by enhancing their intelligence in supporting the building operations and the well-being and comfort of building occupants while minimizing the overall energy consumption. The different level of sophistication has also historical implications as, over the last decades, BEMSs have developed from simple digital systems that perform simple monitoring functions to expert systems that can perform advanced energy optimizations for building operations. These initial considerations highlight how implementations of BEMSs in buildings can vary depending on the desired energy management objectives and on the types of connected energy resources.

In recent years, there has been a wider acceptance and deployment of on-site renewable energy resources to reduce the energy consumptions of buildings. Such strategies have also been embraced in the implementation of net-zero energy building strategies. Considering the stochastic and intermittent nature of on-site renewable energy, there is a need to optimize their utilization, and this imposes new challenges in their management. The functionalities of a BEMS are well suited to handle this task to enable the overall building operations to be optimized for the available renewable energy produced on-site. These BEMSs' capabilities are also usually relied upon when dealing with the external macro energy systems that a building is connected to and that include power grids, gas networks, and heat supply systems. These external energy systems supply power from remote plants to serve the energy demand of buildings through energy transmission and delivery infrastructures.

An example of a BEMS is illustrated in Fig. 5.1 by considering the case of a residential house. By relying on a wireless communication network, the BEMS monitors the operation status of a variety of energy resources in the building and takes actions to optimize their operations. The BEMS is also able to interact with the external grid through the Advanced Metering Infrastructure (AMI). The decision making of a BEMS relies on the information collected from the building environment and, based on this, the BEMS is then able to manage the building's energy production and consumption by applying sophisticated energy optimization techniques that could focus at the operations of one building or multiple building systems.

### 5.2.2 Roles of BEMSs

Depending on their operational tasks and managed equipment, a BEMS can take one or multiple of the roles listed below.

**(1)** *"Brain" of a building*—By embedding advanced algorithms and control logics into BEMSs, BEMSs can provide intelligence to buildings, by making buildings smarter

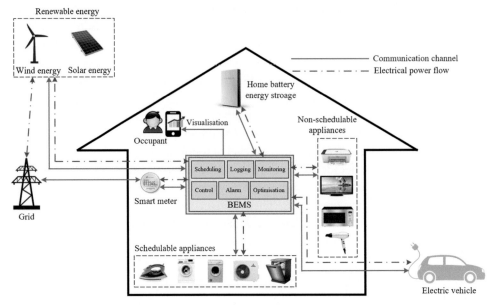

**Fig. 5.1** Example of a building energy management system deployed in a residential building.

in responding to the environment and by improving their operations. Supported by the information infrastructure and data collection and management logics, BEMSs can promptly perceive their real-time operational condition, reduce the wasted energy consumption, and optimize energy use of the building. For example, a BEMS can automatically identify the occupancy state of a room and turn off the light when there are no people in the room.

**(2)** *Assistant of the occupant*—The primary objective of BEMSs is to assist occupants by enhancing their indoor comfort and satisfaction and by reducing their efforts in decision-making processes. A BEMS acts like an intelligent housekeeper. For example, it understands the occupant's comfort and satisfaction requirements in different aspects (e.g., thermal comfort, vision comfort, and appliance usage satisfaction) that could be accommodated by means of strategies ranging from the setting up of simple operational rules for the operations of the energy resources to performing complex occupant behavior and lifestyle analyses. A BEMS can also assist the occupant to deal with the time-varying electricity tariffs. It is challenging for building occupants to evaluate the available time-varying electricity tariffs and to define an advantageous response on their side. An occupant would not be interested to navigate through the details of a time-varying electricity tariff and/or would be reluctant to adjust the appliance usage behaviors to respond to it. The difficulty of this task increases with the consideration that time-varying electricity tariffs vary over the duration of a day and can be updated on a regular basis. For example, if a time-varying electricity tariff requires different energy prices every few hours, it would be difficult for the occupant to manually decide the appliances' operation schedules subjected to such frequently updated prices. In this case, BEMSs can take over the work of automatically arranging the operation of building-side energy resources so as to minimize the energy cost charged by the time-varying energy tariff while preserving the occupant's comfort and satisfaction.

**(3)** *Facilitator between the occupant and the grid*—Modern buildings are able to have active, two-way communications with the smart grid. In this context, BEMSs act as intermediate between the occupant and the grid, e.g., a BEMS arranges the operation of the building-side energy resources with awareness of the occupant's life requirements, or it can automatically assist the grid to improve the grid's operations. For example, it is possible to deploy DSM approaches that aim at re-shaping end customers' electricity consumption profiles that could have significant effects in reducing peak energy demands at the grid level or regulating the voltage at the feeder in which the building connects to the grid. The use of a BEMS to assist the grid operations at a building level is necessary to deal with the complexity of the grid operation requirements and to assist power utility companies to monitor and control each building's operation conditions in a fine-grained manner.

### 5.2.3 Benefits of BEMSs

The benefits of BEMSs can be summarized in the following points:

- *Reduced energy cost*—BEMSs maintains a balance between energy use and the building's operating requirements. BEMSs can help to improve the utilization level of the building's on-site renewable energy and to optimize the building's energy use subjected to dynamic energy tariffs. These will lead to the reduced energy bill to the occupant and/ or the building manager.
- *Better energy conservation*—By intelligently managing and controlling building-side energy resources, BEMSs can optimize the building's energy use and reduce unnecessary energy consumption, and this will enhance the building's ability in conserving energy.
- *Reduced human efforts in decision-making processes and enhanced comfort and satisfaction*—As a kind of expert system, BEMSs can automatically operate energy resources in buildings for the occupant and building manager, and this can lead to saving in their time and efforts. By intelligently regulating the indoor climate and appliance operation, BEMSs can also enhance the occupant's comfort and satisfaction in the building. These can help improving the occupant's productivity and quality of life.
- *Enhanced demand response capability of buildings*—BEMSs can provide computer-aided assistance to the grid in setting up demand response programs. With the help of BEMSs, smart grids can better manage the regional peak power consumption and other grid operational factors (e.g., reverse power flows, nodal voltages, and grid frequency), which will improve the grid's security, reliability, and energy efficiency.

- *Reduced building maintenance cost*—BEMSs can enable evaluation of energy equipment's functionality and early identification of equipment failure. For example, a BEMS can diagnose the mechanical failure of HVAC components and send alarms to the building manager when a component it detected to be close to failure. In this way, BEMSs can support the building manager to promote preventive maintenance practices.

## 5.3 Typical energy resources managed by BEMSs

Building-side energy resources refer to the physical devices deployed in buildings that either generate or consume energy. It is useful to group the most common building-side energy resources considered and connected to BEMSs into the categories of power generation sources, energy storage systems, HVACs, and electric appliances. These are briefly introduced below.

### 5.3.1 Power generation sources

Power generation resources that can be managed by BEMSs mainly consist of distributed renewable energy sources and small-capacity distributed fossil fueled generators that are deployed at the building's site to generate energy for building use. Renewable energy resources are used to serve the daily energy demand of a building and to reduce its dependency on the external power grid supply. Among the most common renewable energy sources, there are wind turbines, PV solar panels, and solar thermal collectors that are typically installed on buildings' envelopes, e.g., roofs and facades.

Distributed fossil fueled generators are usually used as generation backups in critical commercial, industrial, and public buildings (e.g., hospitals and factories) to provide power supply support. They consist mainly of fuel-cell generators and of combined heat and power (CHP) generators. Fuel-cell technologies convert chemical energy into electricity by an electrochemical reaction involving hydrogen and oxygen [3], and the CHP technology generates the concurrent production of electricity and useful thermal energy (heating and/or cooling) from a single fuel source [4].

### 5.3.2 Energy storage systems

Energy storage systems are used to store energy for buildings' use. The most commonly used energy storage system consists of Battery Energy Storage Systems (BESSs). When used in combination with renewable energy sources deployed in a building, BESSs can be charged in time periods when there is an excessive renewable power output and can be discharged to serve the building's energy demand in time periods when there is no renewable power output, i.e., where there is not adequate wind (for wind turbines) or sunlight (for PV solar panels).

In recent years, other energy storage devices, such as electric vehicles (EVs), have been gaining popularity and have found application in building energy management strategies. A typical EV is equipped with a battery that provides electrical power to drive the electric motor of the vehicle for propulsion. When an EV is plugged into a charging slot of a building, it can be regarded as a BESS that can be used by the building. An EV can absorb electricity from the building's on-site energy sources and can provide electricity to the building through charging and discharging actions, respectively. Such an integrated arrangement (between a building and an EV) is usually referred to as *Vehicle-to-Building* (V2B) integration [5]. V2B strategies are gaining importance as a consequence of the wider spread of EVs sold worldwide as highlighted in Fig. 5.2 with the number of electric car registrations in different parts of the world between 2015 and 2020 [6]. The home-used EVs in the current market can be generally categorized into two classes: battery electric vehicles (BEVs) and plug-in hybrid electric vehicles (PHEVs). BEVs rely completely

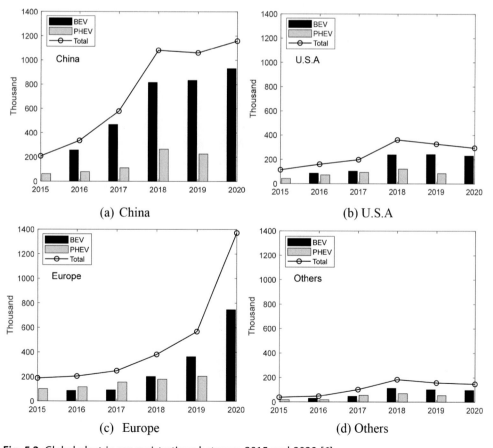

**Fig. 5.2** Global electric car registrations between 2015 and 2020 [6].

on electricity provided by the self-contained battery to operate. Depending on the model, a fully charged BEV is typically able to drive in the range of tens to hundreds of kilometers. A BEV can be charged with two modes that consist of a fast charging and a slow charging mode. The fast charging operation can complete a charging cycle in a short period that can range between tens of minutes and an hour, while slow charging can take up to nearly a full day. The slower the charging rate is the lesser the battery is degraded during the charging process. A PHEV is powered by both a battery and an internal gasoline combustion engine. For the first tens of miles, battery is used to drive the PHEV. After the battery is depleted, the vehicle switches to use the internal combustion engine. The flexibility of PHEVs allows drivers to use electricity as often as possible while taking advantage of the gasoline option when needed.

### 5.3.3 HVAC systems

HVAC systems are an important part of buildings. They are responsible for controlling the temperature, humidity, and air quality in the indoor environment. HVAC systems are large energy consumption sources in buildings. It is estimated that a typical office HVAC system accounts for approximately 40% of the building's total energy consumption [7]. HVACs also dominate buildings' peak power demand in hot summer days.

Due to their significant role in building energy consumption, HVAC systems are important components managed by BEMSs. This usually involves the adoption of certain energy management strategies to reduce HVAC's operation in specific time periods (typically peak demand hours with high electricity tariffs) and to shift its operation to other lower demand time periods. In this process, the indoor thermal comfort requirement of the occupant needs to be adequately preserved. This can be achieved with an effective energy management of the HVAC systems that account for the indoor temperature variations and, when available, for the occupants' comfort satisfaction feedback.

### 5.3.4 Electric appliances

Electric appliances can be regarded as equipment and devices that act as energy consumers in buildings. Modern electric appliances can be equipped with wireless communication interfaces and actuators that enable machine-to-machine communications between electric appliances and with the BEMS. A BEMS can monitor an appliance's operational state and send control signals (e.g., on/off switch commands and parameter settings) to manage its operation.

Electrical appliances can possess different levels of operational flexibility. For example, some appliances can be operated at flexible times, such as washing machines and dish washers. This means that these appliances' operation time can be shifted to a certain extent even if their power consumption remains usually unvaried during operation. Some appliances (such as lighting systems and amplifiers) possess an adjustable operating power and their operating times are determined by the occupant and can hardly be shifted. Some appliances could have the flexibility to vary both their operating time and power consumption.

It should be noted that some appliances are critical for an occupant's basic lifestyle. These are expected to be directly controlled (i.e., manually) by the occupant and are usually not controlled by a BEMS. Examples of these appliances include televisions and computers. The abrupt interruption of the operation of these appliances could cause unacceptable disturbances to an occupant.

## 5.4 Main functional features of BEMSs

BEMSs can be designed and programmed to achieve a wide range of objectives. This section summarizes some of the representative functions of BEMSs, and the list does not want to be exhaustive considering the wide range of functions that are currently being developed for BEMSs.

### 5.4.1 Monitoring of building's power generation and consumption

Monitoring the building's power generation (typically from on-site renewable energy sources) and consumption is the most basic function of BEMSs. Such monitoring plays a fundamental role in building energy management as it enables the BEMS to understand the real-time energy performance of the building and to determine when and how to adjust the building's energy generation and consumption. The accumulation of the monitored data can be used to forecast the building's future power generation and consumption (see Sections 5.4.2 and 5.4.3 below), and it enables the BEMS to monitor the health condition of the energy resources and diagnose possible mechanical failures. The monitored data can also be used to perform advanced tasks, such as analyzing the occupant's behaviors and negotiating with other buildings on energy trading activities (see Section 5.4.5).

Power generation and consumption monitoring can be performed at different levels. Monitoring power generation and consumption of the entire building can be achieved by installing smart meters. Smart meters can measure the electricity at the circuit to which it connects and can calculate the power. They can then automatically send the power reading to the BEMS for analysis. Monitoring power consumption of individual pieces of electric equipment can be achieved in two ways. One way is to integrate submetering devices into the equipment or to install the devices at the power points the equipment connects, and the submetering devices can measure the power consumption of the equipment and automatically send the readings to the BEMS. The main limitation of this approach is the complexity and costs associated with the installation of the submetering devices in the individual pieces of equipment. A second approach relies on Non-intrusive Load Monitoring (NILM, also known as the Non-Intrusive Appliance Load Monitoring (NIALM)) technology that disaggregates a profile of the total power consumption (or voltage or current) of a group of appliances into the power/voltage/current profiles of the individual appliances. NILM is considered to be cost-effective as it does not require the installation of submetering devices in individual appliances. Despite this, the disaggregation accuracy is a major challenge.

## 5.4.2 Forecasting of building's power load

The primary responsibility of a BEMS is to manage the building's energy consumption in an energy-efficient manner. To achieve this, a BEMS often needs to forecast the building's power consumption over a finite future time horizon (e.g., for the subsequent day), so that it can schedule the controllable energy resources to: (1) effectively meet the building's power demand, or (2) reshape the building's power consumption profile to a more desirable pattern (e.g., to reduce the building's peak power consumption). Forecasting building's power load is not only important for energy management of a single building, but it is also important for the grid. In power systems' operations, power load forecasting is a fundamental task, which enables the grid operator to understand the grid's power demand, so that the power generation units can be scheduled ex-ante to meet the demand. A BEMS can send the power load prediction it makes for the building to the grid operator through the AMI. By collecting the power load forecasting results from multiple BEMSs and other energy management systems, the grid operator can estimate the future power load of a certain geographical region consisting of multiple buildings and other energy loads.

A building's power demand depends on the behavior of building occupants, the physical characteristics of the building, the climatic conditions, and it varies over time. Typically, the lifestyle and preferences of occupants influence the operation time and energy consumptions of the building appliances and equipment. For example, the energy consumption of a HVAC system largely depends on the occupant's lifestyle and indoor thermal comfort requirements, the thermal and optical characteristics of the building envelope, and the climatic conditions (e.g., outdoor air temperature and solar radiation). It is common for building power load forecasting to rely on analyses of historical power load data of the building, and this is usually performed by using statistical analysis-based methods [8,9] and machine learning-based methods [10,11]. Statistical analysis-based methods analyze the patterns from the time series data of historical power load readings and generate the future power load values. Machine learning-based methods use historical power load readings to train machine learning models, and the models are then used to generate power load predictions. We introduce basic power load forecasting methods in Chapter 6.

## 5.4.3 Forecasting of on-site renewable power output

Another important task of BEMSs consists of the forecasting of the power output of the building's on-site renewable energy sources. Renewable energy is stochastic in nature. A BEMS often needs to predict the power output of the building's on-site renewable energy sources. Based on this, it specifies operation plans for the energy resources in the building to better utilize the renewable energy and to reduce the amount of energy that the building needs to import from the grid. Forecasting of the renewable power output is not discussed in detail in this book, and some commonly practiced forecasting approaches can be found in dedicated references that are briefly introduced in the following.

The timescale of renewable energy forecasting usually ranges from minutes to weeks. For wind power, its forecasting is usually achieved by forecasting the wind speed first and then converting it to wind power based on a wind turbine's power generation model (as introduced, for example, in Chapter 2). Model-driven forecasting techniques have been successfully applied for wind power forecasting and several of these techniques (e.g., [12–14]) rely on Numerical Weather Prediction (NWP) methods, e.g., [15].

If the forecasting horizon is not too long (i.e., it ranges from few minutes to few hours), the forecasting of wind speed and power can usually be achieved by using time series analysis methods without resorting to NWP models. Time series analysis is a specific way of analyzing a sequence of data points collected with regular time intervals with the aim of extracting meaning statistics and other characteristics of the data. In time series analysis, data points are collected at regular intervals over a certain period of time rather than just being recorded intermittently or randomly. Typical time series analysis methods for wind speed and power forecasting include autoregressive moving average (ARMA) models [16,17], Kalman filtering approaches [18,19], and artificial neural networks (ANNs) [20,21]. With the recent advances of machine learning technologies, deep learning models [22,23] have also been used for wind power forecasting. Hybrid wind power prediction techniques that combine NWP models and time series analysis methods have also been proposed (e.g., [24–27]). For example, data-driven methods have been widely used for such forecasting tasks. These methods use machine learning models (such as artificial neural networks and linear regression models) to analyze the pattern of historically recorded time series data and to generate predictions.

NWP models and time series analysis methods have also been actively applied to solar irradiation and solar energy (including PV solar power and solar thermal energy) forecasting, e.g., [28–34]. It has been recognized that solar resource and solar energy forecasting in very short spatial and time scales (e.g., 0–100 m and 0–30 min, respectively) is still a challenging task, e.g., [35]. For this reason, sky images from ground-based cameras [36] have been widely used in recent years to enhance the forecasting accuracy of solar resource and solar energy. Such images contain information of high spatial and temporal variability of clouds and hence can provide the needed inputs of solar irradiance for numerical models, e.g., [37–39].

## 5.4.4 Scheduling of building-side energy resources

The most critical function of BEMSs consists in the development of operational plans for building-side energy resources (such as those introduced in Section 5.3). By relying on wireless communication and automatic control facilities, BEMSs can send commands to schedule the operations of energy resources.

Energy resource scheduling can be performed with different strategies, ranging from simple rules to sophisticated, algorithmic designs. The objectives of the energy resource scheduling could also vary. For example, the most common objective of energy resource scheduling is to minimize the building's energy cost charged by dynamic energy tariffs [40]. Other typical objectives focus at managing the building's peak power consumption, assisting the grid to regulate the grid's operation [41], and minimizing the disturbance of power outage events to the building [42]. The scheduling strategy also needs to be specifically designed based on the physical model characteristics of the different types of the available energy resources.

Since a building can often be conceptualized as an occupant-centric environment, almost in all of energy resource scheduling strategies, the satisfaction of the occupant's comfort is considered as a primary requirement. Based on this, energy resources are then scheduled to achieve different building energy management objectives. We introduce scheduling strategies for different types of energy resources in Chapters 8, 9, and 10.

### 5.4.5 Sharing energy with other energy entities

Since the beginning of this century, along with the development of smart grids, small renewable energy sources have been increasingly deployed in buildings, by assigning buildings the ability of generating energy. With the integration of renewable energy sources, the energy produced by a building during some time intervals can be larger than its energy demand. For example, at mid-day and under large solar radiation density, a rooftop solar panel may generate solar energy that is larger than the building's energy demand at that time.

In recent years, people have been investigating the mechanism of enabling the surplus energy generated by an energy entity (e.g., a building) to be shared with other energy entities. Typically, this is achieved through the peer-to-peer (P2P) energy trading paradigm [43,44]. With P2P energy trading, for example, the surplus energy generated by the rooftop PV solar panel of a house can be sold to its neighboring houses. In this context, the energy buyer and seller need to negotiate the energy trading amount and price, and this process often needs to be automated by energy management systems of both parties—for buildings, this is performed by BEMSs. To perform P2P energy sharing and trading, a BEMS needs to monitor the building's energy demand and production, understand the occupant's requirements in energy trading (e.g., with what price the occupant would be happy to sell energy to or buy energy from other buildings), and communicate with other energy management systems to determine energy trading transactions. P2P energy trading mechanisms are dealt with in Chapter 11.

## 5.4.6 Human-machine communication

A building is a human-centric environment. BEMSs need to communicate with different stakeholders (e.g., building occupant, the building manager, and/or the building owner) and cowork with them to manage the building's energy performance. Such a human-machine communication can be facilitated by a variety of devices, such as light-emitting diode (LED) screens, personal computers, and smartphones. BEMSs can send different messages to people textually or visually to share the energy performance and operational condition of the buildings. For example, a BEMS can send alarms to the building manager when it detects a mechanical failure risk for an energy resource in the building, or it can also report the energy consumption and energy bill to the occupant visually with charts and tables. The occupants can also input their requirements on the energy resources' operations into the BEMS. Based on the inputted information, the BEMS can schedule the energy resources' operations to better preserve the occupants' comfort and satisfaction. Such a BEMS-occupant communication implicitly underpins the energy management strategy designs introduced in Chapters 8, 9, and 10.

BEMSs can also provide more customized energy-related services to building occupants when enable to learn their lifestyles and habits. For example, a BEMS can intelligently recommend energy-efficient products and energy usage plans that fit the occupant's preferences and lifestyles of the occupants. By adopting the energy-efficient recommendations, the buildings' energy performance can be improved, and to a larger extent, this can also have a beneficial effect on the grid. Such a personalized energy service recommendation scenario is referred to as the "occupant-to-grid integration" in this book and it is dealt with in Chapter 14.

## References

[1] S. Wang, J. Xie, Integrating building management system and facilities management on the Internet, Autom. Constr. 11 (6) (2002) 707–715.
[2] L.G. Mansson, D. McIntyre, A Summary of Annexes 16&17 Building Energy Management Systems, International Energy Agency, Paris, France, 1997.
[3] O.Z. Sharaf, M.F. Orhan, An overview of fuel cell technology: fundamentals and applications, Renew. Sust. Energ. Rev. 32 (2014) 810–853.
[4] M.M. Maghanki, B. Ghobadian, G. Najafi, R.J. Galogah, Micro combined heat and power (MCHP) technologies and applications, Renew. Sust. Energ. Rev. 28 (2013) 510–524.
[5] G. Barone, A. Buonomano, F. Calise, C. Forzano, A. Palombo, Building to vehicle to building concept toward a novel zero energy paradigm: modelling and case studies, Renew. Sust. Energ. Rev. 101 (2019) 625–648.
[6] "Global car registrations and market share 2015–2020." Iea.org https://www.iea.org/data-and-statistics/charts/global-electric-car-registrations-and-market-share-2015-2020 (Accessed 08 March 2022).
[7] HVAC, Factsheet, Department of the Environment and Energy, Australian Government, 2013.
[8] G. Juberias, R. Yunta, J.G. Moreno, C. Mendivil, A new ARIMA model for hourly load forecasting, in: Proc. 1999 IEEE Transmission and Distribution Conference, New Orleans, LA, USA, 1999, April.
[9] H.M. Al-harnadi, S.A. Soliman, Short-term electric load forecasting based on Kalman filtering algorithm with moving window weather and load model, Electr. Power Syst. Res. 68 (1) (2004) 47–59.

[10] W. Kong, Z.Y. Dong, D.J. Hill, F. Luo, Y. Xu, Short-term residential load forecasting based on resident behaviour learning, IEEE Trans. Power Syst. 33 (1) (2017) 1087–1088.

[11] D.L. Marino, K. Amarasinghe, M. Manic, Building energy load forecasting using deep neural networks, in: Proc. of 42nd Annual Conference of the IEEE Industrial Electronics Society, 2016, October.

[12] B. Bochenek, J. Jurasz, A. Jaczewski, G. Stachura, P. Sekula, T. Strzyzewski, M. Wdowikowski, M. Figurski, Day-ahead wind power forecasting in Poland based on numerical weather prediction, Energies 14 (8) (2021) 1–18.

[13] A. Bossavy, R. Girard, G. Kariniotakis, Forecasting ramps of wind power production with numerical weather prediction ensembles, Wind Energy 16 (1) (2013) 51–63.

[14] Y. Liu, Y. Wang, L. Li, S. Han, D. Infield, Numerical weather prediction wind correction methods and its impact on computational fluid dynamics based wind power forecasting, J. Renew. Sustain. Energy 8 (2016).

[15] A.C. Lorenc, Analysis methods for numerical weather prediction, Q. J. R. Meteorol. Soc. 112 (474) (1986) 1177–1194.

[16] S. Rajagopalan, S. Santoso, Wind power forecasting and error analysis using the autoregressive moving average modelling, in: Proc. 2009 IEEE Power & Energy Society General Meeting, Calgary, Canada, 2009, July.

[17] J. Wang, Q. Zhou, X. Zhang, Wind power forecasting based on time series ARMA model, in: Proc. of IOP Conference Series: Earth and Environmental Science, vol. 199 (2), 2018, pp. 1–6.

[18] P. Louka, G. Galanis, N. Siebert, G. Kariniotakis, P. Katsafados, I. Pytharoulis, G. Kallos, Improvements in wind speed forecasts for wind power prediction purposes using Kalman filtering, J. Wind Eng. Ind. Aerodyn. 96 (12) (2008) 2348–2362.

[19] C.D. Zuluaga, M.A. Alvarez, E. Giraldo, Short-term wind speed prediction based on robust Kalman filtering: an experimental comparison, Appl. Energy 156 (2015) 321–330.

[20] G.N. Kariniotakis, G.S. Stavrakakis, E.F. Nogaret, Wind power forecasting using advanced neural networks models, IEEE Trans. Energy Convers. 11 (4) (1996) 762–767.

[21] B. Huang, Y. Liang, X. Qiu, Wind power forecasting using attention-based recurrent neural networks: a comparative study, IEEE Access 9 (2021) 40432–40444.

[22] Y. Wang, R. Zou, F. Liu, L. Zhang, Q. Liu, A review of wind speed and wind power forecasting with deep neural networks, Appl. Energy 304 (2021) 1–24.

[23] Y.Y. Hong, C.L. Paulo, P. Rioflorido, A hybrid deep learning-based neural network for 24-h ahead wind power forecasting, Appl. Energy 250 (2019) 530–539.

[24] Q. Xu, D. He, N. Zhang, C. Kang, Q. Xia, J. Bai, J. Huang, A short-term wind power forecasting approach with adjustment of numerical weather prediction input by data mining, IEEE Trans. Sustain. Energy 6 (4) (2015) 1283–1291.

[25] K. Higashiyama, Y. Fujimoto, Y. Hayashi, Feature extraction of NWP data for wind power forecasting using 3D-convolutional neural networks, Energy Procedia 155 (2018) 350–358.

[26] G. Qu, J. Mei, D. He, Short-term wind power forecasting based on numerical weather prediction adjustment, in: Proc. of 11th IEEE International Conference on Industrial Informatics (INDIN), Bochum, Germany, 2013, July.

[27] F. Cassola, M. Burlando, Wind speed and wind energy forecast through Kalman filtering of numerical weather prediction model output, Appl. Energy 99 (2012) 154–166.

[28] V. Sharma, U. Cali, V. Hagenmeyer, R. Mikut, J.A.G. Ordiano, Numerical weather prediction data free solar power forecasting with neural networks, in: Proc. of 9th International Conference on Future Energy Systems, 2018, pp. 604–609.

[29] R.A. Verzijlbergh, P.W. Heijnen, S.R. de Roode, A. Los, H.J.J. Jonker, Improved model output statistics of numerical weather prediction based irradiance forecasts for solar power applications, Sol. Energy 118 (2015) 634–645.

[30] B. Singh, D. Pozo, A guide to solar power forecasting using ARMA models, in: Proc. of 2019 IEEE PES Innovative Smart Grid Technologies Europe (ISGT-Europe), 2019, October.

[31] S. Suksamosorn, N. Hoochareon, J. Songsiri, Post-processing of NWP forecasts using Kalman filtering with operational constraints for day-ahead solar power forecasting in Thailand, IEEE Access 9 (2021) 105409–105423.

[32] E. Lzgi, A. Oztopal, B. Yerli, M.K. Keymak, A.D. Sahin, Short-mid-term solar power prediction by using artificial neural networks, Sol. Energy 86 (2) (2012) 725–733.

[33] M. Rana, I. Koprinska, V.G. Agelidis, Forecasting solar power generated by grid connected PV systems using ensembles of neural networks, in: Proc. of 2015 International Joint Conference on Neural Networks (IJCNN), 2015, July.

[34] A. Gensler, J. Henze, B. Sick, N. Raabe, Deep learning for solar power forecasting – an approach using AutoEncoder and LSTM neural networks, in: Proc. of 2016 IEEE International Conference on Systems, Man, and Cybernetics (SMC), 2016, October.

[35] G. Kariniotakis (Ed.), Renewable Energy Forecasting, Elsevier, Amsterdam, Netherland, 2017.

[36] S. Dev, F.M. Savoy, Y.H. Lee, S. Winkler, Cloud Imaging Using Ground-Based Sky Cameras (Online), Available from: https://soumyabratadev.files.wordpress.com/2016/08/adsc-phd-day.pdf. (Accessed 11 July 2023).

[37] B. Urquhart, B. Kurtz, E. Dahlin, M. Ghonima, J.E. Shields, J. Kleissl, Development of a sky imaging system for short-term solar power forecasting, Atmos. Measur. Techn. 8 (2) (2015) 875–890.

[38] Z. Zhen, J. Liu, Z. Zhang, F. Wang, H. Chai, Y. Yu, X. Lu, T. Wang, Y. Lin, Deep learning based surface irradiance mapping model for solar PV power forecasting using sky image, IEEE Trans. Ind. Appl. 56 (4) (2020) 3385–3396.

[39] Z. Zhen, Z. Wang, F. Wang, Z. Mi, K. Li, Research on a cloud image forecasting approach for solar power forecasting, Energy Procedia 142 (2017) 362–368.

[40] F. Luo, G. Ranzi, C. Wan, Z.Y. Dong, Z. Xu, A multi-stage home energy management system with residential photovoltaic penetration, IEEE Trans. Ind. Inform. 15 (1) (2019) 116–126.

[41] C. Vivekananthan, Y. Mishra, G. Ledwich, F. Li, Demand response for residential appliances via customer reward scheme, IEEE Trans. Smart Grid 5 (2) (2014) 809–820.

[42] F. Zhang, F. Luo, Y. Liu, G. Ranzi, Z.Y. Dong, Hierarchically resilient energy management scheme for residential communities under grid outages, IET Smart Grid 3 (2) (2020) 174–181.

[43] Y. Zhou, J. Wu, C. Long, W. Ming, State-of-the-art analysis and perspectives for peer-to-peer energy trading, Engineering 6 (7) (2020) 739–753.

[44] M.R. Alam, M. St-Hilaire, T. Kunz, Peer-to-peer energy trading among smart homes, Appl. Energy 238 (2019) 1434–1443.

# CHAPTER 6

# Building power load forecasting

## Contents

## 6.1 Introduction

Performing predictive analysis for energy consumption of buildings refers to forecasting the amount and/or patterns of buildings' future energy consumption based on the buildings' historical energy consumption data and other factors that could influence the buildings' energy consumption (e.g., meteorological data and building structure data). Predictive analyses can be useful for different purposes, for example, to understand the future energy demand of the buildings in a precinct so that urban planners and designers can better plan the construction of the district heating and power generation infrastructures. From an operational viewpoint, building energy management plans can be defined based on the forecasted energy demand of a building.

Predictive analysis for the energy consumption of a building can be carried out at different time scales. Long-term analysis could be estimating the monthly or annual energy consumption amount of a building. Short-term analysis involves the forecasting of the power consumption of the building over a short finite future horizon usually with a duration from several hours to 1 day. Short-term building power load forecasting is important for building energy management because the latter defines operation plans for energy resources in the building. For this reason, this chapter focuses on short-term building power load forecasting, and methodologies to perform long-term forecasting can be found in dedicated references.

Power load forecasting also referred to as simply "load forecasting", is a fundamental task in the operation of power systems and has been studied for decades. Traditionally, load forecasting has not been performed at a single building scale but has focussed on the regional scale to forecast the total power consumption of a geographical area consisting of several buildings and other possible energy loads. This is a consequence of the fact that, traditionally, power measurement devices have been installed in substations to monitor their electrical power which represents the aggregated power consumption of all energy loads connected to the bus where the substation is located. Based on the measured regional power consumption data, load forecasting can be performed to predict the region's future power consumption. With the development of smart grids, smart meters [1] have been increasingly deployed in buildings to record their power consumption Based on the historically recorded power consumption data it is then possible to forecast the power consumption of single buildings.

This chapter introduces some basic short-term load forecasting (STLF) methods that have been applied to generate power consumption predictions from a set of historically recorded time series power data. The methods introduced in this chapter can be used in predicting the power load at both regional and single-building scales. We first introduce two typical statistical analysis methods that can generate future power loads from statistical patterns exhibited in the historical time series power load data. This is then followed by the description of the basic concepts and principles of artificial neural networks

(ANNs) as well as their applications in STLF. In the last part of the chapter, we provide a short overview of deep neural networks-based building power load forecasting techniques.

## 6.2 Explanatory and time series STLF models

There could be a variety of factors affecting the target object of a forecasting task. For example, the power consumption of a building could be affected by factors such as the outdoor climate, the materials used in the building's envelope, the window's characteristics and orientation, and the building occupant's behaviors. Depending on the extent to which these factors are considered, STLF models can be generally classified into two types: *explanatory models* and *time series models*.

### 6.2.1 Explanatory STLF models

Explanatory models perform forecasting by analyzing the impact of the influencing factors. For example, an explanatory model for building power load forecasting can be represented as:

$$\mathbf{L}^f = f(\text{outdoor temperature, time of day, day of week,}$$

$$\text{building's orientation, building's location, occupant's life behavior, error)} \quad (6.1)$$

where $\mathbf{L}^f$ denotes the forecasted power consumption of the building and $f(\cdot)$ denotes the explanatory model for STLF. The "error" term in Eq. (6.1) allows for random variations and effects of the relevant factors not included in the model.

Explanatory models are useful because, ideally, they can comprehensively incorporate the information about the influencing factors of the forecasting target. There are several challenges in the practical implementation of explanatory models for building power load forecasting. Firstly, it could be difficult to enumerate or understand all the factors that affect the target building's power consumption. Secondly, even if the influencing factors can be understood, it would be challenging to quantify how they affect the building's power consumption. Thirdly, to predict the building's future power consumption, it is necessary to predict the future values of the influencing factors and this will introduce additional nontrivial forecasting tasks, therefore significantly increasing the complexity of the STLF task.

The ultimate goal of a forecasting task is to predict what will happen, without necessarily having to know why it will happen [2]. Based on this principle and the challenges of explanatory models mentioned above, time series forecasting models are more commonly used for STLF models.

### 6.2.2 Time series STLF models

*Time series data* is a collection of observations obtained through repeated measurements over time. A building's power consumption recorded by a smart meter over a certain period at a regular sampling interval is time series data. Fig. 6.1 illustrates the power consumption time series data of four residential houses in Sydney over four consecutive days (data sources: the Australian "Smart Grid, Smart City" customer trial dataset [3]).

Without any loss of generality, a time series (denoted as $\boldsymbol{y}$) can be expressed as an array:

$$\boldsymbol{y} = [y_1, y_2, y_3, \ldots] \tag{6.2}$$

where an element $y_t$ ($t = 1, 2, 3, \ldots$) in $\boldsymbol{y}$ represents the data sampled at time point $t$. Eq. (6.2) is written in a general form that can be used to represent any time series data, such as buildings' power load, air temperature, and wind speed.

Time series forecasting models use the time series data of the forecasting target in the past to predict future values and the time series model for STLF can then be represented as:

$$\mathbf{L}^f = f\left(\mathbf{L}^h, \Omega, \text{error}\right) \tag{6.3}$$

where $f(\cdot)$ denotes the time series model for forecasting and $\Omega$ denotes the set of parameters of the model. $\mathbf{L}^h$ denotes the historical time series power consumption data, i.e., the

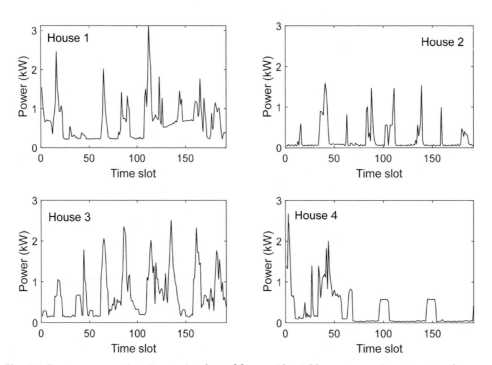

**Fig. 6.1** Power consumption time series data of four residential houses over 4 consecutive days.

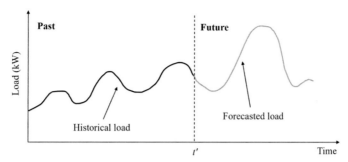

**Fig. 6.2** Illustration of time series analysis based STLF.

observed power consumption data in the past. $\mathbf{L}^f$ denotes the forecasted power consumption data. These vectors are expressed as:

$$\mathbf{L}^h = [y_{t'}, y_{t'-1}, y_{t'-2}, \ldots] \tag{6.4}$$

$$\mathbf{L}^f = [\widehat{y}_{t'+1}, \widehat{y}_{t'+2}, \ldots, \widehat{y}_{t'+K}] \tag{6.5}$$

where the symbol $t'$ denotes the current time slot. In Eq. (6.4), $y_{t'-i}$ denotes the $(i-1)$th historical power load value (also known as a historical observation, in watts or kW) prior to the current time slot. For example, if the historical power load data contains the readings of a smart meter over the past 5 days at a sampling interval of 1 h, the number of data items in $\mathbf{L}^h$ is equal to 120 ($=24 \times 5$). In Eq. (6.5), $K$ represents the number of future power consumption values to be forecasted (also known as the *forecasting window*), and $\widehat{y}_{t'+k}$ ($k=1:K$) denotes the $k$th forecasted power consumption value (in watts or kW).

The values in $\mathbf{L}^h$ (i.e., the observed power consumption data in the past) are called *predictor variables* and the values in $\mathbf{L}^f$ (i.e., the power consumption in the future) are called *forecast variables* [2]. Time series-based STLF can be intuitively illustrated in Fig. 6.2.

Time series models for STLF can be based on statistical analysis methods or machine learning methods. In the next two sections, we will introduce two statistical analysis models that have been widely applied in time series forecasting applications: the exponential smoothing model and the autoregressive integrated moving average (ARIMA) model. Ref. [2] provides a comprehensive introduction to these models. In Section 6.5, we will introduce the basic principles of ANNs and their applications in STLF.

## 6.3 Exponential smoothing-based STLF

### 6.3.1 Overview

The development of the exponential smoothing can be traced back to the late 1950s [4] and, since then, it has been successfully applied to a variety of forecasting tasks including

STLF [5,6]. Exponential smoothing provides forecasts as weighted averages of the past observations with higher weights given to the more recent observations.

In Sections 6.3.2 and 6.3.3, we will introduce two fundamental exponential smoothing models. The first one, denoted as the *simple exponential smoothing*, is the simplest representation of a exponential smoothing model. It is suitable for generating forecasts that possess no clear trends in the observed data. The second model considered is denoted as *damped trend exponential smoothing* model and it is suitable for generating forecasts when the observed data exhibits trends. In Section 6.3.4, we will provide an application example to demonstrate the use of these models when performing STLF.

## 6.3.2 Simple exponential smoothing

In the simple exponential smoothing model, forecasts are calculated using the weighted averages of the historical observation. The one-time step forecast is performed as:

$$\widehat{y}_{t'+1} = \alpha y_{t'} + \alpha(1-\alpha)y_{t'-1} + \alpha(1-\alpha)^2 y_{t'-2} + \alpha(1-\alpha)^3 y_{t'-3} \cdots \quad (6.6)$$

where $0 \leq \alpha \leq 1$ is the smoothing parameter that needs to be prespecified. Based on the fact that in many forecasting applications, more recent observations are usually more meaningful for generating the future values, in the simple exponential smoothing method, the weights of the historical observations (e.g., $\alpha$, $\alpha(1-\alpha)$, $\alpha(1-\alpha)^2$,...) decrease exponentially as the observations are dated back in time, from which the name "exponential smoothing". The smoothing parameter $\alpha$ controls the rate at which the weights decay.

The simple exponential smoothing model is usually represented in the *component form*, which consists of a *forecast equation* and a *smoothing equation* (i.e., the two components of the method). The equations are expressed in a recursion form as follows:

$$\text{Forecast equation}: \widehat{y}_{t+1} = \ell_t \quad (6.7)$$

$$\text{Smoothing equation}: \ell_t = \alpha y_t + (1-\alpha)\ell_{t-1} \quad (6.8)$$

where $\ell_t$ is called the level of the series at time point $t$, and $y_t$ represents the historical load value at time point $t$ ($t=1:t'$, as represented in Eq. (6.4)). For clarity, the expressions of Eqs. (6.7) and (6.8) can be expanded as:

$$\widehat{y}_2 = \ell_1 = \alpha y_1 + (1-\alpha)\ell_0 \quad (6.9)$$

$$\widehat{y}_3 = \ell_2 = \alpha y_2 + (1-\alpha)\ell_1 = \alpha y_2 + \alpha(1-\alpha)y_1 + (1-\alpha)^2 \ell_0 \quad (6.10)$$

$$\widehat{y}_4 = \alpha y_3 + (1-\alpha)\ell_2 = \alpha y_3 + \alpha(1-\alpha)y_2 + \alpha(1-\alpha)^2 y_1 + (1-\alpha)^3 \ell_0 \quad (6.11)$$

$$\cdots$$

$$\widehat{y}_{t'+1} = \sum_{i=0}^{t'-1} \alpha(1-\alpha)^i y_{t'-i} + (1-\alpha)^{t'} \ell_0 \quad (6.12)$$

where $\ell_0$ represents the first estimated value at time point 1, from which the recursive calculation starts, while $y_1$, $y_2$ and $y_3$ depict the historical load values at times $t=1$, $t=2$ and $t=3$, respectively.

Note that when $t < t'$, $\widehat{y}_t$ generated by the simple exponential smoothing model represents the *fitted values* of the observations in the historical dataset. In such a case, we "pretend" that we do not know the actual data value at time point $t$ (i.e., $y_t$), and we use the model to generate a forecasted value $\widehat{y}_t$. When $t = t'$, the value of $\widehat{y}_{t'+1}$ from Eq. (6.12) represents the one-time step forecast as the actual data value $y_{t'+1}$ is unknown, i.e., it has not yet occurred and, therefore, it has not yet been recorded. Eq. (6.12) is also called the *weighted average form* of the simple exponential smoothing model.

Simple exponential smoothing generates "flat" forecasts because all forecasts are with the same value, equal to the last level component $\ell_{t'}$:

$$\widehat{y}_{t'+k} = \widehat{y}_{t'+1} = \ell_{t'}, \quad \forall k = 1 : K \tag{6.13}$$

based on which Eq. (6.12) can also be written as:

$$\widehat{y}_{t'+k} = \ell_{t'} = \sum_{i=0}^{t'-1} \alpha(1-\alpha)^i y_{t'-i} + (1-\alpha)^{t'} \ell_0, \quad \forall k = 1 : K \tag{6.14}$$

### 6.3.3 Damped trend exponential smoothing

With the simple exponential smoothing, all forecasts end up having the same constant value. This approach cannot capture the trend of the observed data if it exists. When dealing with time series data that exhibits trends, the forecasts can be generated with the damped trend exponential smoothing method [2], developed by expanding the simple exponential smoothing. With this method, the forecasts are generated by three components: a *forecast equation*, a *level equation*, and a *trend equation* that can be expressed as:

$$\text{Forecast equation} : \widehat{y}_{t+k} = \ell_t + \left(\emptyset + \emptyset^2 + \ldots + \emptyset^k\right) b_t \tag{6.15}$$

$$\text{Level equation} : \ell_t = \alpha y_t + (1-\alpha)(\ell_{t-1} + \emptyset b_{t-1}) \tag{6.16}$$

$$\text{Trend equation} : b_t = \beta^*(\ell_t - \ell_{t-1}) + \left(1 - \beta^*\right)\emptyset b_{t-1} \tag{6.17}$$

where $b_t$ denotes an estimate of the trend of the series at time point $t$; $0 \leq \emptyset \leq 1$ is the damping parameter; $0 \leq \alpha \leq 1$ is the smoothing parameter for the level; $0 \leq \beta^* \leq 1$ is smoothing parameter for the trend; and $\ell_t$ is calculated following Eq. (6.8). When $t < t'$,

Eq. (6.15) generates the fitted values of the historical observations. With $t=t'$, it generates forecasts for future values. It can be seen that the method no longer generates flat forecasts, i.e., for different values of $k$ the generated forecasts $(\widehat{y}_{t+k})$ differ.

When $\emptyset=0$, the approach degenerates in the simple exponential smoothing method, i.e., the simple exponential smoothing method is a special case of the damped trend exponential smoothing method with $\emptyset=0$. When $\emptyset=1$, Eqs. (6.15)–(6.17) can be simplified as:

$$\text{Forecast equation}: \widehat{y}_{t+k} = \ell_t + kb_t \tag{6.18}$$

$$\text{Level equation}: \ell_t = \alpha L_t^h + (1-\alpha)(\ell_{t-1} + b_{t-1}) \tag{6.19}$$

$$\text{Trend equation}: b_t = \beta^*(\ell_t - \ell_{t-1}) + \left(1 - \beta^*\right)b_{t-1} \tag{6.20}$$

that consists of a special case of the method usually referred to as the Holt's linear trend method [7]. This approach can generate forecasts with a constant trend, i.e., either increasing or decreasing.

For the values of $\emptyset$ decreasing from 1 to 0, the parameter $\emptyset$ "damps" the trend of the forecasts towards constant values sometimes in the future. With $0 < \emptyset < 1$, when $k \to \infty$, $\widehat{y}_{t'+k}$ tends to converge to the value of $\ell_{t'} + \frac{\emptyset b_{t'}}{1-\emptyset}$.

## 6.3.4 Application examples for STLF techniques

We use application examples to demonstrate the exponential smoothing-based STLF. To minimize the complexity of the example, we consider a scenario with the 4 historical load data values shown in Table 6.1 that are sampled at an interval of 15 min. The data is from the Pecan Street energy customer dataset [8], recoding power consumption of a resident in Austin, U.S over 45 min.

### 6.3.4.1 Application example for the simple smoothing-based STLF

We now apply the simple exponential smoothing and the damped trend exponential smoothing methods to the time series in Table 6.1 to generate fitted load values for the time points between 2 and 4 (i.e., $\widehat{y}_2, \widehat{y}_3$, and $\widehat{y}_4$) and to forecast the load values for the two future time points 5 and 6 (denoted as $\widehat{y}_5$ and $\widehat{y}_6$, respectively). For the application

**Table 6.1** Historical load data values used in the application examples.

| Time point 1 | Time point 2 | Time point 3 | Time point 4 |
|---|---|---|---|
| $y_1 = 1.19\,\text{kW}$ | $y_2 = 1.64\,\text{kW}$ | $y_3 = 3.07\,\text{kW}$ | $y_4 = 3.9\,\text{kW}$ |

of the simple exponential smoothing, we initialize $\ell_0 = y_1 = 1.19$ kW and set $\alpha = 0.6$. Based on Eqs. (6.7)–(6.14), we can then calculate:

$$\widehat{y}_2 = \alpha y_1 + (1-\alpha)\ell_0 = 0.6 \times 1.19 + (1-0.6) \times 1.19 = 1.19 \text{ kW}$$

$$\widehat{y}_3 = \alpha y_2 + \alpha(1-\alpha)y_1 + (1-\alpha)^2\ell_0$$

$$= 0.6 \times 1.64 + 0.6 \times (1-0.6) \times 1.19 + (1-0.6)^2 \times 1.19 = 1.46 \text{ kW}$$

$$\widehat{y}_4 = \alpha y_3 + \alpha(1-\alpha)y_2 + \alpha(1-\alpha)^2 y_1 + (1-\alpha)^3\ell_0$$

$$= 0.6 \times 3.07 + 0.6 \times (1-0.6) \times 1.64 + 0.6 \times (1-0.6)^2 \times 1.19$$

$$+ (1-0.6)^3 \times 1.19 = 2.43 \text{ kW}$$

$$\widehat{y}_5 = \alpha y_4 + \alpha(1-\alpha)y_3 + \alpha(1-\alpha)^2 y_2 + \alpha(1-\alpha)^3 y_1 + (1-\alpha)^4\ell_0$$

$$= 0.6 \times 3.9 + 0.6 \times (1-0.6) \times 3.07 + 0.6 \times (1-0.6)^2 \times 1.64 + 0.6$$

$$\times (1-0.6)^3 \times 1.19 + (1-0.6)^4 \times 1.19 = 3.31 \text{ kW}$$

$$\widehat{y}_6 = \widehat{y}_5 = 3.31 \text{ kW}$$

Different values for the parameter $\alpha$ will produce different forecasting results. Table 6.2 shows the load forecasting results for four different values of $\alpha$.

As a demonstration, Fig. 6.3 shows the STLF results by applying the simple exponential smoothing method to a one-day load profile of a resident in the Great Sydney area. The load data is from the Australian "Smart Grid, Smart City" dataset [3]. In the figure, the actual load data is plotted, together with the load forecasting results under two different settings: $\alpha = 0.9$ and $\alpha = 0.5$. In the forecasting curves, each data point is the result of one-step load forecasting generated by the simple exponential smoothing method applied on the load values measured before that time point.

**Table 6.2** Simple exponential smoothing-based load forecasting results with different values of $\alpha$.

| | Time point 1 | Time point 2 | Time point 3 | Time point 4 | Time point 5 | Time point 6 |
|---|---|---|---|---|---|---|
| Historical load | $y_1 = 1.19$[a] | $y_2 = 1.64$ | $y_3 = 3.07$ | $y_4 = 3.9$ | – | – |
| $\alpha = 0.6$ | – | $\widehat{y}_2 = 1.19$ | $\widehat{y}_3 = 1.46$ | $\widehat{y}_4 = 2.43$ | $\widehat{y}_5 = 3.31$ | $\widehat{y}_6 = 3.31$ |
| $\alpha = 0.3$ | – | $\widehat{y}_2 = 1.19$ | $\widehat{y}_3 = 1.33$ | $\widehat{y}_4 = 1.85$ | $\widehat{y}_5 = 2.46$ | $\widehat{y}_6 = 2.46$ |
| $\alpha = 0.1$ | – | $\widehat{y}_2 = 1.19$ | $\widehat{y}_3 = 1.24$ | $\widehat{y}_4 = 1.42$ | $\widehat{y}_5 = 1.67$ | $\widehat{y}_6 = 1.67$ |
| $\alpha = 0.9$ | – | $\widehat{y}_2 = 1.19$ | $\widehat{y}_3 = 1.60$ | $\widehat{y}_4 = 2.92$ | $\widehat{y}_5 = 3.80$ | $\widehat{y}_6 = 3.80$ |

[a]The unit of the data items is kW.

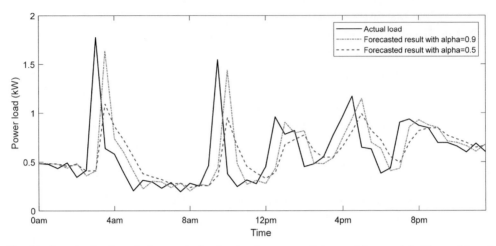

**Fig. 6.3** One-day load prediction results by the simple exponential smoothing model with $\alpha=0.9$ and $\alpha=0.5$.

### 6.3.4.2 Application example for the damped trend exponential smoothing-based STLF

We now apply the damped trend exponential smoothing method to the historical load data in Table 6.1. We set $\alpha=0.6$, $\emptyset=0.5$ and $\beta^{*}=0.7$. $b_0$ is initialized to be 0 and $\ell_0$ is initialized as $\ell_0=y_1=1.19$ kW

**(1)** Calculation of $\widehat{y}_2$:

$$\ell_1 = \alpha y_1 + (1-\alpha)(\ell_0 + \emptyset b_0) = 0.6 \times 1.19 + (1-0.6) \times (1.19 + 0.5 \times 0) = 1.19$$
$$b_1 = \beta^{*}(\ell_1 - \ell_0) + (1-\beta^{*})\emptyset b_0 = 0.7 \times (1.19 - 1.19) + (1-0.7) \times 0.5 \times 0 = 0$$
$$\widehat{y}_2 = \widehat{y}_{1+1} = \ell_1 + \emptyset b_1 = 1.19 + 0.5 \times 0 = 1.19 \text{ kW}$$

**(2)** Calculation of $\widehat{y}_3$:

$$\ell_2 = \alpha y_2 + (1-\alpha)(\ell_1 + \emptyset b_1) = 0.6 \times 1.64 + (1-0.6) \times (1.19 + 0.5 \times 0) = 1.46$$
$$b_2 = \beta^{*}(\ell_2 - \ell_1) + (1-\beta^{*})\emptyset b_1 = 0.7 \times (1.46 - 1.19) + (1-0.7) \times 0.5 \times 0 = 0.19$$
$$\widehat{y}_3 = \widehat{y}_{2+1} = \ell_2 + \emptyset b_2 = 1.46 + 0.5 \times 0.19 = 1.56 \text{ kW}$$

**(3)** Calculation of $\widehat{y}_4$:

$$\ell_3 = \alpha y_3 + (1-\alpha)(\ell_2 + \emptyset b_2) = 0.6 \times 3.07 + (1-0.6) \times (1.46 + 0.5 \times 0.19)$$
$$= 2.46$$
$$b_3 = \beta^{*}(\ell_3 - \ell_2) + (1-\beta^{*})\emptyset b_2 = 0.7 \times (2.46 - 1.46) + (1-0.7) \times 0.5 \times 0.19$$
$$= 0.73$$
$$\widehat{y}_4 = \widehat{y}_{3+1} = \ell_3 + \emptyset b_3 = 2.46 + 0.5 \times 0.73 = 2.83 \text{ kW}$$

**(4)** Calculation of $\widehat{y}_5$:

$$\ell_4 = \alpha y_4 + (1 - \alpha)(\ell_3 + \emptyset b_3) = 0.6 \times 3.9 + (1 - 0.6) \times (2.46 + 0.5 \times 0.73) = 3.47$$
$$b_4 = \beta^*(\ell_4 - \ell_3) + (1 - \beta^*)\emptyset b_3 = 0.7 \times (3.47 - 2.46) + (1 - 0.7) \times 0.5 \times 0.73$$
$$= 0.82$$
$$\widehat{y}_5 = \widehat{y}_{4+1} = \ell_4 + \emptyset b_4 = 3.47 + 0.5 \times 0.82 = 3.88 \text{ kW}$$

**(5)** Calculation of $\widehat{y}_6$:

$$\widehat{y}_6 = \widehat{y}_{4+2} = \ell_4 + (\emptyset + \emptyset^2)b_4 = 3.47 + (0.5 + 0.5^2) \times 0.82 = 4.09 \text{ kW}$$

Table 6.3 shows the load forecasting results for four different combinations of the parameters $\alpha$, $\emptyset$, and $\beta^*$.

Fig. 6.4 shows the load forecasting results by applying the damped trend exponential smoothing method with two parameter settings to the one–day load profile used in the example in Section 6.3.4.1. In the forecasting curves, each data point is a fitted value of the actual load, i.e., each data point is the result of one–step load forecasting generated by the damped trend exponential smoothing method based on the load values occurred prior to the time point (i.e., based on $t$ equal to be the number of time points occurred before the current time point and $k=1$ in Eq. (6.15)).

**Table 6.3** Damped trend exponential smoothing-based load forecasting results with different parameter values.

| | Time point 1 | Time point 2 | Time point 3 | Time point 4 | Time point 5 | Time point 6 |
|---|---|---|---|---|---|---|
| Historical load | $y_1 = 1.19^a$ | $y_2 = 1.64$ | $y_3 = 3.07$ | $y_4 = 3.9$ | – | – |
| $\alpha=0.6$, $\emptyset=0.5$, $\beta^*=0.7$ | – | $\widehat{y}_2 = 1.19$ | $\widehat{y}_3 = 1.56$ | $\widehat{y}_4 = 2.83$ | $\widehat{y}_5 = 3.88$ | $\widehat{y}_6 = 4.09$ |
| $\alpha=0.3$, $\emptyset=0.2$, $\beta^*=0.4$ | – | $\widehat{y}_2 = 1.19$ | $\widehat{y}_3 = 1.34$ | $\widehat{y}_4 = 1.90$ | $\widehat{y}_5 = 2.56$ | $\widehat{y}_6 = 2.57$ |
| $\alpha=0.9$, $\emptyset=0.8$, $\beta^*=0.2$ | – | $\widehat{y}_2 = 1.19$ | $\widehat{y}_3 = 1.66$ | $\widehat{y}_4 = 3.18$ | $\widehat{y}_5 = 4.14$ | $\widehat{y}_6 = 4.38$ |
| $\alpha=0.7$, $\emptyset=0.7$, $\beta^*=0.7$ | – | $\widehat{y}_2 = 1.19$ | $\widehat{y}_3 = 1.66$ | $\widehat{y}_4 = 3.24$ | $\widehat{y}_5 = 4.34$ | $\widehat{y}_6 = 4.79$ |

$^a$The unit of the data items is kW.

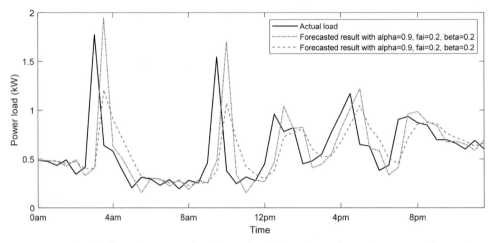

**Fig. 6.4** One-day load prediction results obtained with the damped trend exponential smoothing model with two parameters.

### 6.3.5 Selection of parameters for the exponential smoothing models

As shown in Sections 6.3.2 and 6.3.3 for the simple exponential smoothing model, the parameter $\alpha$ and the initialization variable $\ell_0$ need to be prespecified before the model can be applied, while when using the damped trend exponential smoothing model, the parameters $\alpha$, $\emptyset$, $\beta^*$ and the initialization variable $\ell_0$ need to be initially assigned. As we can see from the examples in Section 6.3.4, different settings of these parameters (here the parameters include the initialization variable $\ell_0$) lead to different forecasting results.

One way to determine the values of these parameters is to rely on previous forecasting experience. An objective and reasonable approach is desirable to be established for this task that can involve the identification of a combination of parameter values that can minimize the *sum of the squared errors (SSE)* obtained from the historical time series dataset (i.e., the observed data):

$$\text{SSE} = \sum_{i=1}^{t'} (y_i - \widehat{y}_i)^2 \tag{6.21}$$

The identification of the parameter values that can minimize the SSE is an optimzsation problem and it can be performed by relying on trial-based approach: try different combinations of the parameter values on the historical dataset and select the combination that produces the smallest SSE. More structured optimization techniques can also be applied for this task (e.g., the evolutionary computation technique introduced in Chapter 7) that consider Eq. (6.21) to be the objective function. For more details about parameter selection in exponential smoothing, reference should be made to dedicated publications, e.g., [2].

## 6.4 ARIMA-based STLF

### 6.4.1 Introduction

The ARIMA model is a statistical analysis model that has been widely used for predicating future trends of time series data. ARIMA model generates forecasting for stationary time series data. Before focussing on the details of the ARIMA model and its application in STLF, we first introduce the concepts of *stationary* and *nonstationary* time series data.

### 6.4.2 Stationary and nonstationary time series data

In a stationary time series, its statistical properties, such as the mean and the standard deviation, do not depend on the time at which the series is observed [2]. In the case of a nonstationary time series, its trends or seasonality influence the variation of the time series at different points in time as, for example, shown for the load profiles in Fig. 6.1.

A typical example of stationary time series is referred to as the *white noise* that consists of a time series in which the variables are independent and identically distributed with finite mean and variance. In this manner, no matter at which time point a white noise is observed, it always looks the same. In particular, a white noise with zero mean is called a Gaussian white noise that is shown in Fig. 6.5.

### 6.4.3 White noise model

We now introduce the mathematical representation of white noise because it will be required in the presentation of the ARIMA model.

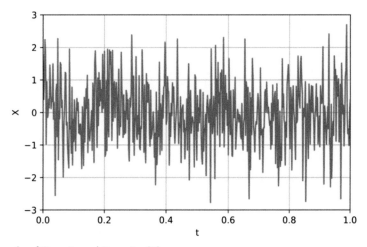

**Fig. 6.5** Example of Gaussian white noise [9].

A Gaussian white noise is denoted as $\varepsilon_t$, where $t$ is the time point index, and it possesses a zero mean and a finite variant:

$$\varepsilon_t \sim N(0, \sigma^2) \qquad (6.22)$$

A white noise with a finite mean can be represented as:

$$y_t = c + \varepsilon_t \qquad (6.23)$$

where $c$ is a constant parameter, representing the mean of the white noise; and $\varepsilon_t$ is a Gaussian white noise data item. When $c=0$, the model of Eq. (6.23) depicts a Gaussian white noise.

A brief interpretation of Eq. (6.23) is provided in this paragraph. We estimate the data value in the time series at time point $t$ as a constant value $c$, and there is a difference between the observed value of the variable (i.e., $y_t$) and the estimated value $c$, which is $\varepsilon_t$. The term $\varepsilon_t$ is denoted as the estimation residual or estimation error of $y_t$. The data values at different time points are independent and do not correlate with each other, i.e., the value of a white noise variable at a time point is not affected by the variables in the past time points. Because of this, we can regard the white noise as unpredictable. For many time series data, a correlation exists among the data values. For example, a building occupant's behavior at one time point will affect his/her behavior in subsequent time points, and this allows correlations to exist among the building's power load values at different time points.

When performing forecasting for a time series, we need to model the correlation among the variables in the time series, and this is what the ARIMA model does. As the name *autoregressive integrated moving average* implies, an ARIMA model simulates the correlation through two submodels: an *autoregressive model* and a *moving average model*.

## 6.4.4 Autoregressive model

An autoregressive model determines a variable's value based on past observations:

$$\widehat{y}_t = c + \phi_1 y_{t-1} + \phi_2 y_{t-2} + \ldots + \phi_p y_{t-p} \qquad (6.24)$$

$$y_t = \widehat{y}_t + \varepsilon_t = c + \varepsilon_t + \phi_1 y_{t-1} + \phi_2 y_{t-2} + \ldots + \phi_p y_{t-p} \qquad (6.25)$$

where $\phi_1, \ldots, \phi_p$ are parameters. $p$ is denoting as the order of the model and the value of $y_t$ is estimated from the previous $p$ observations in the time series. $\widehat{y}_t$ is the fitted/forecasted value for the data at time point $t$, and $y_t$ is the actual observation at time point $t$. The observed value of $y_t$ can be represented as the fitted value plus a white noise error $\varepsilon_t$. For $t < t'$ ($t'$ indicates the current time point), $\widehat{y}_t$ is a fitted value generated by the autoregressive model for the historical observation. For $t \geq t'$, $\widehat{y}_t$ is a forecasted value generated by the model for a future time point, and in this case, since $y_t$ is unknown, we consider $\varepsilon_t = 0$ and $y_t = \widehat{y}_t$.

The autoregressive model of Eq. (6.25) is usually denoted as an AR($p$) model. Obviously, for an AR(1) model: when $\phi_1 = 0$, Eq. (6.25) is equivalent to the white noise model of Eq. (6.23), while, when $c = \phi_1 = 0$, $y_t$ is equivalent to the Gaussian white noise.

### 6.4.5 Moving average model

Different from the autoregressive model that generates the fitted/forecasted value from the past observations in the time series, A moving average model achieves this from the estimation errors of the past values:

$$\widehat{y}_t = c + \theta_1 \varepsilon_{t-1} + \theta_2 \varepsilon_{t-2} + \cdots + \theta_q \varepsilon_{t-q} \tag{6.26}$$

$$y_t = \widehat{y}_t + \varepsilon_t = c + \varepsilon_t + \theta_1 \varepsilon_{t-1} + \theta_2 \varepsilon_{t-2} + \cdots + \theta_q \varepsilon_{t-q} \tag{6.27}$$

where $\theta_1, \ldots, \theta_p$ are parameters; and $q$ is the order of the model that indicates the number of observations in the time series of the estimation errors used to evaluate $\widehat{y}_t$. The moving average model of Eq. (6.27) is usually denoted as the MA($q$) model.

### 6.4.6 Overview of the ARIMA model

The autoregressive model describes the correlation between a variable's value and the past observations, and the moving average model correlates a variable's value to the estimation errors of the past observations. In time series forecasting, these two kinds of correlations need to be considered at the same time. The ARIMA model can be described by combining the autoregressive and the moving average models:

$$\widehat{y}_t = c + \phi_1 y_{t-1} + \phi_2 y_{t-2} + \cdots + \phi_p y_{t-p} + \theta_1 \varepsilon_{t-1} + \theta_2 \varepsilon_{t-2} + \cdots + \theta_q \varepsilon_{t-q} \tag{6.28}$$

$$y_t = \widehat{y}_t + \varepsilon_t = c + \varepsilon_t + \phi_1 y_{t-1} + \phi_2 y_{t-2} + \cdots + \phi_p y_{t-p} + \theta_1 \varepsilon_{t-1} + \theta_2 \varepsilon_{t-2} + \cdots + \theta_q \varepsilon_{t-q} \tag{6.29}$$

that highlight how, in the ARIMA model, the fitted/forecasted value $\widehat{y}_t$ is generated from both of the past $p$ observations and the estimation errors of the past $q$ observations.

That ARIMA model can only be applied to stationary time series. For nonstationary time series, the estimation error $\varepsilon_t$ usually does not appear to be a Gaussian white noise, and therefore it is not suitable for being processed with the ARIMA model. Many real-world time series can be regarded as nonstationary data, such as the buildings' power load data. To apply the ARIMA model, we need to extract a stationary component from the nonstationary time series. This can be achieved by the *differencing operation* described in the following section.

### 6.4.7 Differencing operation on time series data

Differencing is a useful way to generate a stationary time series from a nonstationary time series. The first order differencing operation calculates the difference between two consecutive data values in a time series:

$$y_t^{(1)} = y_t - y_{t-1} \tag{6.30}$$

The first order differenced time series describes the changes of the original time series. Obviously, the first item in the first order differenced time series is $y_2^{(1)}$, calculated from $y_2$ and $y_1$. If there are $t'$ observed data values in the time series, there are $t' - 1$ values in the first order differenced time series.

In many cases, by applying the first order differencing operation to a nonstationary time series it is possible to generate a stationary, differenced time series. Sometimes the first order differenced time series does not appear to be stationary and, in these cases, the differencing operation needs to be re-applied to the first order differenced time series to generate the second order time series as:

$$y_t^{(2)} = y_t^{(1)} - y_{t-1}^{(1)} = (y_t - y_{t-1}) - (y_{t-1} - y_{t-2}) = y_t - 2 \times y_{t-1} + y_{t-2} \tag{6.31}$$

The second order differenced time series describes the "change in the changes" of the original time series. If there are $t'$ observed data values in the time series, there are $t' - 2$ values in the second order differenced time series. In most of the cases, performing the differencing operation twice is sufficient to generate a stationary time series from the original, nonstationary time series.

Once the stationary time series is obtained by applying the differencing operation, the ARIMA model can be applied to the differenced, stationary time series as shown in the following based on Eq. (6.29):

$$\widehat{y}_t^{(d)} = c + \phi_1 y_{t-1}^{(d)} + \phi_2 y_{t-2}^{(d)} + \cdots + \phi_p y_{t-p}^{(d)} + \theta_1 \varepsilon_{t-1} + \theta_2 \varepsilon_{t-2} + \cdots + \theta_q \varepsilon_{t-q} \tag{6.32}$$

$$y_t^{(d)} = \widehat{y}_t^{(d)} + \varepsilon_t = c + \varepsilon_t + \phi_1 y_{t-1}^{(d)} + \phi_2 y_{t-2}^{(d)} + \cdots + \phi_p y_{t-p}^{(d)} + \theta_1 \varepsilon_{t-1} + \theta_2 \varepsilon_{t-2} + \cdots + \theta_q \varepsilon_{t-q} \tag{6.33}$$

where $y_t^{(d)}$ is the differenced time series generated from the $d$th-order differencing operation, and it is the time series to which the ARIMA model applies. Eq. (6.33) is usually denotes as the ARIMA($p$, $d$, $q$) model. The white noise model is represented by ARIMA(0,0,0) and $c = 0$. ARIMA($p$, 0, 0) describes the autoregression model and ARIMA(0,0, $q$) depicts the moving average model.

## 6.4.8 Forecasting generated with the ARIMA model

The ARIMA model can be used to generate the future values of a given time series. The data value in a future time point $t' + k$ (denoted as $y_{t'+k}$) is considered to depend on the data values in the previous $p$ time points and the residuals in the previous $q$ time points.

For a previous time slot $t^* < t' + k$, the forecasted data generated by the ARIMA model is used when $t^* > t'$ and the residual at $t^*$ is set equal to zero. When $t^* < t'$, the observed data value and the corresponding residual in the time series is used. In this manner, the forecasting formula for $y_{t'+k}$ is expressed as:

$$y_{t'+k} = c + \phi_1 y_{t'+k-1} + \phi_2 y_{t'+k-2} + \cdots + \phi_p y_{t'+k-p} + \theta_1 \varepsilon_{t'+k-1}$$
$$+ \theta_2 \varepsilon_{t'+k-2} + \cdots + \theta_q \varepsilon_{t'+k-q} \tag{6.34}$$

$$\varepsilon_{t^*} = 0, \text{if } t^* > t', \forall t^* = t' + k - q : t' + k - 1 \tag{6.35}$$

## 6.4.9 Application example for an ARIMA-based STLF

We can now set up a simple example to demonstrate the application of the ARIMA-based STLF. Let us consider the 6 historical load data values shown in Table 6.4. The data values are from the Pecan Street dataset [8] and are sampled every 15 min.

To generate the nonstationary time series, we apply the differencing operation to the data based on Eq. (6.30) to generate a first order differenced time series:

$$\mathbf{L}^{h(1)} = \left[ y_2^{(1)}, y_3^{(1)}, y_4^{(1)}, y_5^{(1)}, y_6^{(1)} \right] = [y_2 - y_1, y_3 - y_2, y_4 - y_3, y_5 - y_4, y_6 - y_5]$$
$$= [-0.19, -0.28, 0.49, -0.23, -1.93]$$

For ease of notation, we denote $\mathbf{L}^{h(1)}$ as $\mathbf{z}$:

$$\mathbf{z} = [z_1, z_2, z_3, z_4, z_5] = [-0.19, -0.28, 0.49, -0.23, -1.93]$$

where $z_1 = y_2^{(1)}$, $z_2 = y_3^{(1)}$, ..., $z_5 = y_6^{(1)}$.

We then apply an ARIMA(2,1,2) model to the differenced time series $\mathbf{z}$, The parameter $c$ is set to be the mean of $\mathbf{z}$: $c = \frac{z_1 + z_2 + z_3 + z_4 + z_5}{5} = \frac{-0.19 - 0.28 + 0.49 - 0.23 - 1.93}{5} = -0.43$. Since we assigned the parameters $p = 2$ and $q = 2$, the ARIMA model predicts a value based on the last two historical values and the prediction error of the last two historical values.

Table 6.4 6 historical load data values used in the application example.

| Time point 1 | Time point 2 | Time point 3 |
|---|---|---|
| $y_1 = 5.23 \, \text{kW}$ | $y_2 = 5.04 \, \text{kW}$ | $y_3 = 4.76 \, \text{kW}$ |
| Time point 4 | Time point 5 | Time point 6 |
| $y_4 = 5.25 \, \text{kW}$ | $y_5 = 5.02 \, \text{kW}$ | $y_6 = 3.09 \, \text{kW}$ |

We then initialize $\widehat{z}_1 = \widehat{z}_2 = c = -0.43$. $\varepsilon_1$ and $\varepsilon_2$ are then initialized as $\varepsilon_1 = z_1 - \widehat{z}_1 = -0.19 + 0.43 = 0.24$ and $\varepsilon_2 = z_2 - \widehat{z}_2 = -0.28 + 0.43 = 0.15$.

The error items ($\varepsilon_1$, $\varepsilon_2$, ...) represent the prediction errors of the items in the differenced time series $z$, not the prediction errors of the original historical load time series. We then set the parameters: $\phi_1 = -0.4$, $\phi_2 = 0.3$, $\theta_1 = 0.5$, and $\theta_2 = 0.2$.

The ARIMA(2,1,2) model is then applied to generate fitted values for $z_3$, $z_4$, $z_5$ (denoted as $\widehat{z}_3$, $\widehat{z}_4$ and $\widehat{z}_5$, respectively) and to forecast two future load values that are not in the historical load data, i.e., $\widehat{z}_6$ and $\widehat{z}_7$:

**(1)** evaluation of $\widehat{z}_3$ and $\varepsilon_3$:

$$\widehat{z}_3 = \widehat{y}_4^{(1)} = c + \phi_1 z_2 + \phi_2 z_1 + \theta_1 \varepsilon_2 + \theta_2 \varepsilon_1$$

$$= -0.43 + (-0.4) \times (-0.43) + 0.3 \times (-0.43) + 0.5 \times 0.15 + 0.2 \times 0.24$$

$$= -0.26$$

$$\varepsilon_3 = z_3 - \widehat{z}_3 = 0.49 - (-0.26) = 0.75$$

**(2)** evaluation of $\widehat{z}_4$ and $\varepsilon_4$:

$$\widehat{z}_4 = \widehat{y}_5^{(1)} = c + \phi_1 z_3 + \phi_2 z_2 + \theta_1 \varepsilon_3 + \theta_2 \varepsilon_2$$

$$= -0.43 + (-0.4) \times 0.49 + 0.3 \times (-0.28) + 0.5 \times 0.75 + 0.2 \times 0.15$$

$$= -0.31$$

$$\varepsilon_4 = z_4 - \widehat{z}_4 = -0.23 - (-0.31) = 0.08$$

**(3)** evaluation of $\widehat{z}_5$ and $\varepsilon_5$:

$$\widehat{z}_5 = \widehat{y}_6^{(1)} = c + \phi_1 z_4 + \phi_2 z_3 + \theta_1 \varepsilon_4 + \theta_2 \varepsilon_3$$

$$= -0.43 + (-0.4) \times (-0.23) + 0.3 \times 0.49 + 0.5 \times 0.08 + 0.2 \times 0.75 = 0$$

$$\varepsilon_5 = z_5 - \widehat{z}_5 = -1.93 - 0 = -1.93$$

**(4)** evaluation of $\widehat{z}_6$ and $\varepsilon_6$:

$$\widehat{z}_6 = \widehat{y}_7^{(1)} = c + \phi_1 z_5 + \phi_2 z_4 + \theta_1 \varepsilon_5 + \theta_2 \varepsilon_4$$

$$= -0.43 + (-0.4) \times (-1.93) + 0.3 \times (-0.23) + 0.5 \times (-1.93) + 0.2 \times 0.08$$

$$= -0.68$$

$$\varepsilon_6 = 0$$

**(5)** evaluation of $\widehat{z}_7$ and $\varepsilon_7$:

$$\widehat{z}_7 = \widehat{y}_8^{(1)} = c + \phi_1 \widehat{z}_6 + \phi_2 z_5 + \theta_1 \varepsilon_6 + \theta_2 \varepsilon_5$$

$$= -0.43 + (-0.4) \times (-6.68) + 0.3 \times (-1.93) + 0.5 \times 0 + 0.2 \times (-1.93)$$

$$= -1.12$$

$$\varepsilon_7 = 0$$

where $\widehat{z} = [\widehat{z}_1, \dots, \widehat{z}_7]$ is the fitted/forecasted value generated by the ARIMA model for the differenced time series of the historical load. The items in $\widehat{z}$ need to be postprocessed to obtain the fitted/forecasted load values as:

$$\widehat{y}_2 = y_1 + \widehat{y}_2^{(1)} = y_1 + \widehat{z}_1 = 5.23 - 0.43 = 4.8 \text{ kW}$$

$$\widehat{y}_3 = y_2 + \widehat{y}_3^{(1)} = y_2 + \widehat{z}_2 = 5.04 - 0.43 = 4.61 \text{ kW}$$

$$\widehat{y}_4 = y_3 + \widehat{y}_4^{(1)} = y_3 + \widehat{z}_3 = 4.76 - 0.26 = 4.50 \text{ kW}$$

$$\widehat{y}_5 = y_4 + \widehat{y}_5^{(1)} = y_4 + \widehat{z}_4 = 5.25 - 0.31 = 4.94 \text{ kW}$$

$$\widehat{y}_6 = y_5 + \widehat{y}_6^{(1)} = y_5 + \widehat{z}_5 = 5.02 - 0 = 5.02 \text{ kW}$$

$$\widehat{y}_7 = y_6 + \widehat{y}_7^{(1)} = y_6 + \widehat{z}_6 = 3.09 - 0.68 = 2.41 \text{ kW}$$

$$\widehat{y}_8 = \widehat{y}_8 + \widehat{y}_8^{(1)} = \widehat{y}_8 + \widehat{z}_7 = 2.41 - 1.12 = 1.29 \text{ kW}$$

Fig. 6.6 shows the comparison of a 12-h actual load profile (data source: the Australian "Smart Grid, Smart City" customer trial dataset [3]) and the fitted load values generated

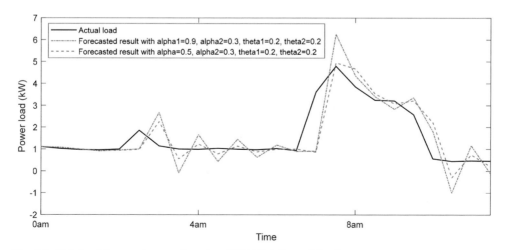

**Fig. 6.6** 12-h load forecasting results using ARIMA(2,1,2) model using two parameter sets.

with ARIMA(2,1,2) for the load profile using the following two parameter sets: (1) $\phi_1 = 0.4$, $\phi_2 = 0.1$, $\theta_1 = 0.6$, and $\theta_2 = -0.1$; and (2) $\phi_1 = 0.2$, $\phi_2 = 0.1$, $\theta_1 = 0.3$, and $\theta_2 = 0.2$.

### 6.4.10 Parameter selection for the ARIMA model

For an ARIMA($p,d,q$) model, there are $p + q$ parameters that need to be prespecified: $\phi_1$, ..., $\phi_p$ and $\theta_1$, ..., $\theta_q$.

When $p = 1$ or 2, the following empirical rules are usually applied to the selection of the parameter $\phi_1$ and $\phi_2$: (i) when $p = 1$, $-1 < \phi_1 < 1$; (ii) when $p = 2$, $-1 < \phi_2 < 1$, $\phi_1 + \phi_2 < 1$, and $\phi_2 - \phi_1 < 1$.

When $q = 1$ or 2, the following empirical rules are usually applied to the selection of the parameter $\theta_1$ and $\theta_2$: (i) when $p = 1$, $-1 < \theta_1 < 1$; (ii) when $p = 2$, $-1 < \theta_2 < 1$, $\theta_1 + \theta_2 > -1$, and $\theta_1 - \theta_2 < 1$.

For the cases of $p \geq 3$ and/or $q \geq 3$, the ranges to be used for the parameter values become more complex, and further information on these can be found in the dedicated references, e.g., [10].

Similarly to the parameter selection for the exponential smoothing models (discussed in Section 6.3.5), the SSE (Eq. 6.21) can be used to evaluate the performance of ARIMA model under different parameter settings, and the optimization techniques can be applied to find the optimal parameter values that can minimize the SSE from the historical observations.

## 6.5 ANN-based STLF

### 6.5.1 Introduction

Apart from statistical analysis-based methods (e.g., exponential smoothing and ARIMA models), ANNs [11,12] can also be used to process time series data and to perform forecasting. ANN-based methods represent the mainstream of STLF in recent years due to the strong ability of ANNs on analyzing the correlation among time series load data. In this section, we first introduce the basic concept of ANN, and we then provide a simple application example to illustrate how ANNs can be used for STLF tasks.

### 6.5.2 Biological background of ANNs

ANNs represent a subset of machine learning [13], a field of training computer systems to learn from data and to improve their performance. As the name "artificial neural networks" implies, ANNs are inspired by human brains and they mimic the way neurons signal to one another in biological neural networks.

The human brain is backboned by a huge number of interconnected cells, called neurons. It is estimated that there are approximately 86 billion neurons in a human brain [14].

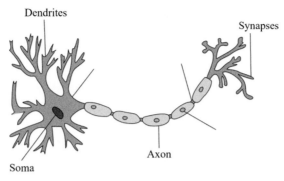

**Fig. 6.7** Schematic of a biological neuron [15].

The interconnected neurons send chemical materials to each other to help the mind conceptualize the environment, interpret things, reason, and make decisions. Fig. 6.7 depicts a biological neuron, which consists of the following basic elements:

- *dendrites*—they are tree-like branches that act as a receiver of the information sent from other neurons it connects to;
- *soma*—it is the body of the neuron and it is responsible for processing the information received by the dendrite;
- *axon*—it is a long, slender projection on a neuron and acts as a "cable" through which neurons can send information; and
- *synapses*—it consists of the connection between the axon and the dendrites of other neurons.

In biological neural networks, when a neuron is excited, it sends chemical materials to other neurons it is connected to, and this will change the electric potential of those neurons. If the electric potential of a neuron exceeds a threshold, the neuron will be activated and it will send chemical materials to other neurons.

### 6.5.3 M-P neuron model

In 1943, inspired by the working mechanism of biological neurons, McCulloch and Pitts established a simple artificial neuron model, called the "M-P neuron model" [16], which consisted of the basic component of ANNs. Fig. 6.8 shows the model of a M-P neuron.

In the M-P neuron model depicted in Fig. 6.8, a neuron (called the "target neuron") connects to $N$ other neurons (called "external neurons") with weighted connections. The target neuron receives the inputs from the $N$ external neurons and generates an output $\hat{y}$. By denoting the input from the $i$th ($i = 1:N$) external neuron as $x_i$ and the weight of the channel connecting the target neuron and the $i$th external neuron as $\omega_i$, the weighted sum of the inputs from the external neurons can be combined with a bias $\theta$, and the sum is

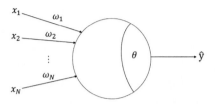

**Fig. 6.8** M-P neuron model.

processed by an *activation function* to determine the output $\widehat{y}$. This process can be expressed as:

$$\widehat{y} = f\left(\sum_{n=1}^{N} x_i \omega_i + \theta\right) \tag{6.36}$$

where the function $f(\cdot)$ is the *activation function*. Ideally, the activation function should be a step function (depicted by sgn($\cdot$)), which maps an input to be a binary value "1" or "0", indicating when the neuron is activated or inhibited as:

$$\text{sgn}(x) = \begin{cases} 1, & \text{if } x \geq 0 \\ 0, & \text{if } x < 0 \end{cases} \tag{6.37}$$

Since the step function has some unfriendly mathematical properties, e.g., discontinuity, the Sigmoid function is usually used in practice as the activation function. The Sigmoid function converts an input into an output value within (0,1):

$$\text{Sigmoid}(x) = \frac{1}{1 + e^{-x}} \tag{6.38}$$

Fig. 6.9 shows the comparison between the step function and the Sigmoid function. There are other functions that can be used as the activation function, such as the Tanh

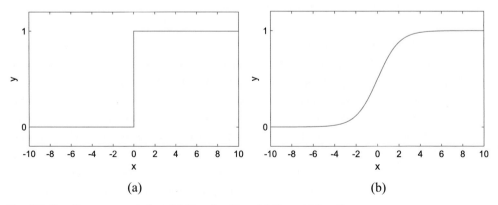

**Fig. 6.9** Function representation. (a) Step function. (b) Sigmoid function.

function, Linear function, Rectified Linear Units (ReLU) function, and Softmax function. Reference [17] provides a detailed introduction to the mathematical forms of different kinds of activation functions. By organizing multiple interconnected neurons as a layered structure, we can build up an ANN.

### 6.5.4 Perceptron

An ANN consists of multiple interconnected neurons that are organized as a layered structure. The simplest layered structure is the 2-layer structure, including an input layer and an output layer. The neurons in the input layer are responsible for sending input data. In the output layer, each neuron is a M-P neuron, and it connects to all the input neurons. Each neuron in the output layers receives the inputs from all the neurons in the input layer and produces an output based on Eq. (6.36).

An ANN with such a 2-layer structure is called a *perceptron*, shown in Fig. 6.10. In the figure, $N$ and $L$ represent the number of neurons in the input layer and in the output layer, respectively. $x_i$ $(i=1:N)$ represents the input from the $i$th neuron in the input layer. $\hat{y}_j$ and $\theta_j$ $(j=1:L)$ represent the output and the bias of the $j$th neuron in the output layer, respectively. $\omega_{i,j}$ $(i=1:N, j=1:L)$ represents the weight of the channel connecting the $i$th neuron in the input layer and the $j$th neuron in the output layer.

In perceptron, only one neuron layer (i.e., the output layer) has the ability of using the activation function to process information. The neurons in the input layer simply deliver input data to the output layer. Because of this, the perceptron is also called the one-layer ANN.

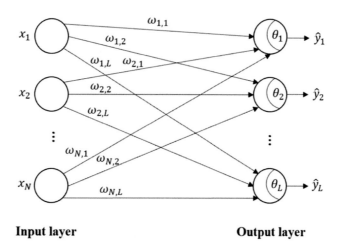

**Fig. 6.10** Structure of the perceptron.

### 6.5.5 Multi-layer feedforward neural networks

The simple structure of a perceptron exhibits limited performance in performing machine learning tasks. A more common way for building up ANNs is to set up one or more neuron layers between the input and output layers. These layers are called *hidden layers*, and such a neuron network has the following features:

**(1)** the neuron network has one input layer, one or more hidden layers, and one output layer;

**(2)** the neurons in the hidden layer(s) and in the output layer are M-P neurons;

**(3)** the neurons in the input layer are only responsible for sending data to the first hidden layer;

**(4)** a neuron does not connect to other neurons in the same layer;

**(5)** a neuron connects to all the neurons in the neighboring layer(s); and

**(6)** a neuron does not connect to any neurons in nonneighboring layers.

An ANN with such a multi-layer structure is called a *multi-layer feedforward neural network*. The word "feedforward" is used to indicate there are no loop connections in the neural network. Fig. 6.11 illustrates the structure of a multi-layer feedforward neural network consisting of three hidden layers. In a multi-layer feedforward neural network, all neurons in the hidden and output layers are associated with biases and activation functions, while the neurons in the input layer are not associated with biases and activation functions.

### 6.5.6 Training of ANNs

The performance of an ANN relies on the setting of the weights of the connection channels and the biases of the neurons. For example, for a building power load forecasting

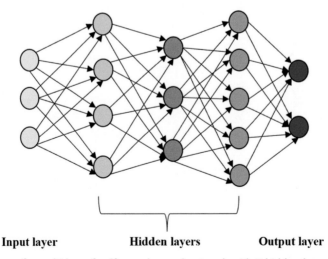

**Input layer        Hidden layers        Output layer**

**Fig. 6.11** Structure of a multi-layer feedforward neural network with 3 hidden layers.

task, it is desirable that, by properly setting the values of the weights and thresholds, the ANN can accurately forecast the load of the considered building in a future time point. This is achieved through a *training* process conducted on a *training dataset*.

A training dataset consists of multiple *samples* and can be represented as:

$$\mathbf{D}^{train} = \{(\boldsymbol{x}_1, \boldsymbol{y}_1), (\boldsymbol{x}_2, \boldsymbol{y}_2), ..., (\boldsymbol{x}_M, \boldsymbol{y}_M)\}, \quad \boldsymbol{x}_m \in \mathbb{R}^N, \boldsymbol{y}_m \in \mathbb{R}^L, m = 1 : M \qquad (6.39)$$

where $\mathbf{D}^{train}$ represents the training dataset; $M$ is the number of samples in the training dataset; $(\boldsymbol{x}_m, \boldsymbol{y}_m)$ represents the $m$th ($i = 1:M$) sample in the training dataset, in which $\boldsymbol{x}_m$ is the input of the sample, which consists of $N$ real values; and $\boldsymbol{y}_m$ is the output of the sample, which consist of $L$ real values. The input and output of a sample can be expressed as a $N$-dimensional vector and a $L$-dimensional vector:

$$\boldsymbol{x}_m = [\boldsymbol{x}_{m,1}, \boldsymbol{x}_{m,2}, ..., \boldsymbol{x}_{m,N}], \quad m = 1 : M \qquad (6.40)$$

$$\boldsymbol{y}_m = [\boldsymbol{y}_{m,1}, \boldsymbol{y}_{m,2}, ..., \boldsymbol{y}_{m,L}], \quad m = 1 : M \qquad (6.41)$$

The samples are used for the training of an ANN. For the ANN to be trained, $N$ and $L$ neurons are set up in the input layer and in the output layer, respectively. For each sample, the input $\boldsymbol{x}_m$ is provided at the input layer, where each dimensional value $x_{m,i}$ ($i = 1: N$) is inputted into a neuron, which delivers it to the neurons in the output layer (for perceptron) or to the neurons in the first hidden layer (for multi-layer feedforward neural networks). Based on the weights of the connection channels and the bias values in the neural network, the $j$th ($j = 1:L$) neuron generates an output of one dimension for the sample, denoted as $\widehat{y}_{m,j}$. In this way, the output generated by the output layer for the $i$th sample can be denoted as $\widehat{\boldsymbol{y}}_m = [\widehat{y}_{m,1}, \widehat{y}_{m,2}, ..., \widehat{y}_{m,L}]$.

The output generated by the ANN $\widehat{\boldsymbol{y}}_m$ is then compared with the "true" output $\boldsymbol{y}_m$, and the difference (called the "error") is used for adjusting the weights and biases of the neurons in the hidden layer(s) and in the output layer. The training process for an ANN uses a training dataset to determine the parameters of the ANN (i.e., weights and biases), and the knowledge "learnt" by an ANN depends on these parameters.

A trained ANN is often tested against a dataset (denoted as *test dataset*) containing samples that are not used for ANN training. The purpose of this operation is to check if it can still generate outputs that are close to the true outputs of the samples in the test dataset.

In the next two sections, we introduce the training methods that are commonly used for the perceptron and multi-layer feedforward neural networks.

## 6.5.7 Training of a perceptron

For a perceptron with $N$ input neurons and $L$ output neurons, there are a total of $N \times L + L$ parameters to be trained. This includes $N \times L$ weights of the channels connecting the input and output neurons (i.e., $\omega_{i,j}$ ($i = 1:N, j = 1:L$)) and the $L$ biases of the output neurons (i.e., $\theta_j$ $(j=1:L)$). Given a training sample $(\boldsymbol{x} \in \mathbb{R}^N, \boldsymbol{y} \in \mathbb{R}^L)$, the parameters of the perceptron are updated as follows:

$$\omega_{i,j} = \omega_{i,j} + \Delta\omega_{i,j}, \quad i = 1:N, \ j = 1:L \tag{6.42}$$

$$\Delta\omega_{i,j} = -\eta\left(y_j - \widehat{y}_j\right)x_i \tag{6.43}$$

$$\theta_j = \theta_j + \Delta\theta_j, \quad j = 1:L \tag{6.44}$$

$$\Delta\theta_j = -\eta\left(y_j - \widehat{y}_j\right) \tag{6.45}$$

where $\eta \in (0,1)$ is a prespecified parameter called the learning rate. Eqs. (6.42)–(6.45) show that if the output of the perceptron is equal to the true output in the sample, then the perceptron's parameters do not change or, otherwise, they will be adjusted based on the error and the learning rate.

## 6.5.8 Training of a multi-layer feedforward neural network

### 6.5.8.1 Overview

Compared to the perceptron, a multi-layer feedforward neural network is more complex since it has one or multiple hidden layers. Without any loss of generality, let us consider a multi-layer feedforward neural network with $N$ neurons in the input layer, $L$ neurons in the output layer, and $K$ hidden layers. This implies that the input data of the neural network has $N$ dimensional values and the output has $L$ dimensional values. We divide the parameters of the multi-layer feedforward neural network into two categories: (i) the parameters associated with the output layer these parameters directly affect the output of the neural network; and (ii) the parameters associated with the hidden layer (s) these parameters indirectly affect the output of the neural network.

We index the hidden layers as 1, 2, ..., $K$ and consider the input layer as the 0th layer and the output layer as the $(K+1)$th layer. The notation used for representing a multi-layer feedforward neural network is collected in Table 6.5.

There are a total of $NI^1 + \sum_{k=2}^{K} I^{k-1}I^k + I^K L$ weights to be trained and these include: (1) the weights of the $NI^1$ channels connecting the $N$ neurons in the input layer and the $I^1$ neurons in the first hidden layer; (2) the weights of the $\sum_{k=2}^{K} I^{k-1}I^k$ channels connecting two neighboring hidden layers; and (3) the weights of the $I^K L$ channels connecting the last hidden layer and the output layer. There is also a total of $\sum_{k=1}^{K} I^k + L$ bias

**Table 6.5** Notation for representing a multi-layer feedforward neural network.

| Notation | Definition |
|---|---|
| $N$ | Number of neurons in the input layer |
| $L$ | Number of neurons in the output layer |
| $K$ | Number of hidden layers |
| $I^k$ | Number of neurons in the $k$th layer ($k=0:K+1$, $I^0=N$, $I^{K+1}=L$) |
| $f_j^{K+1}(\bullet)$ | Activation function used by the $j$th neuron in the output layer ($j=1:L$) |
| $f_i^k(\bullet)$ | Activation function used by the $i$th neuron in the $k$th hidden layer ($k=1:K$, $i=I^k$) |
| $\eta$ | Learning rate |
| $\theta_j^{K+1}$ | Bias of the $j$th neuron in the output layer ($j=1:L$) |
| $\theta_i^k$ | Bias of the $i$th neuron in the $k$th hidden layer ($k=1:K$, $i=1:I^k$) |
| $\omega_{i,j}^{K+1}$ | Weight of the channel connecting the $i$th neuron in the last hidden layer and the $j$th neuron in the output layer ($i=1:I^K$, $j=1:L$) |
| $\omega_{z,i}^k$ | Weight of the channel connecting the $z$th neuron in the $(k-1)$th layer and the $i$th neuron in the $k$th layer ($k=1:K$, $z=1:I^{k-1}$, $i=1:I^k$) |
| $\boldsymbol{x}$ | Input of the training sample |
| $\boldsymbol{y}$ | Output of the training sample |
| $x_i$ | $i$th dimensional value in the input of the training sample $\boldsymbol{x}=[x_1,\cdots,x_N]$. It is also the input into the $i$th neuron in the input layer ($i=1:N$) |
| $y_j$ | $j$th dimensional value in the output of the training sample $\boldsymbol{y}=[y_1,\cdots,y_L]$ ($j=1:L$) |
| $b_i^k$ | Input into the activation function used by the $i$th neuron in the $k$th hidden layer ($k=1:K$, $i=1:I^k$) |
| $b_j^{K+1}$ | Input into the activation function used by the $j$th neuron in the output layer ($j=1:L$) |
| $\widehat{y}_j^{K+1}$ | Output generated by the $j$th neuron in the output layer ($j=1:L$) |
| $\widehat{y}_i^k$ | Output generated by the $i$th neuron in the $k$th layer ($k=0:K$, $i=1:I^k$). For $k=0$, $\widehat{y}_i^0 = x_i$ ($i=1:N$) |

values to be trained that consist of: (1) the biases of the $\sum_{k=1}^{K} I^k$ neurons in the $K$ hidden layers; and (2) the biases of $L$ neurons in the output layer. Therefore, for a multi-layer feedforward neural network, there are $NI^1 + \sum_{k=2}^{K} I^{k-1} I^k + I^K L + \sum_{k=1}^{K} I^k + L$ parameters to be trained.

The most widely used method for training multi-layer feedforward neural networks is called the *error backpropagation algorithm*, also known as the "BP" algorithm. Given a training sample ($\boldsymbol{x} \in \mathbb{R}^N$, $\boldsymbol{y} \in \mathbb{R}^L$), the BP algorithm starts by evaluating the *sum of the squared errors* (denoted as $E$ here in short) between the neural network's output and the true output in the sample:

$$E = \frac{\sum_{j=1}^{L}\left(\widehat{y}_j^{K+1} - y_j\right)}{2} \tag{6.46}$$

based on which the BP algorithm updates the neural network's parameters as follows:

$$\omega_{z,i}^k = \omega_{z,i}^k + \Delta\omega_{z,i}^k, \quad k = 1:K+1, z = 1:I^{k-1}, i = 1:I^k \tag{6.47}$$

$$\theta_i^k = \theta_i^k + \Delta\theta_i^k, \quad k = 1:K+1, i = 1:I^k \tag{6.48}$$

As the name "error backpropagation" implies, BP algorithm considers the sum of the squared errors as an error signal that affects the update of the parameters of the output layer and "propagates" backwards to affect the parameters' updates of the hidden layers. The BP algorithm uses a gradient based method to determine the updated magnitudes of the parameters, i.e., $\Delta\omega_{z,i}^k$ and $\Delta\theta_i^k$ ($k=1:K+1$, $i=1:I^k$). Based on this error backpropagation principle, we introduce the update methodology for the output layer in the next section, followed by the update methodology for the hidden layers.

### 6.5.8.2 Parameter update for an output layer

To determine $\Delta\omega_{i,j}^{K+1}$ and $\Delta\theta_j^{K+1}$ ($i=1:I^K$, $j=1:L$), the BP algorithm calculates the gradient of the sum of the squared errors with respect to $\omega_{i,j}^{K+1}$ and $\theta_j^{K+1}$. For the weight $\omega_{i,j}^{K+1}$, the gradient is calculated as:

$$\frac{\partial E}{\partial \omega_{i,j}^{K+1}} = \frac{\partial \sum_{j=1}^{L}\frac{\left(\widehat{\gamma}_j^{K+1}-\gamma_j\right)^2}{2}}{\partial \omega_{i,j}^{K+1}} = \left(\widehat{\gamma}_j^{K+1}-\gamma_j\right)\frac{\partial\left(\widehat{\gamma}_j^{K+1}-\gamma_j\right)}{\partial \omega_{i,j}^{K+1}} \tag{6.49}$$

Since $\gamma_j$ does not depend on $\omega_{i,j}^{K+1}$ and $\widehat{\gamma}_j^{K+1} = f_j^{K+1}\left(b_j^{K+1}\right)$, Eq. (6.49) can be re-written as:

$$\frac{\partial E}{\partial \omega_{i,j}^{K+1}} = \left(\widehat{\gamma}_j^{K+1}-\gamma_j\right)\frac{\partial\left(\widehat{\gamma}_j^{K+1}-\gamma_j\right)}{\partial \omega_{i,j}^{K+1}} = \left(\widehat{\gamma}_j^{K+1}-\gamma_j\right)\frac{\partial\widehat{\gamma}_j^{K+1}}{\partial \omega_{i,j}^{K+1}}$$

$$= \left(\widehat{\gamma}_j^{K+1}-\gamma_j\right)\frac{\partial f\left(b_j^{K+1}\right)}{\partial \omega_{i,j}^{K+1}}$$

$$= \left(\widehat{\gamma}_j^{K+1}-\gamma_j\right)f_j^{K+1'}\left(b_j^{K+1}\right)\frac{\partial b_j^{K+1}}{\partial \omega_{i,j}^{K+1}} \tag{6.50}$$

$b_j^{K+1}$ is calculated as:

$$b_j^{K+1} = \sum_{i=1}^{I^K}\omega_{i,j}^{K+1}\widehat{\gamma}_i^K + \theta_j^{K+1} = \sum_{i=1}^{I^K}\omega_{i,j}^{K+1}f_i^K\left(b_i^K\right) + \theta_j^{K+1} \tag{6.51}$$

and:

$$\frac{\partial b_j^{K+1}}{\partial \omega_{i,j}^{K+1}} = f_i^K\left(b_i^K\right) = \widehat{\gamma}_i^K \tag{6.52}$$

By substituting Eq. (6.52) into Eq. (6.50), we obtain:

$$\frac{\partial E}{\partial \omega_{i,j}^{K+1}} = \left(\widehat{y}_j^{K+1} - y_j\right) f_j^{K+1\prime}\left(b_j^{K+1}\right) \frac{\partial b_j^{K+1}}{\partial \omega_{i,j}^{K+1}} = \left(\widehat{y}_j^{K+1} - y_j\right) f_j^{K+1\prime}\left(b_j^{K+1}\right) \widehat{y}_i^{K} \quad (6.53)$$

We now define $\delta_j^{K+1}$ to denote the error signal at the $j$th neuron in the output layer, which includes all the items covering the subscript $j$ in Eq. (6.53):

$$\delta_j^{K+1} = \left(\widehat{y}_j^{K+1} - y_j\right) f_j^{K+1\prime}\left(b_j^{K+1}\right) \quad (6.54)$$

that can be re-written as:

$$\frac{\partial E}{\partial \omega_{i,j}^{K+1}} = \delta_j^{K+1} \widehat{y}_i^{K} \quad (6.55)$$

Based on the gradient, $\Delta \omega_{i,\,j}^{K+1}$ is determined as:

$$\Delta \omega_{i,j}^{K+1} = -\eta \frac{\partial E}{\partial \omega_{i,j}^{K+1}} = -\eta \left(\widehat{y}_j^{K+1} - y_j\right) f_j^{K+1\prime}\left(b_j^{K+1}\right) \widehat{y}_i^{K} = -\eta \delta_j^{K+1} \widehat{y}_i^{K} \quad (6.56)$$

$\omega_{i,\,j}^{K+1}$ is then updated as:

$$\omega_{i,j}^{K+1} = \omega_{i,j}^{K+1} + \Delta \omega_{i,j}^{K+1} = \omega_{i,j}^{K+1} - \eta \delta_j^{K+1} \widehat{y}_i^{K}, \quad i = 1 : I^K, j = 1 : L \quad (6.57)$$

We then derive the update method for the biases $\theta_j^{K+1}$ ($j$=1:$L$). Similarly to the weights, we first calculate the gradient:

$$\frac{\partial E}{\partial \theta_j^{K+1}} = \left(\widehat{y}_j^{K+1} - y_j\right) \frac{\partial \left(\widehat{y}_j^{K+1} - y_j\right)}{\partial \theta_j^{K+1}} = \left(\widehat{y}_j^{K+1} - y_j\right) f_j^{K+1\prime}\left(b_j^{K+1}\right) \frac{\partial b_j^{K+1}}{\partial \theta_j^{K+1}} \quad (6.58)$$

Based on the expanded expression of $b_j^{K+1}$ in Eq. (6.51), we can express:

$$\frac{\partial b_j^{K+1}}{\partial \theta_j^{K+1}} = \frac{\partial \left[\sum\limits_{i=1}^{I^K} \omega_{i,j}^{K+1} f_i^{K}\left(b_i^{K}\right) + \theta_j^{K+1}\right]}{\partial \theta_j^{K+1}} = 1 \quad (6.59)$$

and:

$$\frac{\partial E}{\partial \theta_j^{K+1}} = \left(\widehat{y}_j^{K+1} - y_j\right) f_j^{K+1\prime}\left(b_j^{K+1}\right) = \delta_j^{K+1} \quad (6.60)$$

Based on the gradient, $\Delta\theta_j^{K+1}$ is calculated as:

$$\Delta\theta_j^{K+1} = -\eta \frac{\partial E}{\partial \theta_j^{K+1}} = -\eta\left(\widehat{y}_j^{K+1} - y_j\right)f_j^{K+1'}\left(b_j^{K+1}\right) = -\eta\delta_j^{K+1} \tag{6.61}$$

Based on Eq. (6.48), $\theta_j^{K+1}$ is then updated as:

$$\theta_j^{K+1} = \theta_j^{K+1} - \eta\delta_j^{K+1}, \quad j = 1:L \tag{6.62}$$

### 6.5.8.3 Parameter update for the last hidden layer

In this section, we introduce the update process for the parameters of the hidden layers (i.e., $\omega_{z,i}^k$ and $\theta_i^k$ ($k=1:K$, $z=1:I^{k-1}$, $i=1:I^k$). Let us consider the parameter update for the last hidden layer first (i.e., the $K$th hidden layer, whose outputs will be fed into the output layer), and then we will generalize the methodology to the parameter update for an arbitrary hidden layer in the next section.

Similarly to the parameter update process applied to the output layer, the BP algorithm calculates the gradient of the sum of the squared errors of the neural network's output with respect to $\omega_{z,i}^K$ and $\theta_i^K$. For $\omega_{z,i}^K$, the gradient is calculated as:

$$\frac{\partial E}{\partial \omega_{z,i}^K} = \frac{\partial \frac{\sum_{j=1}^{L}\left(\widehat{y}_j^{K+1} - y_j\right)^2}{2}}{\partial \omega_{z,i}^K} = \sum_{j=1}^{L}\left(\widehat{y}_j^{K+1} - y_j\right)\frac{\partial \widehat{y}_j^{K+1}}{\partial \omega_{z,i}^K} \tag{6.63}$$

Since $\widehat{y}_j^{K+1} = f_j^{K+1}\left(b_j^{K+1}\right)$, Eq. (6.63) can be re-written as:

$$\frac{\partial E}{\partial \omega_{z,i}^K} = \sum_{j=1}^{L}\left(\widehat{y}_j^{K+1} - y_j\right)\frac{\partial \widehat{y}_j^{K+1}}{\partial \omega_{z,i}^K} = \sum_{j=1}^{L}\left(\widehat{y}_j^{K+1} - y_j\right)\frac{\partial f_j^{K+1}\left(b_j^{K+1}\right)}{\partial \omega_{z,i}^K}$$
$$= \sum_{j=1}^{L}\left(\widehat{y}_j^{K+1} - y_j\right)f_j^{K+1'}\left(b_j^{K+1}\right)\frac{\partial b_j^{K+1}}{\partial \omega_{z,i}^K} \tag{6.64}$$

In Eq. (6.64), we need to calculate the partial derivative $\frac{\partial b_j^{K+1}}{\partial \omega_{z,i}^K}$ and $b_j^{K+1}$ is the input into the activation function used by the $j$th neuron in the output layer that depends on the outputs of the neurons in the $K$th hidden layer. The outputs of the neurons in the $K$th hidden layer depend also on the inputs of the activation functions used by those neurons. Therefore, $b_j^{K+1}$ can be expanded as:

$$b_j^{K+1} = \sum_{i=1}^{I^K}\omega_{i,j}^{K+1}\widehat{y}_i^K + \theta_j^{K+1} = \sum_{i=1}^{I^K}\omega_{i,j}^{K+1}f_i^K\left(b_i^K\right) + \theta_j^{K+1}$$
$$= \sum_{i=1}^{I^K}\omega_{i,j}^{K+1}f_i^K\left(\sum_{z=1}^{I^{K-1}}\omega_{z,i}^K\widehat{y}_z^{K-1} + \theta_i^K\right) + \theta_j^{K+1} \tag{6.65}$$

that shows how $b_j^{K+1}$ indirectly depends on $\omega_{z,i}^K$. Since $g_j = \sum_{i=1}^{I^K} v_{i,j} z_i^K + \theta_j$, $z_i^K = f(b_i^K)$, and $b_i^K = \sum_{z=1}^{I^{K-1}} \omega_{z,i}^K z_z^{K-1} + \gamma_i^K$, the following expression can be obtained by applying the chain rule:

$$\frac{\partial b_j^{K+1}}{\partial \omega_{z,i}^K} = \frac{\partial b_j^{K+1}}{\partial \widehat{y}_i^K} \frac{\partial \widehat{y}_i^K}{\partial \omega_{z,i}^K} = \frac{\partial \sum_{i=1}^{I^K} \omega_{i,j}^{K+1} \widehat{y}_i^K + \theta_j^{K+1}}{\partial \widehat{y}_i^K} \frac{\partial \widehat{y}_i^K}{\partial \omega_{z,i}^K} = \omega_{i,j}^{K+1} \frac{\partial \widehat{y}_i^K}{\partial \omega_{z,i}^K} = \omega_{i,j}^{K+1} \frac{\partial f_i^K(b_i^K)}{\partial \omega_{z,i}^K}$$

$$= \omega_{i,j}^{K+1} f_i^{K\prime}(b_i^K) \frac{\partial b_i^K}{\partial \omega_{z,i}^K} = \omega_{i,j}^{K+1} f_i^{K\prime}(b_i^K) \frac{\partial \left( \sum_{z=1}^{I^{K-1}} \omega_{z,i}^K \widehat{y}_z^{K-1} + \theta_i^K \right)}{\partial \omega_{z,i}^K}$$

$$= \omega_{i,j}^{K+1} f_i^{K\prime}(b_i^K) \widehat{y}_z^{K-1}$$

$$(6.66)$$

We substitute Eq. (6.66) into Eq. (6.64) and define the error signal that has backpropagated to the $i$th neuron in the last hidden layer as:

$$\delta_i^K = \sum_{j=1}^L \left( \widehat{y}_j^{K+1} - y_j \right) f_j^{K+1\prime} \left( b_j^{K+1} \right) \omega_{i,j}^{K+1} f_i^{K\prime}(b_i^K) = \sum_{j=1}^L \delta_j^{K+1} \omega_{i,j}^{K+1} f_i^{K\prime}(b_i^K) \quad (6.67)$$

and:

$$\frac{\partial E}{\partial \omega_{z,i}^K} = \sum_{j=1}^L \delta_j^{K+1} \omega_{i,j}^{K+1} f_i^{K\prime}(b_i^K) \widehat{y}_z^{K-1} = \delta_i^K \widehat{y}_z^{K-1} \qquad (6.68)$$

From the expression $\frac{\partial E}{\partial \omega_{z,i}^K} = \delta_i^K \widehat{y}_z^{K-1}$, we can see that to calculate the gradient of the weight between the $z$th neuron in the second last hidden layer and the $i$th neuron in the last hidden layer, we need to calculate the error signal that has moved backwards to the $i$th neuron from the output layer ($\delta_i^K$) and the output generated from the $z$th neuron in the last hidden layer ($\widehat{y}_z^{K-1}$).

For the biases in the last hidden layer $\theta_i^K$ ($k=1:I^K$), its update process can be obtained in a similar manner as the weights. Firstly, the gradient is calculated as:

$$\frac{\partial E}{\partial \theta_i^K} = \frac{\partial \frac{\sum_{j=1}^L \left( \widehat{y}_j^{K+1} - y_j \right)^2}{2}}{\partial \theta_i^K} = \sum_{j=1}^L \left( \widehat{y}_j^{K+1} - y_j \right) \frac{\partial \widehat{y}_j^{K+1}}{\partial \theta_i^K}$$

$$= \sum_{j=1}^L \left( \widehat{y}_j^{K+1} - y_j \right) f_j^{K+1\prime} \left( b_j^{K+1} \right) \frac{\partial b_j^{K+1}}{\partial \theta_i^K} \qquad (6.69)$$

and, by applying the chain rule, $\frac{\partial b_j^{K+1}}{\partial \theta_i^K}$ can be calculated as:

$$\frac{\partial b_j^{K+1}}{\partial \theta_i^K} = \frac{\partial b_j^{K+1}}{\partial \widehat{\gamma}_i^K}\frac{\partial \widehat{\gamma}_i^K}{\partial \theta_i^K} = \frac{\partial \sum_{i=1}^{I^K}\omega_{i,j}^{K+1}\widehat{\gamma}_i^K + \theta_j^{K+1}}{\partial \widehat{\gamma}_i^K}\frac{\partial \widehat{\gamma}_i^K}{\partial \theta_i^K} = \omega_{i,j}^{K+1}\frac{\partial \widehat{\gamma}_i^K}{\partial \theta_i^K} = \omega_{i,j}^{K+1}\frac{\partial f_i^K(b_i^K)}{\partial \theta_i^K}$$

$$= \omega_{i,j}^{K+1}f_i^{K'}(b_i^K)\frac{\partial b_i^K}{\partial \theta_i^K} = \omega_{i,j}^{K+1}f_i^{K'}(b_i^K)\frac{\partial \left(\sum_{z=1}^{I^{K-1}}\omega_{z,i}^K\widehat{\gamma}_z^{K-1} + \theta_i^K\right)}{\partial \theta_i^K}$$

$$= \omega_{i,j}^{K+1}f_i^{K'}(b_i^K)$$

(6.70)

By substituting Eq. (6.70) into Eq. (6.69), we can get:

$$\frac{\partial E}{\partial \theta_i^K} = \sum_{j=1}^{L}\left(\widehat{\gamma}_j^{K+1} - \gamma_j\right)f_j^{K+1'}(b_j^{K+1})\omega_{i,j}^{K+1}f_i^{K'}(b_i^K)$$

$$= f_i^{K'}(b_i^K)\sum_{j=1}^{L}\delta_j^{K+1}\omega_{i,j}^{K+1} = \delta_i^K$$

(6.71)

The parameters of the last hidden layer are then updated as:

$$\omega_{z,i}^K = \omega_{z,i}^K - \eta\delta_i^K\widehat{\gamma}_z^{K-1}, \quad i=1:I^K, z=1:I^{K-1}$$

(6.72)

$$\theta_i^K = \theta_i^K - \eta\delta_i^K, \quad i=1:I^K$$

(6.73)

### 6.5.8.4 Parameter update for an arbitrary hidden layer

Eqs. (6.72) and (6.73) can be generalized to describe the calculation of the weight gradients and bias gradients for an arbitrary $k$th hidden layer in a multi-layer feedback neuron network as:

$$\frac{\partial E}{\partial \omega_{z,i}^k} = \delta_i^k\widehat{\gamma}_z^{k-1}, \quad k=1:K, i=1:I^k, z=1:I^{k-1}$$

(6.74)

$$\frac{\partial E}{\partial \theta_i^k} = \delta_i^k, \quad k=1:K, i=1:I^k$$

(6.75)

where $\delta_i^k = \sum_{z=1}^{I^{k+1}}\delta_z^{k+1}\omega_{i,z}^{k+1}f_i^{k'}(b_i^k)$.

Eq. (6.74) suggests that for an arbitrary hidden layer $k$, the weight gradient $\frac{\partial E}{\partial \omega_{z,i}^k}$ is calculated as the product of two terms: (i) $\delta_i^k$ – the error that has backpropagated to the $i$th neuron in the $k$th hidden layer from "below layers" (i.e., from the output layer to the $k$th hidden layer); and (ii) $\widehat{\gamma}_z^{k-1}$—the output generated from the $z$th neuron in the "above layer" (i.e., the $(k-1)$ th hidden layer or the input layer, and or $k=1$ the above layer represents the input layer).

Eq. (6.75) shows that for an arbitrary hidden layer $k$, the bias gradient $\frac{\partial E}{\partial \theta_i^k}$ is $\delta_i^k$. The weights and biases at an arbitrary $k$th hidden layer can then be updated as:

$$\omega_{z,i}^k = \omega_{z,i}^k - \eta \delta_i^k \, \widehat{y}_z^{k-1}, \quad k = 1:K, i = 1:I^k, z = 1:I^{k-1} \tag{6.76}$$

$$\theta_i^k = \theta_i^k - \eta \delta_i^k, \quad k = 1:K, i = 1:I^k \tag{6.77}$$

### 6.5.8.5 Overall workflow of training of a multi-layer feedforward neural network using BP algorithm

For convenience purposes, Table 6.6 summarizes the parameter update expressions for a multi-layer feedforward neural network which have been derived in the previous sections.

It is important to note that the equations in Table 6.6 represent recursive calculation processes. This implies that the calculation of $\delta_i^k$ starts from the calculation of $\delta_i^{K+1}$; based on $\delta_i^{K+1}$, $\delta_i^K$ can be calculated, and then $\delta_i^{K-1}$ and so on. This process continues until the value for $\delta_i^k$ is obtained that denotes the principle of error backpropagation. For the evaluation of $\widehat{y}_i^k$, we need to calculate the output of the neurons in the $(k-1)$th layer, i.e., $\widehat{y}_i^{k-1}$, which depends on $b_i^{k-1}$, and $b_i^{k-1}$ further depends on $\widehat{y}_i^{k-2}$ and so on until the input layer is reached. This implies that, to calculate $\widehat{y}_i^k$, we need to start from calculating $b_i^1$ $(i = 1:I^1)$ using the input of the training sample $x$; then we can calculate $\widehat{y}_i^1$ $(i = 1:I^1)$

**Table 6.6** Parameter update expressions for a multi-layer feedforward neural network.

| Layer | Expression |
|---|---|
| Output layer | $\delta_j^{K+1} = \left( \widehat{y}_j^{K+1} - y_j \right) f_j^{K+1\prime} \left( b_j^{K+1} \right)$ |
| | $\dfrac{\partial E}{\partial \omega_{i,j}^{K+1}} = \delta_j^{K+1} \widehat{y}_i^K$ |
| | $\dfrac{\partial E}{\partial \theta_j} = \delta_j^{K+1}$ |
| | $\omega_{i,j}^{K+1} = \omega_{i,j}^{K+1} - \eta \delta_j^{K+1} \widehat{y}_i^K$ |
| | $\theta_j = \theta_j - \eta \delta_j^{K+1}$ |
| $k$th hidden layer | $\delta_i^k = \sum\limits_{z=1}^{I^{k+1}} \delta_z^{k+1} \omega_{i,z}^{k+1} f_i^{k\prime} \left( b_i^k \right)$ |
| | $\dfrac{\partial E}{\partial \omega_{z,i}^k} = \delta_i^k \widehat{y}_i^{k-1}$ |
| | $\dfrac{\partial E}{\partial \theta_i^k} = \delta_i^k$ |
| | $\omega_{z,i}^k = \omega_{z,i}^k - \eta \delta_i^k \widehat{y}_z^{k-1}$ |
| | $\theta_i^k = \theta_i^k - \eta \delta_i^k$ |

based on $b_i^1$; then can determine $b_i^2$ $(i=1:I^2)$ based on $\widehat{y}_i^1$ and $\widehat{y}_i^2$ from $b_i^2$ until $\widehat{y}_i^k$ is evaluated from $b_i^k$ $(i=1:I^k)$.

Algorithm 6.1 shows the overall procedure for the training of a multi-layer feedforward neural network.

Algorithm 6.1 starts by preparing the training dataset (Line 1). The neural network structure is then set up and the activation functions and the learning rate are specified (Lines 2–4). The training process can now start. The algorithm initializes all the weights and biases of the neural network as random values in the range of [0, 1] (Line 5). It then adjusts the weights and biases subjected to every sample included in the training dataset. For each training data sample, it calculates the neural network's output based on the current values of the weights and biases in the network (Line 7) and it then updates the weights and biases of the output layer (Lines 8 and 9). The algorithm updates the weights and biases of the hidden layers layer-by-layer in a backward manner, starting from the last hidden layer (Lines 10–16). After the parameter update process has been applied to all training data samples, the neural network training process is completed.

---

### ALGORITHM 6.1 Procedure for the training of a multi-layer feedforward neural network using the BP algorithm.

**Start**

1. Input the training dataset $\mathbf{D}^{train}$;
2. Input the structural parameters of the neural network, i.e., $N$, $L$, $K$, $I^1$, ..., $I^K$);
3. Specify the activation function for each neuron in the hidden and output layers;
4. Specify the learning rate $\eta$;
5. Randomly initialize $\omega_{z,i}^k$ and $\theta_i^k$ $(k=1:K+1, z=1:I^{k-1}, i=1:I^k)$ within [0,1];
6. **For** each training data sample $(\mathbf{x}_m, \mathbf{y}_m)$ in $\mathbf{D}^{train}$;
7.     Calculate $\widehat{y}_m^{K+1}$ based on the current values of the weights and biases;
8.     Calculate $\dfrac{\partial E}{\partial \omega_{i,j}^{K+1}}$ and $\dfrac{\partial E}{\partial \theta_i^{K+1}}$ following Eqs. (6.55) and (6.60);
9.     Update $\omega_{i,j}^{K+1}$ and $\theta_i^{K+1}$ following Eqs. (6.57) and (6.62);
10.     Set $k=K$;
11.     Calculate $\dfrac{\partial E}{\partial \omega_{z,i}^k}$ and $\dfrac{\partial E}{\partial \theta_i^k}$ following the recursive processes described by Eqs. (6.74) and (6.75);
12.     Update $\omega_{z,i}^k$ and $\theta_i^k$ following Eqs. (6.76) and (6.77);
13.     Set $k=k-1$;
14.     **If** $k>0$
15.       **Go to** Line 11;
16.     **End If**
17. **End For**

**End**

### 6.5.8.6 Derivative of the activation function

Note that in the training process of a multi-layer feedforward neural network introduced earlier, we need to calculate the terms representing the error signal backpropagation, i.e., $\delta_i^k$ ($k=1:K+1$, $i=1:I^k$), which involve the calculation of the derivative of the activation function (denoted as $f'(\bullet)$). The explicit form of the derivative depends on the mathematical form of the used activation function. For example, if the Sigmoid function is used as the activation function, it has the following mathematical representation:

$$f'(x) = \text{Sigmoid}'(x) = \text{Sigmoid}(x)(1 - \text{Sigmoid}(x)) = \frac{1}{1 + e^{-x}} \times \left(1 - \frac{1}{1 + e^{-x}}\right)$$

(6.78)

The derivative in Eq. (6.78) needs to be substituted into the equations collected in Table 6.2 to calculate $\delta_i^k$. Reference [17] provides mathematical expressions for derivatives of other common activation functions.

## 6.5.9 Application example for an ANN-based STLF

We now use a simple example to demonstrate how ANNs can be used to perform STLF tasks.

### 6.5.9.1 Training load dataset

In this simple example, we assume the current date to be Wednesday and to have a historical load dataset containing only two samples, which represent the power load values of a building between 1:30pm and 2:20pm in the previous 2 days (Monday and Tuesday), respectively. The load values are sampled every 10min, i.e., each load value represents the building's power consumption over a period of 10min.

We want to use this dataset to train a multi-layer feedforward neural network, so that in future days (e.g., Wednesday, Thursday and Friday), once we know our building's power consumption between 1:30pm and 2pm, we can use the trained neural network to predict its power consumption in the future two time intervals, i.e., between 2 and 2:10pm and 2:10–2:20pm. For each sample in the training dataset, we treat the first 3 load values, which represent the building's power consumption between 1:30 and 2pm, as the input of the neural network. We then consider the last 2 load values that depict the building's power consumption between 2pm and 2:20pm, as the output of the neural network. Table 6.7 shows the 2-sample based historical load dataset used in this example.

### 6.5.9.2 Configuration of the neural network

For simplicity purposes, we consider a multi-layer feedforward neural network formed by a very simple structure to perform this STLF task. The neural network consists of one input layer, one hidden layer, and one output layer. According to the training dataset in Table 6.7, we have 3 input items and 2 output items, and the number of the neurons in the input layer and output layer are 3 and 2, respectively (i.e., $N=3$ and $L=2$). We

**Table 6.7** Historical load dataset used in the application example.

| | Input | | | Output | |
|---|---|---|---|---|---|
| | 1:30–1:40 pm | 1:40–1:50 pm | 1:50–2 pm | 2–2:10 pm | 2:10–2:20 pm |
| Sample 1 (Monday) | 4.3[a] | 3.8 | 3.5 | 3.5 | 3.2 |
| Sample 2 (Tuesday) | 5.8 | 4.2 | 6.1 | 6.4 | 7.7 |

[a]The unit of the data in the table is kW.

consider that there are only 2 neurons in the hidden layer. The structure of the neural network used in this application is illustrated in Fig. 6.12. The learning rate $\eta$ of the neural network is taken as 0.1.

When applying ANNs for predictive tasks (e.g., the STLF task in this example), activation functions such as Sigmoid are not applicable to the neurons of the output layer, because they will "normalize" the output of a neuron into the range of [0, 1], which obviously might not fit the value ranges of the outputs in the training dataset, e.g., the outputs in Table 6.7. Therefore, for predictive tasks, we usually assume the neurons in the output layer outputs the same value as the input. We use the *linear function* for the neurons in the output layer:

$$\text{Linear}(x) = x \tag{6.79}$$

As shown in Eq. (6.79), the linear activation function does not change the input of the neuron as the derivative of the linear activation function remains equal to 1:

$$\text{Linear}'(x) = 1 \tag{6.80}$$

In this example, we apply the Sigmoid function to the two neurons in the hidden layer and apply linear function to the two neurons in the output layer.

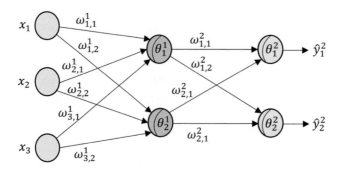

**Input layer**          **Hidden layer**          **Output layer**

**Fig. 6.12** Structure of the 1-hidden layer feedforward neural network used in the application example.

### 6.5.9.3 Initialization of weights and biases

The neural network needs a set of initial parameter values to start working. At the beginning, we randomly initialize the weights and biases of the neural network within $(0, 1)$, as shown in Tables 6.8 and 6.9.

### 6.5.9.4 Training of the neural network using the training load dataset

After the neural network is initialized, the two samples in Table 6.7 are used to train the neural network sequentially. For the first sample, the first 3 load values are fed into the two neurons in the hidden layer from the input layer. We denote the first sample as $(\boldsymbol{x}, \boldsymbol{y})$, and the outputs of the two neurons of the hidden layer are calculated:

$$\widehat{y}_1^1 = \text{Sigmoid}(b_1^1) = \text{Sigmoid}(\omega_{1,1}^1 x_1 + \omega_{2,1}^1 x_2 + \omega_{3,1}^1 x_3 + \theta_1^1)$$
$$= \text{Sigmoid}(0.15 \times 4.3 + 0.83 \times 3.8 + 0.54 \times 3.5 + 0.13) = \text{Sigmoid}(5.82)$$
$$= \frac{1}{1 + e^{-5.69}} = 0.9970$$

$$\widehat{y}_2^1 = \text{Sigmoid}(b_2^1) = \text{Sigmoid}(\omega_{1,2}^1 x_1 + \omega_{2,2}^1 x_2 + \omega_{3,2}^1 x_3 + \theta_2^1)$$
$$= \text{Sigmoid}(0.29 \times 4.3 + 0.08 \times 3.8 + 0.14 \times 3.5 + 0.28) = \text{Sigmoid}(2.32)$$
$$= \frac{1}{1 + e^{-5.69}} = 0.9105$$

The outputs $\widehat{y}_1^1$ and $\widehat{y}_2^1$ are then fed into the two neurons of the output layer to determine the outputs of the neural network:

$$\widehat{y}_1^2 = \text{Linear}(b_1^2) = b_1^2 = \omega_{1,1}^2 \widehat{y}_1^1 + \omega_{2,1}^2 \widehat{y}_2^1 + \theta_1^2 = 0.11 \times 0.997 + 0.96 \times 0.9105 + 0.5$$
$$= 1.4838$$

$$\widehat{y}_2^2 = \text{Linear}(b_2^2) = b_2^2 = \omega_{1,2}^2 \widehat{y}_1^1 + \omega_{2,2}^2 \widehat{y}_2^1 + \theta_2^2 = 0.02 \times 0.997 + 0.77 \times 0.9105 + 0.72$$
$$= 1.4410$$

**Table 6.8** Initialization of weights of the neural network.

| Hidden layer | | | | | | Output layer | | | |
|---|---|---|---|---|---|---|---|---|---|
| $\omega_{1,1}^1$ | $\omega_{2,1}^1$ | $\omega_{3,1}^1$ | $\omega_{1,2}^1$ | $\omega_{2,2}^1$ | $\omega_{3,2}^1$ | $\omega_{1,1}^2$ | $\omega_{2,1}^2$ | $\omega_{1,2}^2$ | $\omega_{2,2}^2$ |
| 0.15 | 0.83 | 0.54 | 0.29 | 0.08 | 0.14 | 0.11 | 0.96 | 0.02 | 0.77 |

**Table 6.9** Initialization of biases of the neural network.

| Hidden layer | | Output layer | |
|---|---|---|---|
| $\theta_1^1$ | $\theta_2^1$ | $\theta_1^2$ | $\theta_2^2$ |
| 0.13 | 0.28 | 0.50 | 0.72 |

Based on the neural network's outputs $\widehat{\gamma}_1$ and $\widehat{\gamma}_2$ and the true outputs in the first sample (i.e., [3.5, 3.2]), the weights and biases are updated following the recursive formulas summarized in Table 6.6. For the output layer, the weights and biases are updated as:

$$\delta_1^2 = \left(\widehat{\gamma}_1^2 - \gamma_1\right)\text{Linear}'\left(b_1^2\right) = \left(\widehat{\gamma}_1^2 - \gamma_1\right) = 1.4838 - 3.5 = -2.0162$$

$$\delta_2^2 = \left(\widehat{\gamma}_2^2 - \gamma_2\right)\text{Linear}'\left(b_2^2\right) = \left(\widehat{\gamma}_2^2 - \gamma_2\right) = 1.441 - 3.2 = -1.7590$$

$$\omega_{1,1}^2 = \omega_{1,1}^2 - \eta\delta_1^2\widehat{\gamma}_1^1 = 0.11 - 0.1 \times (-2.0162) \times 0.997 = 0.31$$

$$\omega_{2,1}^2 = \omega_{2,1}^2 - \eta\delta_1^2\widehat{\gamma}_2^1 = 0.96 - 0.1 \times (-2.0162) \times 0.9105 = 1.14$$

$$\omega_{1,2}^2 = \omega_{1,2}^2 - \eta\delta_2^2\widehat{\gamma}_1^1 = 0.02 - 0.1 \times (-1.759) \times 0.997 = 0.20$$

$$\omega_{2,2}^2 = \omega_{2,2}^2 - \eta\delta_2^2\widehat{\gamma}_2^1 = 0.77 - 0.1 \times (-1.759) \times 0.9105 = 0.93$$

$$\theta_1^2 = \theta_1^2 - \eta\delta_1^2 = 0.5 - 0.1 \times (-2.0162) = 0.70$$

$$\theta_2^2 = \theta_2^2 - \eta\delta_2^2 = 0.72 - 0.1 \times (-1.759) = 0.90$$

For the hidden layer, the weights and biases are updated as follows:

$$\text{Sigmoid}'\left(b_1^1\right) = \text{Sigmoid}'\left(\omega_{1,1}^1 x_1 + \omega_{2,1}^1 x_2 + \omega_{3,1}^1 x_3 + \theta_1^1\right)$$
$$= \text{Sigmoid}'(0.15 \times 4.3 + 0.83 \times 3.8 + 0.54 \times 3.5 + 0.13)$$
$$= \text{Sigmoid}'(5.82) = \frac{1}{1 + e^{-5.82}} \times \left(1 - \frac{1}{1 + e^{-5.82}}\right) = 0.0030$$

$$\text{Sigmoid}'\left(b_2^1\right) = \text{Sigmoid}'\left(\omega_{1,2}^1 x_1 + \omega_{2,2}^1 x_2 + \omega_{3,2}^1 x_3 + \theta_2^1\right)$$
$$= \text{Sigmoid}'(0.29 \times 4.3 + 0.08 \times 3.8 + 0.14 \times 3.5 + 0.28)$$
$$= \text{Sigmoid}'(2.32) = \frac{1}{1 + e^{-2.32}} \times \left(1 - \frac{1}{1 + e^{-2.32}}\right) = 0.0815$$

$$\delta_1^1 = \left(\delta_1^2\omega_{1,1}^2 + \delta_2^2\omega_{1,2}^2\right)\text{Sigmoid}'\left(b_1^1\right) = [-2.0162 \times 0.11 + (-1.759) \times 0.02] \times 0.003 = 0$$

$$\delta_2^1 = \left(\delta_1^2\omega_{2,1}^2 + \delta_2^2\omega_{2,2}^2\right)\text{Sigmoid}'\left(b_2^1\right) = [-2.0162 \times 0.96 + (-1.759) \times 0.77] \times 0.0815$$
$$= -0.27$$

$$\omega_{1,1}^1 = \omega_{1,1}^1 - \eta\delta_1^1 x_1 = 0.15 - 0 = 0.15$$

$$\omega_{2,1}^1 = \omega_{2,1}^1 - \eta\delta_1^1 x_2 = 0.83 - 0 = 0.83$$

$$\omega_{3,1}^1 = \omega_{3,1}^1 - \eta\delta_1^1 x_3 = 0.54 - 0 = 0.54$$

$$\omega_{1,2}^1 = \omega_{1,2}^1 - \eta\delta_2^1 x_1 = 0.29 - 0.1 \times (-0.27) \times 4.3 = 0.41$$

$$\omega_{2,2}^1 = \omega_{2,2}^1 - \eta\delta_2^1 x_2 = 0.08 - 0.1 \times (-0.27) \times 3.8 = 0.18$$

$$\omega_{3,2}^1 = \omega_{3,2}^1 - \eta\delta_2^1 x_3 = 0.14 - 0.1 \times (-0.27) \times 3.5 = 0.23$$

$$\theta_1^1 = \theta_1^1 - \eta\delta_1^1 = 0.13 - 0.1 \times 0 = 0.13$$
$$\theta_2^1 = \theta_2^1 - \eta\delta_2^1 = 0.28 - 0.1 \times (-0.27) = 0.31$$

Similarly, we can use the second training sample to further train the neural network (calculation details are not provided for this second sample). Tables 6.10 and 6.11 show the values of the weights and biases at the beginning (i.e., before training), after training by the first sample, and after the training by the second sample.

Table 6.12 shows the fitted load values for the outputs of the training samples, which are generated before training and after training by the two samples. It can be seen that after training, the neural network can produce outputs that are closer to the outputs of the training samples, demonstrating an enhanced load forecasting ability.

**Table 6.10** Values of the weights before and after training.

| | Hidden layer | | | | | | Output layer | | | |
|---|---|---|---|---|---|---|---|---|---|---|
| | $\omega_{1,1}^1$ | $\omega_{2,1}^1$ | $\omega_{3,1}^1$ | $\omega_{1,2}^1$ | $\omega_{2,2}^1$ | $\omega_{3,2}^1$ | $\omega_{1,1}^2$ | $\omega_{2,1}^2$ | $\omega_{1,2}^2$ | $\omega_{2,2}^2$ |
| Before training | 0.15 | 0.83 | 0.54 | 0.29 | 0.08 | 0.14 | 0.11 | 0.96 | 0.02 | 0.77 |
| Trained by 1st sample | 0.15 | 0.83 | 0.54 | 0.41 | 0.18 | 0.23 | 0.31 | 1.14 | 0.20 | 0.93 |
| Trained by 2nd sample | 0.15 | 0.83 | 0.54 | 0.45 | 0.21 | 0.28 | 0.74 | 1.56 | 0.76 | 1.49 |

**Table 6.11** Values of the biases before and after training.

| | Hidden layer | | Output layer | |
|---|---|---|---|---|
| | $\theta_1^1$ | $\theta_2^1$ | $\theta_1^2$ | $\theta_2^2$ |
| Before training | 0.13 | 0.28 | 0.50 | 0.72 |
| Trained by 1st sample | 0.13 | 0.31 | 0.70 | 0.90 |
| Trained by 2nd sample | 0.13 | 0.31 | 1.13 | 1.46 |

**Table 6.12** ANN-generated fitted load values for the outputs of the 2 training samples.

| | | 2–2:10 pm | 2:10–2:20 pm |
|---|---|---|---|
| First sample | True output | 3.50[a] | 3.20 |
| | Fitted value by ANN before training | 1.48 | 1.44 |
| | Fitted value by ANN after training | 3.40 | 3.69 |
| Second sample | True output | 6.40 | 7.70 |
| | Fitted value by ANN before training | 1.53 | 1.48 |
| | Fitted value by ANN after training | 3.42 | 3.71 |

[a]The units of the data in the table are kW.

## 6.6 Introduction to deep neural network-based STLF

## 6.6.1 Introduction to deep neural networks

In the application example of Section 6.5.9, we have used a simple ANN with only one hidden layer to demonstrate how it can be used for STLF. Such a one-hidden layer structure has limited learning ability, especially when dealing with large training data. Intuitively, one straightforward way of enhancing an ANN's learning ability is to increase the number of hidden layers and/or increase the number of neurons in each hidden layer to better mimic the structure of human brains, which include an enormous number of interconnected neurons.

If we set up multiple hidden layers in an ANN, we then obtain a *deep neural* network. In simple terms, a deep neural network is an ANN with multiple hidden layers. Deep neural networks are the core of deep learning [18,19] and have been extensively applied in different fields, such as image recognition, natural language processing, and load forecasting. As the term "deep" suggests, the input data is processed by multiple layers of neurons to produce the output. The neurons in each hidden layer builds upon the previous layer to refine the output. Such a multi-hidden layer structure can significantly enhance the performance of the neuron network in performing different tasks such as forecasting and classification.

Because of the increased numbers of hidden layers and neurons, the training of deep neural networks is a nontrivial and usually compute-intensive task. Deep neural networks are usually trained with large number of training samples and, because of the multiple hidden layers, they require more computing power and time to complete the training process. The error backpropagation algorithm we introduced in Section 6.5.8 is widely used in training deep neural networks. In this process, stochastic gradient descents [20] are usually used rather than the strict gradient we introduced in Section 6.5.8. The advances of computing hardware have supported the training of deep neural networks that are now trained on workstations, supercomputers, and even at data centers using massive training data.

## 6.6.2 Long short-term memory networks-based STLF

In Section 6.5, we focused on feedforward neural networks, in which the layers are sequentially connected and no loops existed. There are ANNs that are designed with other structures. One kind of deep neural network is called the Long Short-Term Memory (LSTM) network [21]. Unlike feedforward neural networks, LSTM networks have feedback connections, allowing them to process entire sequences of data and not just individual data points. As a result, LSTM networks have been extensively used in time series forecasting tasks and have shown superior performance than other types of deep neural networks. Refs. [22,23] present the use of LSTM networks to do building power load forecasting.

# References

[1] J. Zheng, W. Gao, L. Lin, Smart meters in smart grid: an overview, in: Proceedings of 2013 IEEE Green Technologies Conference, Denver, 2013, April.

[2] G. Athanasopoulos, R.J. Hyndman, Forecasting: Principles and Practice, third ed., OTexts, 2021.

[3] Smart Grid, Smart City Customer Trial Data (Online). Available from: https://data.gov.au/dataset/ds-dga-4e21dea3-9b87-4610-94c7-15a8a77907ef/details (Accessed 03 July 2023).

[4] R.G. Brown, Statistical Forecasting for Inventory Control, McGraw/Hill, 1959.

[5] W.R. Christiaanse, Short-term load forecasting using general exponential smoothing, IEEE Trans. Power Appar. Syst. PAS-90 (2) (1971) 900–911.

[6] P. Ji, D. Xiong, P. Wang, J. Chen, A study on exponential smoothing model for load forecasting, in: Proceedings of 2012 Asia-Pacific Power and Energy Engineering Conference, Shanghai, China, 2012, March.

[7] C.C. Holt, Forecasting Seasonals and Trends by Exponentially Weighted Average, Carnegie Institute of Technology, Pittsburgh, USA, 1957.

[8] Pecan Street Dataport (Online). Available from: https://www.pecanstreet.org/dataport/ (Accessed 03 July 2023).

[9] This file is made available and licensed under the Creative Commons Attribution-Share Alike 3.0 unported license. Attribution: Morn. https://commons.wikimedia.org/wiki/File:White_noise.svg.

[10] D.C. Montgomery, C.L. Jennings, M. Kulahci, Introduction to Time Series Analysis and Forecasting, second ed., John Wiley & Sons, 2011.

[11] A. Krogh, What are artificial neural networks? Nat. Biotechnol. 26 (2008) 195–197.

[12] M.H. Hassoun, Fundamentals of Artificial Neural Networks, The MIT Press, Cambridge, MA, USA, 1995.

[13] Z. Zhou, Machine Learning, Springer Nature, 2021.

[14] B. Voytek, Brain Metrics (Online). Available from: https://www.nature.com/scitable/blog/brain-metrics/are_there_really_as_many/#:~:text=Approximately%2086%20billion%20neurons%20in, between%20200%20and%20400%20billion (Accessed 05 July 2023).

[15] This file is made available and licensed under the Creative Commons Attribution-Share Alike 3.0 unported license. Attribution: Quasar Jarosz. https://commons.wikimedia.org/wiki/File:Neuron_Hand-tuned.svg.

[16] W.S. McCulloch, W. Pitts, A logical calculus of the ideas immanent in nervous activity, Bull. Math. Biophys. 5 (4) (1943) 115–133.

[17] L. Panneerselvam, Activation Functions and Their Derivatives – A Quick & Complete Guide (Online). Available from: https://www.analyticsvidhya.com/blog/2021/04/activation-functions-and-their-derivatives-a-quick-complete-guide/ (Accessed 05 July 2023).

[18] I. Goodfellow, Deep Learning, Random House, Manhattan, USA, 2016.

[19] N. Buduma, N. Locascio, Fundamentals of Deep Learning, O'Reilly Media, Sebastopol, California, USA, 2017.

[20] S. Amari, Backpropagation and stochastic gradient descent method, Neurocomputing 5 (4–5) (1993) 185–196.

[21] A. Sherstinsky, Fundamentals of recurrent neural network (RNN) and long short-term memory (LSTM) network, Phys. D: Nonlinear Phenom. 404 (2020) 132306.

[22] S. Muzaffar, A. Afshari, Short-term load forecasts using LSTM networks, Energy Procedia 158 (2019) 2922–2927.

[23] W. Kong, Z.Y. Dong, Y. Jia, D.J. Hill, Y. Xu, Y. Zhang, Shor-term residential load forecasting based on LSTM recurrent neural network, IEEE Trans. Smart Grid 10 (1) (2019) 841–851.

## CHAPTER 7

# Evolutionary optimization

## Contents

## 7.1 Introduction

Optimization techniques have found extensive applicability in a wide range of disciplines because they enable to identification of advantageous solutions while considering the relevant problem constraints. We perform optimization processes regularly to make decisions in our daily lives even if these optimizations are not performed formally or coded in a programming language. Optimization is here defined as the process that leads to the selection of the best option from a set of candidate options, subject to certain criteria. Optimization is important for building energy management, as BEMSs often need to specify operation plans for building-side energy resources.

In this chapter, we introduce the basic concepts of optimization to establish an overarching framework that can be used to code different algorithms to be deployed for the

*Building Energy Management Systems and Techniques*
https://doi.org/10.1016/B978-0-323-96107-3.00008-4

identification of optimized solutions. An illustrative example is first introduced to provide an overview of the overall process involved in an optimization problem and its numerical solution. After presenting a possible classification for optimization algorithms, particular attention is devoted to evolutionary algorithms because they lend themselves to addressing the complexities embedded in optimization problems typically encountered in the design of Building Energy Management Systems (BEMSs). Three representative evolutionary algorithms are considered and presented in the following. These consist of the Genetic Algorithm (GA), the Particle Swarm Optimization (PSO), and the Differential Evolution (DE) algorithm.

Benchmark functions are then considered to highlight the ability of evolutionary algorithms to identify global minima of functions that possess multiple local minima, therefore representing challenging optimization scenarios. While the focus in this chapter is on minimization problems it is worth highlighting that a maximization problem can be easily treated as a minimization problem by changing the sign of the objective function. Critical remarks are provided to highlight the features of evolutionary optimization. Pseudocodes of all algorithms considered in the chapter have been provided to highlight the details of the solution procedures.

## 7.2 Illustrative optimization example

### 7.2.1 Description of the illustrative example

The illustrative example considers two power generation units that provide energy to a precinct. For simplicity, we consider a specific instant in time and assume that the power load (energy demand) of the building at this specific point in time is 100 kW. A BEMS is considered to drive the optimization process that involves the scheduling of the power outputs from the two generation units while minimizing the overall energy cost. For ease of notation, the power outputs produced by the two units are referred to as $x_1$ and $x_2$, and we assume these variables to be integers. The power generation costs of the two power generation units are associated with the fuel consumption and are assumed to be expressed as follows:

$$C_1 = 0.05 \times x_1^2 \tag{7.1}$$

$$C_2 = 4 \times x_2 \tag{7.2}$$

where $C_1$ and $C_2$ denote the cost of the power generation for units 1 and 2, respectively (in \$). The scheduling of the two units by the BEMS is defined by identifying the values for $x_1$ and $x_2$ that (a) minimize the total power generation cost of the two units, and (b) satisfy the power load of the building, that is, $x_1 + x_2 = 100$ kW.

## 7.2.2 Decision variables, objective function, and constraints

The above example represents a typical optimization problem that is usually defined once the following three key elements are established.

- *Decision variables*—every optimization problem involves one or more variables whose values need to be determined. These variables are denoted as decision variables (or control variables), and, once their values are determined, they represent a solution to the optimization problem. The number of decision variables sets the dimension of the optimization problem, that is, with $N$ decision variables we have a $N$-dimensional optimization problem. The decision variables can be of different types depending on the problem considered, for example, a decision variable can be an integer, a real number, or a binary variable.

In the proposed illustrative example, the decision variables consist of the power output assigned to each generation unit and these have been previously denoted as $x_1$ and $x_2$. When defining the decision variables, it is also important to select the most suitable types of variables. In this case, we consider integer values. Because we have identified two decision variables the size of the optimization problem is two-dimensional. It is convenient to collect the decision variables in a vector as follows:

$$\boldsymbol{x} = [x_1, x_2], x_1, x_2 \in \mathbb{N}_+ \tag{7.3}$$

in which $\mathbb{N}_+$ depicts the set of positive integers.

- *Objective function*—the objective function represents the criteria that we are trying to optimize while varying the decision variables. In setting up an optimization problem, we are usually trying to either maximize or minimize the objective functions. From a practical and implementation viewpoint, we tend to focus on minimization problems. A maximization problem can be transformed into a minimization by simply multiplying the optimization function by $-1$ to invert its sign.

In the proposed example, the objective function aims at minimizing the cost of supplying the energy produced by the two generators and this can be expressed as:

$$\min F = C_1 + C_2 = 0.05 \times x_1^2 + 4 \times x_2 \tag{7.4}$$

- *Constraints*—when dealing with optimization problems we expect the decision variables to vary within preferred ranges of values to ensure the validity and usefulness of the solution. This range of values could reflect a physical requirement, for example, a decision variable has to be positive because represents a physical quantity, or the sought values are required to fall within certain intervals or above certain limits to be meaningful.

In the adopted example, we assume that each generation unit has a capacity of $100\,\text{kW}$ and we need to ensure that the total power production generated by the two units has to match the power load of the building (equal to 100 kW). These requirements can be expressed mathematically with the following constraints:

$$0 \le x_1 \le 100 \text{ kW} \tag{7.5}$$

$$0 \le x_2 \le 100 \text{ kW} \tag{7.6}$$

$$x_1 + x_2 = 100 \text{ kW} \tag{7.7}$$

### 7.2.3 Solving the optimization problem

Once the optimization problem is defined (for example carried out by establishing the decision variables, the objective function, and the required constraints), we need to find suitable solving strategies to identify the optimal solution. The complexity of an optimization problem is usually associated with the dimension of the problem, the types of the decision variables, and the forms of the objective function and constraints.

For the optimization problem represented in Eqs. (7.3)–(7.7), we can use a simple solution approach that enumerates all possible scenarios, that is, writing out all possible combinations of $x_1$ and $x_2$. To do so, we can start by listing the integer values within [0, 100] for $x_1$ based on the range specified in the constraint of Eq. (7.5). We can then determine the corresponding value for $x_2$ from the constraint of Eq. (7.7) as equal to $100 - x_1$. For each combination, we can calculate the objective function value from Eq. (7.4). Fig. 7.1 summarizes the variation of the objective function for different values for $x_1$. In this case, the minimum value of the objective function is \$320.05 which corresponds to the combination of decision variables equal to $[x_1 = 41, x_2 = 59]$ (in kW).

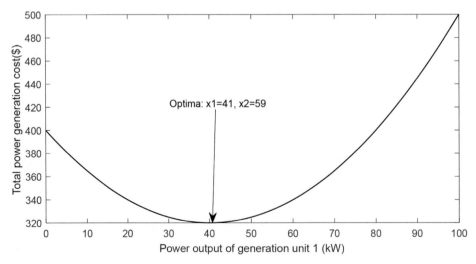

**Fig. 7.1** Variation of the objective function for values of the energy produced by the generation unit ($x_1$) in the illustrative example.

While the enumeration method can be effectively applied to solve the optimization problem of the illustrative example, it would be impractical or prohibitive to use such an approach for more complex problems. For example, in the above illustrative example, if we consider the power outputs of the two generation units are positive real numbers instead of positive integers (i.e., $x_1, x_2 \in \mathbb{R}_+$, where $\mathbb{R}_+$ denotes the set of positive real numbers) and constraints (7.5)–(7.7) to be still applicable, then it is impossible to enumerate all the possible combinations of $x_1$ and $x_2$ and determine the optimal one. Because of this, we need to adopt suitable algorithms to identify minimum solutions in general optimization problems.

## 7.2.4 Solving optimization problems

There is a wide range of numerical techniques that can be applied to identify solutions to optimization problems. Optimization algorithms can be grouped into different categories. For the purpose of this chapter, we distinguish them according to the operations that they require to be carried out on the objective functions when seeking the solution in the following categories:

- *zero-order algorithms* make use of the objective function in the solution process;
- *first-order algorithms* rely on the determination of the first derivative of the objective function; and
- *second-order algorithms* are based on the evaluation of the second derivative of the objective function and, therefore, have a higher computational demand when compared to the previous categories.

The selection of the optimization solution category to be implemented for a particular problem depends on the nature of the objective function and its constraints as well as on the number and types of decision variables. Mathematical programming methods have been developed for solving optimization problems with specific forms. For example, a linear optimization problem (also known as a linear programming problem) that usually consists of an optimization problem with real-valued decision variables, a linear objective function, and linear inequality, can be solved by linear programming techniques such as the simplex method [1]. For some specific forms of optimization problems, mathematical programming methods can generate a deterministic solution that can be mathematically proved to be the global optimal solution in the problem space defined by the decision variables, objective function, and constraints. More information on mathematical programming approaches can be found in dedicated textbooks, for example, [1–4].

There are optimization problems that possess complex forms and can hardly be solved by mathematical programming methods. For example, for a class of optimization problems in which the decision variables are mixed with different types (i.e., real values and integers) and the objective function and/or constraints include nonlinear expressions (usually these problems are referred to as mixed integer, nonlinear programming

(MINLP) problems), mathematical programming approaches can find difficulties in identifying the global minima. This is a consequence of the fact that it is usually difficult to mathematically derive the deterministic optimal solution of a given problem.

Optimization problems in the energy domain, such as those that could exist in typical building energy management applications, are complex in nature (e.g., in the form of MINLP) and these are usually efficiently dealt with zero-order algorithms, such as those implemented with the *evolutionary computation* techniques.

Let us consider an objective function of the following form to better highlight the difficulties in establishing the global minima of complex expressions:

$$f(x) = \sum_{i=1}^{D} \left( x_i^2 - 10\cos(2\pi x_i) + 10 \right) \tag{7.8}$$

where $x_i$ denotes the $i$th decision variable and $D$ denotes the number of decision variables.

The function of Eq. (7.8) is denoted as the Rastrigin's function and defines a challenging objective function commonly used to test optimization solution strategies in the literature. By considering a $D$ value of two in Eq. (7.8), the variation of the objective function in the range of $[-5,5]$ for both decision variables $x_1$ and $x_2$ is illustrated in Fig. 7.2. The plotted surface highlights the many local minima possessed by the function that could be hard to distinguish for minimization algorithms that follow and rely on the shape of the surface, for example, on its slopes, to seek the global minima. An effective optimization algorithm has to be able to distinguish the local minima from the global ones and, to do so, avoid being "trapped" in the local minima locations.

For example, a gradient-descent based optimization method (i.e., a first-order mathematical programming algorithm that relies on the first derivatives of the objective

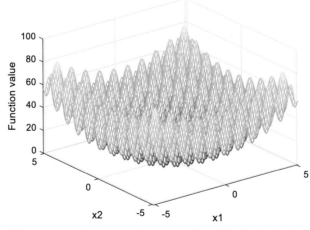

**Fig. 7.2** Variation of the Rastrigin's function in the range of $[-5,5]$ for both decision variables $x_1$ and $x_2$.

function) would start its solution process with an initial guess for the decision variables and continue by seeking the global minima in the direction of the steepest descent as defined by the negative of the gradient (i.e., first derivatives of the objective function). In such a situation, the gradient-descent based optimization method would easily converge to a local minima because of looking for locations of zero gradients. For such a problem, this solution method would not be desirable because not able to "escape" from the local minima.

Optimization problems such as those depicted in Fig. 7.2 can be dealt with by relying on evolutionary algorithms [5]. As implied by the name, evolutionary algorithms rely on strategies encountered in biological intelligence. With such an approach, the optimization is set up as an evolutionary process where only the strongest (and best) solutions can continue to survive until certain conditions are met. With such an approach the process mimics the "survival of the fittest" typically encountered in biological evolution. More details on solution strategies based on evolutionary algorithms are presented in the following section.

## 7.3 Introduction to evolutionary algorithms

### 7.3.1 Background

Evolutionary computation builds on the concepts and dynamics that underpin biological evolution from which it found its inspiration. An interpretation of this approach is provided by the theory of evolution that was first formulated by Charles Darwin in 1859 [6]. This theory provides insight into how organisms change over time as a result of changes in their heritable physical or behavioral traits. The underlying idea is that some of the changes enable organisms to better adapt to their environment and, therefore, provide them with greater chances to survive. Modern genetics and biology have highlighted how the evolution of organisms is a complex process and that it is affected by numerous factors. Despite this, it is possible to identify three fundamental processes that play a key role in driving the evolution process. These consist of genetic recombination, genetic mutation, and natural selection which can be described as follows.

- *Genetic recombination*—biological characters are encoded in genes. Genetic recombination refers to the rearrangement of DNA sequences by a combination of the breakage, re-joining, and copying of chromosomes or chromosome segments. Genetic recombination is a programmed feature of meiosis in most sexual organisms. An offspring carries recombined chromosomes that are inherited from the parents.
- *Gene mutation*—a gene mutation is a permanent alteration in the DNA sequence that makes up a gene. Mutations can range in size. For example, they can affect a single DNA building block up to large segments of chromosomes that include multiple genes. Mutations can produce new characters and are thus an important drive for the organisms' change.

- *Natural selection*—individuals with characters that help them adapt to the environment have greater chances to survive than individuals who do not possess such adaptive capabilities. This process is usually referred to as natural selection. The surviving individuals can then pass their heritable characters to their offspring.

The above three elements in biological evolution are reflected in evolutionary algorithms. In the following, we will introduce several evolutionary algorithms that are useful for the optimization problems relevant to the design of BEMSs. In this process, we also present various examples to demonstrate the effectiveness of the solution strategies of the selected complex optimization problems.

## 7.3.2  Basic concepts in evolutionary computation

Some evolutionary algorithms have been developed to date that mimic biological evolution processes. Even if some evolutionary algorithms do not explicitly relate their underlying mechanisms to biological evolution, they all share the same overarching approach that consists in trying to find the global, or near-global, minima of the given optimization problem by repeatedly generating and trialing different sets of candidate solutions.

The basic concepts at the basis of an evolutionary algorithm are defined in the following.

- *Individuals*—an individual in an evolutionary computation plays the role of an organism in a natural evolution. An individual is encoded as a vector containing the decision variables. It represents a possible candidate solution for the given optimization problem. Each individual is represented based on the following vector:

$$\boldsymbol{x} = [x_1, x_2, \dots x_D] \tag{7.9}$$

where $D$ is the number of the decision variables encoded in the individual (that corresponds to the dimension of the given problem).

- *Fitness degree*—every individual can be evaluated with a fitness degree (also known as *fitness value*) that is analogous to an animal's fitness degree in being able to survive in an environment. The fitness value is obtained through the calculation of the objective function of the given problem.

As mentioned earlier, although an optimization problem can be defined as a maximization or a minimization problem, it is usually convenient to express optimization problems as minimization problems. Based on this, it is common to assume that, the smaller the value calculated with the objective function, the higher the fitness degree of the individual.

- *Population*—it consists of a group of individuals. It is analogous to a population or a swarm of animals. A population is said to have a $N$ size when it contains $N$ individuals.
- *Individual update*—evolutionary algorithms need to update existing individuals, generate new variants of the individuals, and evaluate their fitness degrees. They use this process to seek better solutions. Generating new individuals is analogous to animals' reproduction and mutation, in which chromosome exchange and mutation occur.

- *Selection*—evolutionary algorithms include, explicitly or implicitly, a selection process. This is analogous to the biological selection process. The solutions with high fitness degrees are kept at the expense of the solutions with low fitness degrees that are abandoned.

### 7.3.3 Basic workflow of an evolutionary algorithm

The underlying structure of evolutionary algorithms can be summarized in the following overarching key steps even if the specific details in which these are implemented can vary.

- *Initialization*—at the beginning of the optimization process, an evolutionary algorithm generates a population of individuals. For each individual, the decision variables are generated accounting for the available constraints.
- *Evolution and individual evaluation*—the fitness degree of each individual in the initialized population is evaluated by calculating the value for the objective function for each individual. The best individual, that is, the individual with the smallest objective function value, is stored as the *historically recorded best individual*.

The individuals in the population are then updated by means of predefined strategies that vary among evolutionary algorithms. The updated population is then re-evaluated. The best individual in the updated population is then compared with the historically recorded best individual, and their values are stored.

- *Selection*—selection strategies are applied to eliminate one or more individuals with low fitness degrees. The individuals with high fitness degrees, that is, those with low objective function values, are kept and considered in the subsequent evolution round.
- *Iteration and termination*—the steps associated with the evolution and the selection are repeatedly performed until a certain termination criterion is met. Each round of executing steps is denoted as a generation or iteration. In evolutionary algorithms, the termination criterion is usually met when the maximum generation number is reached and the algorithm outputs the historically recorded best identified individual.

The steps introduced in the proposed basic structure can be described in the pseudocode of Algorithm 7.1. Evolutionary algorithms usually follow the workflow of Algorithm 7.1 even if they can still vary in the implementation of specific steps. For example, different evolutionary algorithms often use different strategies to create the variants of the individuals and to update the population.

   In the remaining of the chapter, we introduce three representative evolutionary algorithms that have been widely used in engineering for solving optimization problems.

### 7.4 Genetic algorithm

The concept of the GA was first proposed by John Holland in the 1960s [7] and it mimics Darwin's evolution theory. In the GA, each individual that represents a candidate solution to the given problem is denoted as a chromosome. Each decision variable in a chromosome is referred to as a gene. The conventional GA adopts a binary encoding scheme where each

---

**ALGORITHM 7.1 Basic workflow of an evolutionary algorithm.**
**Start**
1  Initialize the population;
2  Evaluate the individuals' objective function values and record the individual with the smallest objective as the historically recorded best individual;
3  Repeat
4     Update the population;
5     Perform selection;
6     Update the historically recorded best individual;
7  Until termination criteria are met
8  Output the historically recorded best individual.
**End**

---

chromosome is depicted by a string of binary variables. Each gene is a binary bit, that is, equal to either "0" or "1." This implies that all decision variables, including those of integer and decimal types, are encoded as binary strings. For clarity, a graphical representation of a gene, chromosome, and population in the GA is presented in Fig. 7.3.

In the process of generating new individuals, the GA randomly combines pairs of chromosomes to reproduce two new chromosomes. This involves both gene crossover and mutation operations. For gene crossover, the GA randomly generates a crossover point from the parent chromosomes' genes and it then creates new individuals by exchanging the genes of the parents among themselves until the crossover point is reached as shown graphically in Fig. 7.4.

The crossover operation can be mathematically expressed as:

$$x_i^{new} = \begin{cases} x_i^1 & \text{if } i \leq Cr \\ x_i^2 & \text{otherwise} \end{cases} \quad i = 1 : D \tag{7.10}$$

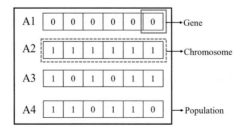

**Fig. 7.3** Graphical representation of a gene, chromosome and population in a GA.

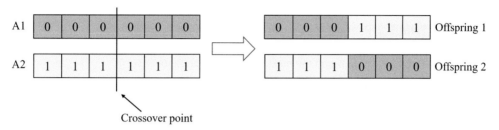

**Fig. 7.4** Graphical representation of a crossover in a GA.

where $D$ is the number of genes; $Cr$ is the crossover factor in the range of $[1, D]$; $x_i^{\text{new}}$ is the value of the $i$th gene in the new chromosome; $x_i^1$ and $x_i^2$ represents the value of the $i$th gene of the two parent chromosomes $\boldsymbol{x}^1$ and $\boldsymbol{x}^2$, respectively.

For the gene mutation, the GA randomly chooses some genes from the new individuals and performs a flip operation of the bits to swap "0" and "1" (i.e., to change "1" into "0" and to change "0" into "1"):

$$x_i^{new} = \sim x_i^{new} \text{ if } i \in \Theta, i = 1 : D \tag{7.11}$$

where $\Theta$ is the set of the randomly generated mutation points and "$\sim$" represents the bit flip operation.

The pseudocode describing the steps involved in the GA is presented in Algorithm 7.2.

In Algorithm 7.2, the function $minByObj(\Theta, n)$ returns $n$ individuals with minimum objective function values from a population of individuals $\Theta$. The returned individuals are sorted by ascending order of their objective function values. The function $obj(\boldsymbol{x})$ returns the objective function value of the individual $\boldsymbol{x}$.

## 7.5 Particle swarm optimization

The PSO is another biological intelligence inspired evolutionary algorithm that was proposed by Kennedy and Eberhart in 1995 [8]. Unlike GA which mimics the theory of evolution, the underlying principles of the PSO are based on the social behavior of animal groups, such as bird flocks and fish swarms. The animal swarms usually have a 'leader-follower' structure, in which a follower's movement is driven by the swarm leader's movement and by its own behavior. For example, when a flock of birds flies together to seek food or perches, one bird acts as the swarm leader and the other birds are followers. A bird would then adjust its fly direction to follow the leader and the adopted flying adjustments would also account for the bird's memory of its past flying trajectory.

---

**ALGORITHM 7.2 Pseudocodes describing the steps of the GA.**

**Start**

1   Set the population size $N$ and the maximum generation number $G$;
2   Initialize the population $\Theta$ as the set of $N$ chromosomes;
3   Set the generation index $g=1$;
4   Evaluate each chromosome's objective function value;
5   Set the historically recorded a best individual $\boldsymbol{x}^*=minByObj(\Theta,1)$;
6   **While** $g<G$
7   Initialize the selected population to be empty by setting $\widetilde{\boldsymbol{P}}=\varnothing$;
8       **While** $(|\widetilde{\boldsymbol{P}}|<N)$
9           Randomly select $\boldsymbol{x}_i^1$ and $\boldsymbol{x}_i^2$ from $\Theta$;
10          Randomly generate the crossover factor $Cr$;
11          Generate two new chromosomes $\boldsymbol{x}^{new1}$ and $\boldsymbol{x}^{new2}$ from $\boldsymbol{x}_i^1$ and $\boldsymbol{x}_i^2$ according to Eq. (7.10);
12          Randomly generate the mutation point set $\boldsymbol{\psi}$;
13          Perform mutations to $\boldsymbol{x}^{new1}$ and $\boldsymbol{x}^{new2}$ according to Eq. (7.11);
14          Set $\Omega=\boldsymbol{x}^1\cup\boldsymbol{x}^2\cup\boldsymbol{x}^{new1}\cup\boldsymbol{x}^{new2}$;
15          $[\boldsymbol{x}^{min1},\boldsymbol{x}^{min2}]=minByObj(\Omega,2)$;
16          Set $\widetilde{\boldsymbol{P}}=\widetilde{\boldsymbol{P}}\cup\boldsymbol{x}^{min1}\cup\boldsymbol{x}^{min2}$;
17          **If** $obj(\boldsymbol{x}^{min1})<obj(\boldsymbol{x}^*)$
18              Set $\boldsymbol{x}^*=\boldsymbol{x}^{min1}$;
19          **End If**
20      **End While**
21      Set $g=g+1$;
22  **End While**
23  Output $\boldsymbol{x}^*$.

**End**

---

The PSO generates a population of particles. When using PSO, the population is referred to as a swarm. Each particle has two attributes: a position and a velocity that is represented by a vector that consists of $D$ variables as follows:

$$\boldsymbol{p}_i=[p_{i,1},p_{i,2},...,p_{i,D}] \tag{7.12}$$

$$\boldsymbol{v}_i=[v_{i,1},v_{i,2},...,v_{i,D}] \tag{7.13}$$

where $i$ is the index of the particle in the swarm; and $\boldsymbol{p}_i$ and $\boldsymbol{v}_i$ are the position and velocity vectors, respectively. At the beginning of the algorithm, PSO randomly initializes each dimensional value of $\boldsymbol{p}_i$ and $\boldsymbol{v}_i$ within the predefined bounds.

With this representation the decision variables are represented by the position of the particles, that is, by vector $\boldsymbol{p}_i$. Each particle's position update is influenced by two factors: (1) the historically recorded best position that the whole swarm has reached, denoted as $\boldsymbol{p}^*=[p_1^*, ..., p_D^*]$ (this is equivalent to vector $\boldsymbol{x}^*$ in the pseudo code of Algorithm 7.2); and (2) the best position that the particle has ever visited, referred to as $\boldsymbol{p}_i^{*,local}=[p_{i,1}^{*,local}, ..., p_{i,D}^{*,local}]$.

In each generation, PSO updates each particle's velocity based on the following expression:

$$v_{i,d}^{new} = w \cdot v_{i,d} + c_p \cdot rnd() \cdot \left(p_{i,d}^{*,local} - p_{i,d}\right) + c_g \cdot rnd() \cdot \left(p_d^* - p_{i,d}\right) \tag{7.14}$$

where the function $rnd()$ generates a random real number in the range of $(0,1)$; $w$ is a parameter that represents an inertia weight; and $c_p$ and $c_g$ are parameters denoted as the cognitive coefficient and the social coefficient, respectively. These three parameters mimic the flying of a bird, whose velocity is balanced by its current velocity, its memory of its historical location $(p_i^{*,local})$, and the global memory denoting the best position reached by the whole swarm $(p^*)$. Based on the updated velocity value, each particle's position is updated as follows:

$$p_i^{new} = \left[p_{i,1}^{new}, p_{i,2}^{new}, \ldots, p_{i,D}^{new}\right] \tag{7.15}$$

$$p_{i,d}^{new} = p_{i,d} + v_{i,d}^{new}, d = 1:D \tag{7.16}$$

in which $p_i^{new}$ depicts the new position that is evaluated and compared with that of its current position $p_i$. If $p_i^{new}$ has a larger fitness degree, that is, a smaller objective function value, then the particle will move to the new position by replacing $p_i$ with $p_i^{new}$ in the following iteration. Alternatively, the particle maintains its current position when entering the subsequent iteration.

The pseudocode associated to the PSO is presented in Algorithm 7.3. The $minByObj(\Theta^p, n)$ function returns the first $n$ individuals with minimum objective function values (i.e., fitness values) from the inputted population $\Theta^p$, and the $obj(x)$ function returns the objective function value of the inputted individual $x$.

## 7.6 Differential evolution algorithm

The DE algorithm was proposed by Storn and Price in 1997 [9]. Even if the DE algorithm does not build on a biological intelligence framework, as done by the GA and PSO, it still relies on the basic structure of evolutionary algorithms introduced in Section 7.3 In each iteration, the DE algorithm uses simple but efficient rules to generate the new individuals for subsequent iterations.

Let us consider that there are $N$ individuals in the population. In each generation, for each individual $x_i$ (with $i = 1,2, \ldots, N$), the DE algorithm first generates a mutant of the individual by randomly choosing two other individuals from the population and by adopting a heuristic rule as follows:

$$m_i = x_{r1} + F \cdot (x_{r2} - x_{r3}) \tag{7.17}$$

where $m_i$ is the mutant for $x_i$. Subscripts $r1$, $r2$, and $r3$ denote three different random integers in the range of $[1,N]$ with $r1 \neq r2 \neq r3 \neq i$. $F$ is a constant parameter denoted as the mutation factor.

---

**ALGORITHM 7.3 Pseudocode describing the steps of the PSO.**

**Start**

1  Set the population size $N$ and the maximum generation number $G$;

2  Set parameters $w$, $c_p$, and $c_g$;

3  Initialize the positions of a swarm of particles $\Theta^P = \{\boldsymbol{p}_1, \boldsymbol{p}_2, ..., \boldsymbol{p}_N\}$;

4  Initialize the velocities of a swarm of particles $\Theta^V = \{\boldsymbol{v}_1, \boldsymbol{v}_2, ..., \boldsymbol{v}_N\}$;

5  Set the generation index $g = 1$;

6  Evaluate each particle's objective function $c_i = obj(\boldsymbol{p}_i)$;

7  Set the historically recorded best individual $\boldsymbol{p}^* = minByObj(\Theta^P, 1)$;

8  Set the local best position of each individual $\boldsymbol{p}_i^{*,\ local} = \boldsymbol{p}_i$;

9  **While** $g < G$

10    **For** $i = 1:N$

11        Set $r_1 = rnd()$;

12        Set $r_2 = rnd()$;

13        Set $\boldsymbol{p}_i^{new} = \boldsymbol{p}_i$;

14        **For** $j = 1:D$

15            Calculate $v_{i,d}^{new}$ according to Eq. (7.14);

16            Set $v_{i,\ d} = v_{i,\ d}^{new}$;

17            Calculate $p_{i,\ d}^{new}$ according to Eq. (7.16);

18        **End For**

19        **If** $obj(\boldsymbol{p}_i^{new}) < obj(\boldsymbol{p}_i)$

20            Set $\boldsymbol{p}_i = \boldsymbol{p}_i^{new}$;

21            **If** $obj(\boldsymbol{p}_i^{new}) < obj(\boldsymbol{p}_i^{*,\ local})$

22                Set $\boldsymbol{p}_i^{*,\ local} = \boldsymbol{p}_i^{new}$;

23                **If** $obj(\boldsymbol{p}_i^{new}) < obj(\boldsymbol{p}^*)$

24                    Set $\boldsymbol{p}^* = \boldsymbol{p}_i^{new}$;

25                **End If**

26            **End If**

27        **End If**

28    **End For**

29    Set $g = g + 1$;

30 **End While**

31 Output $\boldsymbol{p}^*$.

**End**

---

Based on $\boldsymbol{x}_i$ and $\boldsymbol{m}_i$, the DE algorithm then generates a new trial individual $\boldsymbol{u}_i = [u_1, ..., u_D]$ for $\boldsymbol{x}_i$ by performing crossover on $\boldsymbol{x}_i$ and $\boldsymbol{m}_i$:

$$u_{i,j} = \begin{cases} m_{i,j} & \text{if } (rand() \leq Cr) \text{ or } j = a_i \\ x_{i,j} & \text{if } (rand() > Cr) \text{ and } j \neq a_i \end{cases} \tag{7.18}$$

where $Cr$ is a constant parameter denoted as the crossover factor, whose value is in the range of $(0,1)$; and $a_i$ is a random number generated for individual $i$ that falls in the range of $[1, D]$.

---

**ALGORITHM 7.4 Pseudocodes describing the steps of the DE algorithm.**
**Start**
1  Set the population size $N$ and the maximum generation number $G$;
2  Set parameters $F$ and $Cr$;
3  Initialize the individuals $\Theta = \{x_1, x_2, \ldots, x_N\}$;
4  Set the generation index $g = 1$;
5  Set $c_i = obj(x_i)$;
6  Set the historically recorded best individual $x^* = minByObj(\Theta, 1)$;
7  **While** $g < G$
8      **For** $i = 1{:}N$
9          **While** $(r1 == i)$  $r1 = rndInt(1, N)$ **End While**
10         **While** $(r2 == i$ or $r2 == r1)$  $r2 = rndInt(1, N)$ **End While**
11         **While** $(r3 == i$ or $r3 == r1$ or $r3 == r2)$  $r3 = rndInt(1, N)$ **End While**
12         Generate the trial individual $u_i$ by applying Eqs. (7.17) and (7.18);
13         Evaluate the $i$th individual's objective function value $c_i = obj(x_i)$;
14         Evaluate the trial individual's objective function value $c_i^{tr} = obj(u_i)$;
15         **If** $c_i^{tr} < c_i$
16             Set $x_i = u_i$;
17             **If** $(c_i^{tr} < obj(x^*))$
18                 Set $x^* = u_i$;
19             **End If**
20         **End If**
21     **End For**
22     Set $g = g + 1$;
23 **End While**
24 Output $x^*$.
**End**

---

In the crossover operation described in Eq. (7.18), the condition $j = a_i$ ensures that at least one dimension of the trial individual is taken from the mutant. After generating $u_i$, the DE algorithm performs a selection between $x_i$ and $u_i$. If the objective function value of $u_i$ is smaller than $x_i$, then $x_i$ is replaced by $u_i$. Alternatively, $x_i$ is maintained unchanged when entering the subsequent generation.

The pseudocode of the DE algorithm is outlined in Algorithm 7.4.

## 7.7 Control parameters of evolutionary algorithms

An evolutionary algorithm includes one or multiple control parameters, whose values need to be prespecified before running the algorithm. The selection of the control parameters affects the computation time of the algorithm and the solution performance of the algorithm (reflected in the algorithm's ability to generate high-quality individuals, that is, the individuals with low objective function values). There are no general rules for the setting of the values of the control parameters. For a given optimization problem, it is common to evaluate good control parameter values by running different trials.

There are two basic control parameters that exist in most evolutionary algorithms: the population size ($N$) and the maximum generation number ($G$). As previously introduced, the population size specifies the number of individuals maintained in the population, and the maximum generation number specifies the number of rounds in which the algorithm performs the evolution (steps 4–6 in Algorithm 7.1). With large values for these two parameters, the algorithm will be able to try more individual variants and will have a larger chance to find the global optimal solution even if the computation time and used memory will increase accordingly. Therefore, the setting of $N$ and $G$ is a compromise between the computational time and available space and the quality of the final solution obtained from the minimization process.

In addition to $N$ and $G$, algorithms have different control parameters that are related to the algorithms' logics. The setting of these parameters affects the processes of generating individual variants and of updating the population. Tables 7.1–7.4 summarize the control parameters and their empirical value ranges for the three evolutionary algorithms previously introduced, that is, GA, PSO, and DE.

**Table 7.1** Control parameters of GA.

| Parameter | Meaning | Value Range |
|---|---|---|
| $N$ | Size of the population | An integer larger than 4 |
| $G$ | Maximum generation time | An integer larger than 2 |
| $Cr$ | Crossover factor | $[1, D]$ |
| $\psi$ | Set of mutation points | $\psi \in \{1, 2, ..., D\}$. It is usually set so that $\frac{|\psi|}{D} < 0.1$. |

**Table 7.2** Control parameters of PSO.

| Parameter | Meaning | Value range |
|---|---|---|
| $N$ | Size of the population | An integer larger than 4 |
| $G$ | Maximum generation time | An integer larger than 2 |
| $w$ | Inertia weight | $[-1.1, 1.1]$ |
| $c_p$ | Cognitive coefficient | $c_p + c_g \in [0.1, 4.4]$ |
| $c_g$ | | Social coefficient |

**Table 7.3** Control parameters of DE.

| Parameter | Meaning | Value range |
|---|---|---|
| $N$ | Size of the population | An integer larger than 4 |
| $G$ | Maximum generation time | An integer larger than 2 |
| $F$ | Mutation factor | $(0, 2]$ |
| $Cr$ | Crossover factor | $(0, 1)$ |

**Table 7.4** Random integers generated for the mutation operation in the first generation.

|            | r1 | r2 | r3 |
|------------|----|----|----|
| for $x_1$  | 3  | 4  | 2  |
| for $x_2$  | 1  | 3  | 4  |
| for $x_3$  | 4  | 1  | 2  |
| for $x_4$  | 3  | 2  | 1  |

## 7.8 Evolutionary algorithms applied to benchmark functions

In this section we consider two well-known functions used for the benchmarking of optimization problems solved using evolutionary algorithms and that consist of the Ackley's and the Eggholder functions. For illustrative purposes, we apply one of the evolutionary algorithms, that is, the DE algorithm, to evaluate the global minima of the two functions.

### 7.8.1 DE applied to the Ackley's function

The Ackley's function is expressed as follows:

$$f_{ackley}(x) = -20 \cdot \exp\left(-0.2 \cdot \sqrt{D^{-1} \cdot \sum_{i=1}^{D} x_i^2}\right)$$
$$- \exp\left(D^{-1} \cdot \sum_{i=1}^{D} \cos(2\pi \cdot x_i)\right) + 20 + \exp(1) \tag{7.19}$$

To better visualize the optimization results, we take $D=2$ which sets the number of decision variables to be equal to 2. The global minima of the Ackley's function occurs at $[x_1=0, x_2=0]$ with a function value of $f_{ackley}(0,0)=0$ and the variation of this expression is plotted in Fig. 7.5 for $D=2$ and $x_i \in [-50, 50]$.

We now illustrate how the DE works in searching for better solutions for the given objective function (in this example, the Ackley function shown in Eq. (7.19)) in the generation loop. For this purpose, we only consider one generation and a small population consisting of four individuals, that is, $G=1$ and $N=4$. The parameters $F$ and $Cr$ are set to be 0.5 and 0.3, respectively. With these settings, the solving process can be divided into the following two steps: (1) the initialization step, and (2) the first generation step.

#### 7.8.1.1 Initialization

At the beginning of running the DE to solve the problem with the Ackley's function as the objective function, the individuals are randomly generated within the decision variables' value ranges (i.e., $\in [-50, 50]$) to form the initial population (Line 3 in Algorithm 7.4). Consider the four individuals are randomly generated as follows:

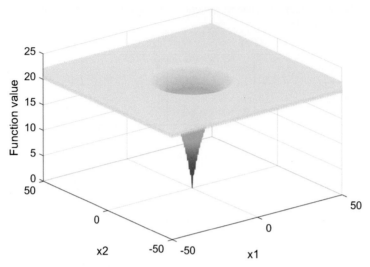

**Fig. 7.5** Variation of the Ackley's function for $D=2$ and $x_i \in [-50, 50]$ (global minimum occurs at $[x_1=0, x_2=0]$ with a function value of $f_{ackley}(0,0)=0$).

$$x_1 = [-20.3, 32.5] \quad x_2 = [47.2, 6.0]$$
$$x_3 = [-13.9, -14.8] \quad x_4 = [20.4, -12.5]$$

DE evaluates the objective function values of the individuals by calculating the value of the Ackley' function under the four individuals. For example, the objective function value of the first individual is:

$$
\begin{aligned}
f_{ackley}(\boldsymbol{x}_1) &= -20 \cdot \exp\left(-0.2 \cdot \sqrt{D^{-1} \cdot \sum_{i=1}^{D} x_{1,i}^2}\right) \\
&\quad - \exp\left(D^{-1} \cdot \sum_{i=1}^{D} \cos(2\pi \cdot x_{1,i})\right) + 20 + \exp(1) \\
&= -20 \cdot \exp\left\{-0.2 \cdot \sqrt{\frac{1}{2} \cdot [(-20.3^2) + 32.5^2]}\right\} \\
&\quad - \exp\left\{\frac{1}{2} \cdot [\cos(-2\pi \cdot 20.3) + \cos(2\pi \cdot 32.5)]\right\} + 20 + \exp(1) \\
&= 22.1
\end{aligned}
$$

In the same way, we can calculate: $f_{ackley}(\boldsymbol{x}_2)=20.8$, $f_{ackley}(\boldsymbol{x}_3)=19.8$, and $f_{ackley}(\boldsymbol{x}_4)=21.6$. By comparing the individuals' objective function values, it can be seen that $f_{ackley}(\boldsymbol{x}_3)$ is the minimum among the four. Therefore, the global best individual during the initialization is set to be $\boldsymbol{x}^* =[-13.9, -14.8]$ (Line 6 in Algorithm 7.4).

### 7.8.1.2 The first generation

After generating the initial population, DE generates one trial individual for each of the four individuals in the initial population following a mutation operation (Eq. (7.17)) and a crossover operation (Eq. (7.18)). The random integers used for the mutation operation are given in Table 7.4:

The following four mutants are then generated based on Eq. (7.17):

$$\boldsymbol{m}_1 = [-27.3, -24.1] \quad \boldsymbol{m}_2 = [-37.5, 31.4]$$
$$\boldsymbol{m}_3 = [-13.4, 0.8] \quad \boldsymbol{m}_4 = [19.9, -28.1]$$

The random parameters generated for the crossover operation are collected in Table 7.5:

Based on the individuals and the generated mutants, the trial individuals are determined based on Eq. (7.18). Each trial individual is generated for an individual currently present in the population (Line 12 in Algorithm 7.4):

$$\boldsymbol{u}_1 = [-27.3, -24.1] \quad \boldsymbol{u}_2 = [47.2, 31.4]$$
$$\boldsymbol{u}_3 = [-13.4, 0.8] \quad \boldsymbol{u}_4 = [19.9, -12.5]$$

By substituting each trial individual in Ackley's function Eq. (7.19), the objective function values of the trial individuals can be calculated: $f_{ackley}(\boldsymbol{u}_1) = 21.2$, $f_{ackley}(\boldsymbol{u}_2) = 21.8$, $f_{ackley}(\boldsymbol{u}_3) = 19.0$, and $f_{ackley}(\boldsymbol{u}_4) = 21.2$.

By comparing the objective function values of each individual and its trial individual, the selection can be performed (Lines 15 and 16 in Algorithm 7.4):

**(1)** $\boldsymbol{u}_1$ is selected between $\boldsymbol{x}_1$ and $\boldsymbol{u}_1$, because $f_{ackley}(\boldsymbol{x}_1) = 22.1 > f_{ackley}(\boldsymbol{u}_1) = 21.2$.
**(2)** $\boldsymbol{x}_2$ is selected between $\boldsymbol{x}_2$ and $\boldsymbol{u}_2$, because $f_{ackley}(\boldsymbol{x}_2) = 20.8 < f_{ackley}(\boldsymbol{u}_2) = 21.8$.
**(3)** $\boldsymbol{u}_3$ is selected between $\boldsymbol{x}_3$ and $\boldsymbol{u}_3$, because $f_{ackley}(\boldsymbol{x}_3) = 19.8 > f_{ackley}(\boldsymbol{u}_3) = 19.0$.
**(4)** $\boldsymbol{u}_4$ is selected between $\boldsymbol{x}_4$ and $\boldsymbol{u}_4$, because $f_{ackley}(\boldsymbol{x}_4) = 21.6 > f_{ackley}(\boldsymbol{u}_4) = 21.2$.

**Table 7.5** Random parameters generated for the crossover operation in the first generation.

|  | $a_i$ |  | rand() |
| --- | --- | --- | --- |
| $a_1$ (for $\boldsymbol{x}_1$) | 1 | for $x_{1,1}$ | 0.45 |
|  |  | for $x_{1,2}$ | 0.14 |
| $a_2$ (for $\boldsymbol{x}_2$) | 2 | for $x_{2,1}$ | 0.86 |
|  |  | for $x_{2,2}$ | 0.30 |
| $a_3$ (for $\boldsymbol{x}_3$) | 1 | for $x_{3,1}$ | 0.27 |
|  |  | for $x_{3,2}$ | 0.05 |
| $a_4$ (for $\boldsymbol{x}_4$) | 1 | for $\boldsymbol{x}_{4,1}$ | 0.60 |
|  |  | for $\boldsymbol{x}_{4,2}$ | 0.71 |

After performing the selection, the population is updated to include $\boldsymbol{u}_1 = [-27.3, -24.1]$, $\boldsymbol{x}_2 = [47.2, 6.0]$, $\boldsymbol{u}_3 = [-13.4, 0.8]$, and $\boldsymbol{u}_4 = [19.9, -12.5]$. For ease of notation, the four selected individuals in the population are renamed from $\boldsymbol{x}_1$ to $\boldsymbol{x}_4$:

$$\boldsymbol{x}_1 = [-27.3, -24.1 \quad \boldsymbol{x}_2 = [47.2, 6.0]$$
$$\boldsymbol{x}_3 = [-13.4, 0.8] \quad \boldsymbol{x}_4 = [19.9, -12.5]$$

In the updated population, $\boldsymbol{x}_3$ relates to the minimum objective function value, that is, 19.0. Since this value (19.0) is smaller than the objective function value of the recorded best individual the algorithm has found (i.e., $\boldsymbol{x}^* = [-13.9, -14.8]$ associated to an objective function value of 19.8), the recorded best individual is updated to be $\boldsymbol{x}^* = [-13.4, 0.8]$ (Lines 17–19 in Algorithm 7.4). In this example we have $G = 1$ and DE terminates after the 1st generation with the optimal solution that it has found, that is, $\boldsymbol{x}^* = [-13.4, 0.8]$.

If $G$ is set to be larger than 1 (that is usually the case), the above procedures is performed based on the updated population in subsequent generations. The procedures are performed for $G$ rounds, after which the algorithm terminates and outputs $\boldsymbol{x}^*$. Ideally, after a number of generations, the global optima $\boldsymbol{x}^* = [0, 0]$ can be found by the algorithm. For example, Fig. 7.6 shows the best individual recorded by DE in each generation with the following settings: $N = 30$, $G = 100$, $F = 0.9$, and $Cr = 0.1$. The curve in Fig. 7.6 is also denoted as the "convergence curve," meaning that the best solution found by the algorithm for the given problem converges to an optimal solution. From the figure, it can be seen that in early generations, the quality of the best solution found by DE

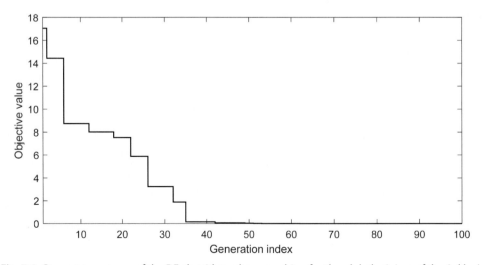

**Fig. 7.6** Convergence curve of the DE algorithm when searching for the global minima of the Ackley's function (with $D = 2$).

is not very good (with objective function values larger than 10). As the generations evolve, the individuals are updated through the DE algorithm's evolution mechanism that enables the best solution found by the algorithm to be continuously improved. This is reflected in the convergence curve, where the objective function value of the current best individual gradually decreases over subsequent generations. After about 50 generations, the objective value of the best solution achieves the minimum value (0), therefore indicating that the algorithm has found the global minima of the Ackley's function.

Fig. 7.7 illustrates the updating process of the individuals in DE's population when solving the Ackley's function. The figures show the distribution of the 30 individuals in the population on the two-dimensional contour of the Ackley's function at the

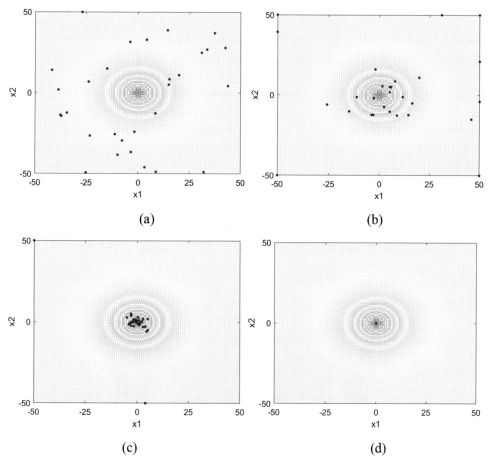

**Fig. 7.7** Finding the global minima of the Ackley's function: distribution of the DE algorithm's population in a two-dimensional space. (A) At the beginning of the generation. (B) After the 10th generation. (C) After the 30th generation. (D) After the 80th generation.

beginning of the evolution, after the 10th generation, after the 30th generation, and after the 80th generation. It can be seen that at the beginning, the 30 individuals are randomly scattered in the two-dimensional problem space. As the optimization progresses, the individuals are gradually updated towards the global minima. After 80 generations, the global minima (0, 0) is identified by the DE algorithm.

## 7.8.2 DE applied to the Eggholder function

The Eggholder function takes two decision variables, and it is expressed as:

$$f_{eh}(x) = -(x_2 + 47)\sin\left(\sqrt{\left|x_2 + \frac{x_1}{2} + 47\right|}\right) - x_1 \sin\left(\sqrt{|x_1 - (x_2 + 47)|}\right) \quad (7.20)$$

The Eggholder function's global minima is difficult to identify because of the large number of local minima. The global minimum of the Eggholder function occurs at $[x_1 = 512.0, \; x_2 = 404.23]$ with corresponding function value of $f_{eh}(512.0, 404.23) = -959.64$. Fig. 7.8 shows the variation of the Eggholder function for $x_1, x_2 \in [-512, 512]$.

We apply the DE algorithm to identify the global minima of the Eggholder function. Figs. 7.9 and 7.10 illustrate the convergence curve and the distribution of the population when applying the DE algorithm. In particular, Fig. 7.9 shows that the DE algorithm finds the global optima within less than ten generations, while Fig. 7.10 depicts the solution process at different generations in a two-dimensional domain. In these calculations, the individuals are initially randomly distributed in the two-dimensional problem space. They then gradually evolve from generation towards the global minima. After 100

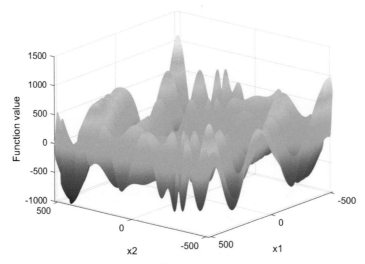

**Fig. 7.8** Variation of the Eggholder function for $x_1, x_2 \in [-512, 512]$ (minimum occurs at $[x_1 = 512.0, x_2 = 404.23]$ with function value of $f_{eh}(512.0, 404.23) = -959.64$).

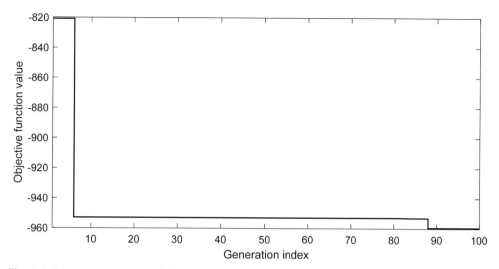

**Fig. 7.9** Convergence curve of the DE algorithm when searching for the global minima of the Eggholder function.

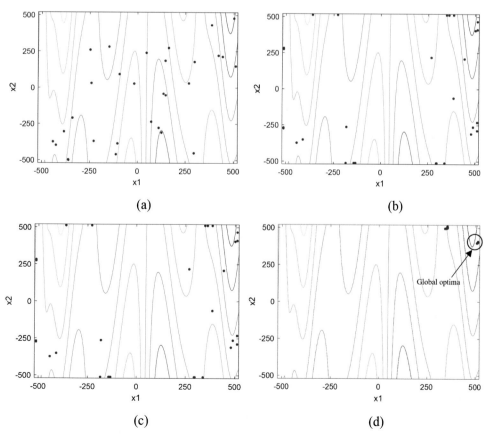

**Fig. 7.10** Finding the global minima of the Eggholder function: distribution of the DE algorithm's population in a two-dimensional space. (A) At the beginning of the generation. (B) After the 10th generation. (C) After the 40th generation. (D) After the 100th generation.

generations, it can be seen that the individuals evolve into two clusters (in Fig. 7.10D): one cluster depicting the global optima (512.0, 404.23), as marked in the figure; and a second cluster located on the left of the global optima and consisting of a local optima. As long as at least one individual is able to identify the global optima, the algorithm is regarded as successfully completing its task. This example shows that the DE algorithm has successfully identified the global optima of the Eggholder function after 100 generations.

### 7.8.3 Evolutionary algorithms applied to high dimensional problems

It is worth noting that, as a heuristic searching technique, an evolutionary algorithm does not guarantee that it always finds the global optima of a given minimization problem. When the dimension of the problem is large and the shapes of the objective function and constraints are complex, the difficulty in determining the global minima of the problem increases. When solving high dimensional problems, it is usually useful to set the parameters $N$ and $G$ to equal relatively large values to enhance the algorithm's ability in identifying the global optimal solution. As discussed in Section 7.7, parameters $N$ and $G$ have a direct impact on the performance of an evolutionary algorithm and they must be selected accounting for the trade-off between the quality of the final best solution and the used computing resource. For a larger value of $N$, more candidate solutions are generated and this enables the algorithm to explore more solutions, with a larger probability of finding better solutions. This expected improved performance comes at the expense of a higher computational demand. Similarly, a larger value of $G$ enables individuals to experience more iterations of their evolutions and, therefore, to have a greater probability of evolving to better solutions. As expected, this improved performance comes at the expense of a larger computational time to complete the optimization.

We now use an example to illustrate the above considerations by applying the DE algorithm to determine the global minima of the Ackley's function with a higher dimension, that is, $D=15$. In this case we have 15 decision variables, that is, $(x_1, \ldots, x_{15})$ and, with the settings of $N=30$, $G=100$, $F=0.9$, and $Cr=0.1$, the calculated convergence curve is plotted in Fig. 7.11. In this trial, the final best solution found by the DE after the 100 generations is:

$$x^* = [-0.9979, 1.1117, -50, -32.3109, -0.2347, -14.0164, 5.3982, 27.8517, 1.4518,$$

$$-30.6967, 0.6010, -0.2205, -26.1755, 50, 4.2946]$$

$$(7.21)$$

and the objective function value of the best solution is 4.1121. It can be seen that the DE algorithm fails to identify the global optima in this case (expected to be $x_1 = x_2 = \ldots = x_{15} = 0$). This is a consequence of the fact that with the increase of the problem's

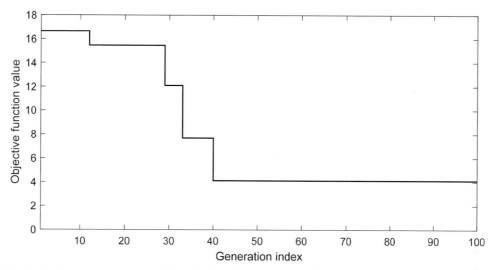

**Fig. 7.11** Convergence curve of the DE algorithm when searching for the global minimum of the Ackley's function (with $D=15$)—solution obtained with $N=30$ and $G=100$.

dimensionality, the searching space of possible solutions becomes large and it becomes more difficult to identify the global optima.

For problems with larger dimensions, an approach for enhancing the evolutionary algorithm's ability in searching for global optima is to increase the population size ($N$) and the generation time ($G$). A larger value of $N$ allows the algorithm to create more individuals in each generation, and a larger value of $G$ allows the algorithm to create a larger number of individual variations. The more individuals the algorithm can trial, the larger chance it can find the global optima. For example, we now increase the values of $N$ and $G$ to be $N=100$ and $G=200$ and re-run the case. This time DE successfully finds the global optimal solution $x_1=x_2=\ldots=x_{15}=0$, which results in an objective function value of 0. Fig. 7.12 depicts the convergence curve obtained with the DE algorithm when solving a 15-dimensional Ackley's function with the settings of $N=100$ and $G=200$.

## 7.9 Remarks for evolutionary optimization

### 7.9.1 Heuristic and stochastic searching-based optimization

As shown in previous sections, evolutionary algorithms do not rely on mathematical derivations to solve an optimization problem. Instead, they use heuristic rules to generate possible solutions for a given problem and try to find the global optima of the problem. When the problem is too complex to be solved by mathematical programming methods, we have to use heuristic strategies to search in the problem space (i.e., to generate and

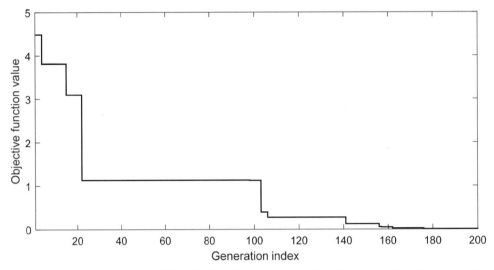

**Fig. 7.12** Convergence curve of the DE algorithm when searching for the global minimum of the Ackley's function (with $D=15$)—solution obtained with $N=100$ and $G=200$.

evaluate possible solutions) and hope the strategies can help to find the global optima, or at least a high-quality solution of the problem.

Many evolutionary algorithms are inspired by biological intelligence. When creating variants of the possible solutions (i.e., the individuals), an important principle is to maintain the "diversity" of the population. An effective evolutionary algorithm should be capable of generating a wide diversity of individuals (reflected in the values of the decision variables forming an individual). In this way, there is a greater chance to generate high-quality individual variants from the diverse individuals in the population. This is analogous to the biological world: the biological diversity of an ecosystem is important for it to remain viable and to generate beneficial traits to make the creatures' offspring better adapt to the environment.

To achieve the individual diversity, evolutionary algorithms usually use stochastic based mechanisms to create the variants of the individuals. By taking DE as an example, when generating the mutant and the trial individual, Eqs. (7.17) and (7.18) define the model rules and include random parameters. Such stochastic individual variant generation mechanisms can also be found in GA (see Section 7.4), PSO (see Section 7.5) and other evolutionary algorithms. These parameters add randomness to the generated individual variants, and prevent them from being deterministically generated by the rules. As a result, such randomness enables the algorithm to be capable of generating and trialing new individual variants that may significantly differ from the current individuals in the population, which therefore provides a chance to find better solutions for the given problem.

It is worth mentioning that because of their stochastic searching nature, for complex problems whose global optima are difficult to find (e.g., the problems with a large number of decision variables and highly nonlinear objective functions and constraints), the output of an evolutionary algorithm over a finite number of generations would be different in different trials of the algorithm. By running an evolutionary algorithm multiple times with the same parameter setting for solving a given problem, it is possible to get different solutions. This is because when the problem is too complex, an evolutionary algorithm cannot guarantee to find the global optima of the problem (see Figs. 7.11 and 7.12 as an example), and with the stochastic mechanism of individual variant generation, the individuals generated by the algorithm in each generation cannot be deterministically predicted. As a result, the individual that is recorded as the best individual at the end of the algorithm execution cannot be deterministically predicted as well.

## 7.9.2 Balance between exploration and exploitation

Evolutionary algorithms are based on heuristic and stochastic based searching approaches for the global optimal solution in the problem's space. There are two terms that well depict the characteristics of the searching process and that are significantly related to an evolutionary algorithm's performance: *exploration* and *exploitation*. Exploration refers to as the process of visiting entirely new regions of a problem space, whilst exploitation is the process of visiting those regions of a problem space within the neighborhood of a previously visited solution [10]. Exploration and exploitation are two cornerstones of evolutionary optimization and both affect an evolutionary algorithm's problem-solving performance. Without exploration, the algorithm would easily converge to local optima, as the individual variants would be trapped around the optima and would lack the ability to visit other regions of the problem space. Without exploitation, the search would be degraded to be a chaos search and would lack a direction (i.e., searching the regions of the problem space in a completely random manner), while exploitation can guide the searching towards the regions where better solutions are likely to be located. It is therefore important to find an effective balance between the exploration and exploitation capabilities when running evolutionary algorithms. Such a balance is achieved through the design of the evolution mechanisms and the selection of the control parameters.

## References

[1] S. Boyd, L. Vandenberghe, Convex Optimization, Cambridge University Press, Cambridge, UK, 2004.
[2] R.K. Arora, Optimization: Algorithms and Applications, CRC Press, Boca Raton, FL, USA, 2015.
[3] J. Lee, S. Layfer (Eds.), Mixed Integer Nonlinear Programming, Springer International Publishing, NY, USA, 2012.
[4] V. Buljak, G. Ranzi, Constitutive Modelling of Engineering Materials – Theory, Computer Implementation and Parameter Identification, Academic Press, Cambridge, MA, USA, 2021.

[5]  A.E. Eiben, J.E. Smith, Introduction to Evolutionary Computation, Springer International Publishing, New York, USA, 2003.

[6]  C. Darwin, On the Origin of Species, sixth ed., Cambridge University Press, Cambridge, UK, 2009.

[7]  D. Whitley, A genetic algorithm tutorial, Stat. Comput. 4 (1994) 65–85.

[8]  J. Kennedy, R. Eberhart, Particle swarm optimization, in: Proceedings of the International Conference on Neural Networks, 1995.

[9]  R. Storn, K. Price, Differential evolution – a simple and efficient heuristic for global optimization over continuous spaces, J. Glob. Optim. 11 (1997) 341–359.

[10]  M. Crepinsek, S. Liu, M. Mernik, Exploration and exploitation in evolutionary algorithms: a survey, ACM Comput. Surv. 45 (3) (2013) 1–33.

# CHAPTER 8

# Energy management of building-integrated battery energy storage systems

## Contents

## 8.1 Introduction

A Battery Energy Storage System (BESS) [1] refers to a system that consists of one or more chemical batteries that absorb and release energy on demand. BESSs are relatively easy to install when compared to other energy storage technologies such as flywheel energy storage [2] and supercapacitor energy storage [3]. In recent years, BESSs have been increasingly deployed in buildings to provide on-site power supply support to building occupants. For example, the home BESS product manufactured by Tesla Energy, that is, the "Tesla Powerwall," can be mounted on the wall of a residential apartment or house to discharge electricity for household use [4].

There are three fundamental uses (or application scenarios) of BESSs in buildings and these consist of:

- *Emergency power backup*: A BESS can be used to provide power to the building in emergencies, for example, in an unexpected power outage event. In this way, it can reduce the disturbance during the emergency period and maintain critical operations within the building;
- *Adapt to time-varying energy tariffs*: Time-varying energy tariffs apply different energy rates to different hours or different periods during a day (see Chapter 4). A BESS can be used to reduce the energy bill for building occupants under the time-varying energy tariff environment. The occupants can charge the BESS in the low-energy price period, and then, in the hours with a high energy price, the BESS can discharge electricity for the building to use. In this way, the amount of energy imported from the grid in high-energy price hours can be reduced and, consequently, building's energy bills can be reduced as well; and.
- *Accommodate the building-side renewable energy and increase utilization of renewable energy*: Renewable energy (such as wind power and solar power) is intermittent, and the power output of a building's on-site renewable energy source (RES) does not usually match the building's power demand. This is depicted in Fig. 8.1, which shows typical

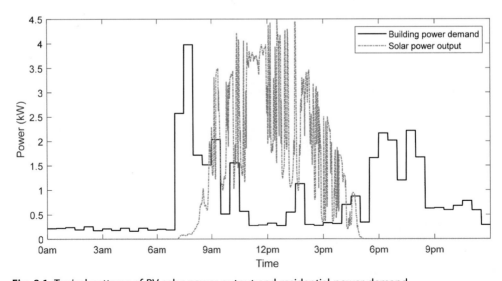

**Fig. 8.1** Typical patterns of PV solar power output and residential power demand.

patterns of the power output from a rooftop photovoltaic (PV) solar panel and the power demand of a residential house. Significant solar power is available at noon and in the afternoon when the power demand is quite low because occupants are usually not home during the daytime. In the evening hours, since occupants are back home and running appliances, the house's power consumption is high (and at this time there is no solar power available). In such a scenario, solar power cannot be utilized to serve the power demand of the house. With the integration of a BESS, the surplus renewable energy can be absorbed in the BESS and discharged later as required. This can improve the utilization of the building's on-site renewable energy.

This chapter introduces some basic energy management strategies for BESSs that are integrated into buildings. In the first part, we briefly review the main types of BESSs that are currently available. The mathematical model of a BESS that can be used for energy management simulations is then presented, followed by energy management strategies for BESSs in the above three application scenarios. An optimization-based BESS energy management approach is also introduced. In the last part of the chapter, we briefly discuss the scenarios of using the battery of Electric Vehicles (EVs) to power buildings using Vehicle-to-Building (V2B) and Vehicle-to-Home (V2H) integration.

## 8.2 Types of BESSs in building applications

Two main types of batteries are commonly used as BESSs in building systems. These consist of lead-acid and lithium-ion batteries and are briefly described in the following:

Lead-acid batteries (see Fig. 8.2) are the most widely used battery type and have been widely used in photovoltaic systems to store solar energy. Compared to other battery types, lead-acid batteries are relatively low cost and possess a long lifetime, and therefore

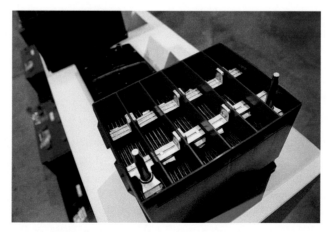

**Fig. 8.2** Components of a lead-acid battery [5].

**Fig. 8.3** Example of a lithium-ion battery [6].

they can be regarded as good energy backup candidates for supporting buildings' operation in an off-grid manner. During operation, lead–acid batteries are slow to charge, and they cannot be fully discharged.

Lithium-ion batteries have a high energy density, that is, they are capable of providing a high voltage and charging storage per unit mass (or per unit volume), and are suitable for high-power applications. Lithium-ion batteries require relatively low maintenance costs. Fig. 8.3 shows a lithium–ion BESS. Lithium-ion batteries tend to overheat and can be damaged at high voltages. This aspect can become a safety concern for several applications. These batteries usually require safety mechanisms to limit voltage and internal pressures, even if the inclusion of these safety mechanisms increases their weight and limits their energy storage performance. Lithium-ion batteries experience aging because, after being used for several years, they can lose capacity or fail. When compared to other types of batteries, the cost of lithium-ion batteries is relatively high.

## 8.3 Model of a BESS

### 8.3.1 Properties of a BESS

The modeling of a BESS in energy management simulations requires account for a set of critical properties that are common to different types of batteries. These properties are briefly outlined in the following (with the relevant notation specified in Table 8.1).

*Charging and discharging power capacities* of a BESS ($P^{bess,rc}$ and $P^{bess,rd}$), also referred to as the rated charging and discharging powers, respectively, are typically measured in kilowatt (kW). The BESS' manufacturer specifies these values. When a BESS is charged or discharged, the charging and discharging power cannot exceed the rated charging and discharging power capacity to avoid overloading the BESS.

*Energy capacity* of a BESS ($E^{bess,r}$) is typically measured in kilowatt-hours (kWh). The energy capacity, also denoted as the rated energy capacity, provides a representation of the amount of energy that can be stored in a BESS and, when considered in combination with the charging and discharging power capacities, it provides insight into the time required to fully charge/discharge a BESS from its empty/full state.

*Charging and discharging efficiencies* of a BESS ($\eta^c$ and $\eta^d$) indicate the efficiency of converting electricity into chemical energy when charging and the efficiency of converting chemical energy in the battery into electricity when discharging, respectively. In the charging and discharging processes, there is a proportion of energy disseminated into the ambient environment as heat, leading to charging and discharging efficiencies lower than 100%.

*State-of-charge* and *depth-of-discharge* of a BESS, commonly denoted as SOC and DOD, are numerical values that vary between 0 and 1. SOC indicates the percentage of energy stored in the battery. A SOC of 0% means the battery is empty and 100% means it is fully charged. DOD indicates the percentage of the battery that has been discharged relative to the overall capacity of the battery.

*Cycle life* is the number of charge/discharge cycles that a battery can sustain in its lifetime. The BESS's cycle life depends on the DOD. If the battery is regularly discharged with a low value of DOD, it is expected to be able to undergo more cycles during its lifetime when compared to a battery that is frequently drained to its maximum DOD. Fig. 8.4 illustrates the dependency of the battery's cycle life on the DOD.

Information on the properties of a BESS, such as those collected in Table 8.1, is usually provided by the manufacturer.

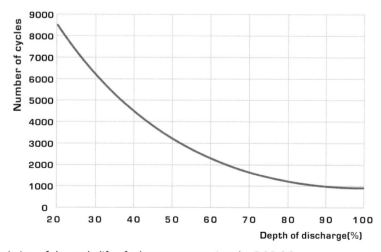

**Fig. 8.4** Variation of the cycle life of a battery concerning the DOD [7].

**Table 8.1** Properties defining the characteristics and model representation of a BESS.

| Notation | Definition |
|---|---|
| $P^{bess,rc}$ | Charging power capacity (in kW) |
| $P^{bess,rd}$ | Discharging power capacity (in kW) |
| $\eta^c$ | Charging efficiency (in %) |
| $\eta^d$ | Discharging efficiency (in %) |
| $E^{bess,r}$ | Energy capacity of the BESS (in kWh) |
| $SOC^{lower}$ | Lower allowable SOC limit (in %) |
| $SOC^{upper}$ | Upper allowable SOC limit (in %) |

## 8.3.2 Worked example

Let us consider a BESS with a charging power capacity of 3 kW and an energy capacity of 15 kWh. The charging and discharging efficiencies are considered to be 95% and 92%, respectively. Given a charging power of 2.5 kW, determine

(1) the time required for the BESS to be fully charged (under continuous charging) from an empty state; and

(2) the time required for the BESS to be fully discharged (under continuous discharging) from a full state.

**Solution**

(1) The charging power applied to the BESS is denoted as $P^{bess,c}$ ($P^{bess,c} \leq P^{bess,rc}$), and the term $P^{bess,c} \times \eta^c$ represents the power that is charged into the BESS. Therefore, the time required for the BESS to be fully charged (denoted by $\Delta t^c$) is

$$\Delta t^c = \frac{E^{bess,r}}{P^{bess,c} \times \eta^c} = \frac{15}{2.5 \times 0.95} = 6.32 \text{ h}$$

(2) The discharging power is referred to as $P^{bess,d}$ ($P^{bess,d} \leq P^{bess,rd}$), and the term $\frac{P^{bess,d}}{\eta^d}$ represents the power taken from (i.e., discharged by) the BESS. With this notation, the value of $P^{bess,d}$ represents the power available for use. The time required for the BESS to be fully discharged (denoted by $\Delta t^d$) can then be estimated as

$$\Delta t^d = \frac{E^{bess,r}}{\frac{P^{bess,d}}{\eta^d}} = \frac{15 \times 0.92}{2.5} = 5.52 \text{ h}$$

## 8.3.3 Worked example

Let us consider a BESS with an energy capacity of 20 kWh, assumed to be fully charged. A 2.4 kW discharging power is applied to the BESS for 180 min. The discharging efficiency is considered to be 100%. Determine the SOC and DOD of the BESS at the end of the discharging operation.

**Solution**

After 180 min of operation and with a discharging power of 1.2 kW, the energy drawn from the battery is $2.4 \times \frac{180}{60} = 7.2$ kWh. The amount of remaining energy in the battery is $20 - 7.2 = 12.8$ kWh. Therefore, the SOC and DOD are

$$\text{SOC} = \frac{12.8}{E^{bess,r}} = \frac{12.8}{20} = 64.0\%$$

$$\text{DOD} = \frac{7.2}{E^{bess,r}} = \frac{7.2}{20} = 36.0\%$$

A BESS needs to avoid over-charging and over-discharging, as this can affect the battery's condition and lifetime. The negative impact of over-discharging (i.e., with high values of DOD) is illustrated in Fig. 8.4. Over-charging can lead to gassing, water loss, and heat generation inside the battery, which are harmful to the battery's cycle life. For example, in lead acid batteries, gassing starts before full charge (i.e., SOC = 100%) is reached and increases as the charging progresses. Because of this, an allowable SOC range is usually set for a BESS, with the lower and upper boundary of the range denoted as $SOC^{lower}$ and $SOC^{upper}$, respectively, (see Table 8.1). The SOC of a BESS should always be kept in the range of $[SOC^{lower}, SOC^{upper}]$ to avoid over-charging and over-discharging that can compromise the long-term performance and condition of the BESS.

## 8.3.4 Operational state variables of a BESS

From the perspective of energy management, a BESS's operational state at a particular time slot can be represented by the three state variables included in Table 8.2. In the adopted notation, the subscript $t$ is used as the index for the time slot. The charging/discharging power ($P_t^{bess}$) is the power applied to draw energy from the BESS or to charge the BESS. $P_t^{bess}$ depicts the charging/discharging power of the BESS that occurs at time slot $t$. For ease of notation, we use a positive value of $P_t^{bess}$ to indicate a charging condition and a negative value to denote discharging. Based on the adopted notation, we can express the power constraint condition of the BESS as follows:

$$\begin{cases} -P_t^{bess} \leq P^{bess,rd}, & \text{if } P_t^{bess} < 0 \\ P_t^{bess} \leq P^{bess,rc}, & \text{if } P_{,t}^{bess} \geq 0 \end{cases} \tag{8.1}$$

**Table 8.2** Operational state variables of a BESS.

| Notation | Definition |
|---|---|
| $P_t^{bess}$ | Charging/discharging power of the BESS at time slot $t$ (in kW) |
| $SOC_t$ | SOC of the BESS at time slot $t$ (in %) |
| $E_t^{bess}$ | Energy stored in the BESS at time slot $t$ (in kWh) |

that ensures that the charging/discharging power of the BESS remains between its rated charging and discharging power limits. The value of $P_t^{bess}$ can be determined from the energy management strategy and predictions.

At time slot $t$, the SOC of the BESS is calculated as

$$SOC_t = \frac{E_t^{bess}}{E^{bess,r}} \qquad (8.2)$$

while the variation of the value of energy stored in the BESS is evaluated as

$$E_{t+1}^{bess} = \begin{cases} E_t^{bess} + \dfrac{P_t^{bess}}{\eta^d} \times \Delta t, \text{ if } P_t^{bess} < 0 \\ E_t^{bess} + P_t^{bess} \times \eta^c \times \Delta t, \text{ if } P_t^{bess} \geq 0 \end{cases}, \qquad (8.3)$$

where $E_t^{bess}$ and $E_{t+1}^{bess}$ represent the energy capacities stored in the BESS at times slots $t$ and $(t+1)$, respectively (in kWh); and $\Delta t$ depicts the duration of a time slot (in hours). When the BESS operates in a discharging mode (i.e., $P_t^{bess}$ is negative) it reduces the energy stored and, therefore, $E_{t+1}^{bess}$ will be smaller than $E_t^{bess}$. Similarly, when the BESS operates in a charging mode (i.e., $P_t^{bess}$ is positive), $E_{t+1}^{bess}$ will be larger than $E_t^{bess}$.

As discussed in Section 8.3.1, it is common to restrict the range of values to be adopted for the SOC of the BESS to extend the lifetime of the BESS. With this consideration, it is not possible to completely charge or discharge the BESS and these limits can be expressed by the additional constraints

$$SOC^{lower} \leq SOC_t \leq SOC^{upper}. \qquad (8.4)$$

### 8.3.5 Worked example for a BESS operational condition

Let us consider a BESS with an energy capacity of 32 kWh and with 14 kWh currently stored. The lower and upper allowable SOC limits of the BESS are 0.1 and 0.8, respectively. Determine the following variables:
(1) The maximum amount of energy that can be charged into the BESS; and
(2) The maximum amount of energy that can be discharged from the BESS.

**Solution**
(1) The maximum amount of energy that can be stored in the BESS is $32 \times 0.8 = 25.6$ kWh. Therefore, the maximum amount of energy that can be charged into the BESS is $25.6 - 14 = 11.6$ kWh.
(2) The minimum amount of energy that can be stored is $32 \times 0.1 = 3.2$ kWh. As a consequence, Therefore, the maximum amount of energy that can be discharged from the BESS is: $14 - 3.2 = 10.8$ kWh.

**Table 8.3** Values of the charging/discharging power and the durations of the three time periods.

|  | Time period 1 | Time period 2 | Time period 3 |
|---|---|---|---|
| Charging/discharging power | −3.2 kW | 4.0 kW | −1.5 kW |
| Duration | 50 min | 30 min | 80 min |

## 8.3.6  Worked example for a BESS operational condition

Let us consider a BESS with an energy capacity of 20 kWh. The charging and discharging efficiencies of the BESS are 80% and 90%, respectively. Let us assume that the BESS is charged and discharged over three time periods. The values of the charging/discharging power and the time durations are shown in Table 8.3. At the beginning of the analysis, the energy stored in the BESS is 15 kWh.

Determine the SOC of the battery at the end of each period.

**Solution**

Let us denote the energy stored in the BESS at the beginning as $E_0^{bess}$, that is, $E_0^{bess} = 15$ kWh, and the charging/discharging power in the three time periods as $P_1^{bess}$, $P_2^{bess}$, and $P_3^{bess}$. The durations of the three time periods are depicted by $\Delta t_1$, $\Delta t_2$, and $\Delta t_3$. The SOC variation of the BESS can then be calculated as shown below.

(1) Discharging of the BESS occurs during time period 1 and, at the end of the time period, the energy stored in the BESS (denoted as $E_1^{bess}$) is

$$E_1^{bess} = E_0^{bess} + \frac{P_1^{bess}}{\eta^d} \times \Delta t_1 = 15 + \frac{-3.2}{0.9} \times \frac{50}{60} = 12.04 \text{ kWh}$$

and the corresponding SOC (denoted as $SOC_1$) is evaluated as

$$SOC_1 = \frac{E_1^{bess}}{E^{bess,r}} = \frac{12.04}{20} = 60.2\%.$$

(2) The BESS is charged during the time period 2. At the end of the time period, the energy stored in the BESS (denoted as $E_2^{bess}$) and the SOC (denoted as $SOC_2$) are

$$E_2^{bess} = E_1^{bess} + P_2^{bess} \times \eta^c \times \Delta t_2 = 12.04 + 4 \times 0.8 \times \frac{30}{60} = 13.64 \text{ kWh}$$

$$SOC_2 = \frac{E_2^{bess}}{E^{bess,r}} = \frac{13.64}{20} = 68.2\%.$$

(3) In time period 3, the BESS is discharged. At the end of the time period, the energy stored in the BESS (denoted as $E_3^{bess}$) and the SOC (denoted as $SOC_3$) become

$$E_3^{bess} = E_2^{bess} + \frac{P_3^{bess}}{\eta^d} \times \Delta t_3 = 13.64 + \frac{-1.5}{0.9} \times \frac{80}{60} = 11.42 \text{ kWh}$$

$$SOC_3 = \frac{E_3^{bess}}{E^{bess,r}} = \frac{11.42}{20} = 57.1\%.$$

## 8.4 BESS integration for emergency power backup for buildings

The technology of a BESS is well suited to deal with power outage events where the BESS can provide energy backup services to buildings and other energy consumers. With the use of a BESS, the disturbance for an occupant caused by a power outage event can be reduced by supporting critical activities within a building. In this section, we design a simple BEMS that discharges a BESS to power a building in a power outage event.

### 8.4.1 Energy management strategy

Let us consider the energy management of a BESS that provides an emergency power supply to a building in a power outage event. The properties of the BESS are depicted by the variable introduced in Table 8.1. Let us consider a power outage event lasting for $T$ time slots with the duration of each time slot equal to $\Delta t$ (in hours). Discharging actions are applied to the BESS to draw energy from the BESS and to serve the building. It is assumed that the building's power demand and the discharging power of the BESS are constant over the duration of a time slot. The power demand profile of the building over the outage period (denoted as $L$) can be represented by the following vector:

$$L = [L_1, ..., L_t, ..., L_T],  \tag{8.5}$$

where $L_t$ is the power demand of the building at time slot $t$ (in kW) with $t = 1:T$. In practice, the value of $L_t$ can be obtained through load forecasting techniques (such as the ones introduced in Chapter 6).

The BESS energy management strategy considered in the following assumes that, at the beginning of each time slot, the BESS is discharged to meet the power demand of the building in that time slot subjected to the amount of energy stored in the BESS and the BESS' operational constraints (8.1) and (8.4). The details of this energy management strategy are described in Algorithm 8.1.

In Line 4 of Algorithm 8.1, the BEMS performs control actions on the BESS in each time slot by starting from the first time slot. The BEMS forecasts the building's power load at the current time slot (Line 4) and calculates the amount of energy that can be discharged from the BESS to cover the power load (Line 5). It then determines the discharging power of the BESS as the minimum one among of the following three values (Line 6): (i) The building's power demand, (ii) the BESS's discharging power capacity, and (iii) the maximum discharging power of the BESS. In Line 6, the function $\min(\cdot)$ returns the minimum number among the inputted numbers. At the end of each time slot, the amount of energy stored in the BESS is updated (Line 7). This control logic is repeated at each time slot until the entire energy management horizon is analyzed, that is, until time slot $T$ is reached (Lines 8–13).

---

### ALGORITHM 8.1 Energy management strategy for a BESS to provide emergency power backup service to a building.

**Start**

1. Input the BESS model ($P^{bess,rd}, E^{bess,r}, \eta^d$, fand $SOC^{lower}$);
2. Input the initial energy stored in the BESS (denoted as $E_0$) and the number of time slots $T$;
3. Start from the first time slot by setting the time index $t=1$;
4. Forecast the value of $L_t$;
5. Calculate the current dischargeable energy in the BESS as $E_t^{dcb} = E_{t-1}^{bess} - E^{bess,\, re} \times SOC^{lower}$;
6. Set $P_t^{bess} = -1 \times min \left( \frac{E_t^{dcb} \times \eta^d}{\Delta t}, L_t, P^{bess,rd} \right)$;
7. Update the energy stored in the BESS as $E_t^{bess} = E_{t-1}^{bess} + \frac{P_t^{bess}}{\eta^d} \times \Delta t$;
8. **If** $t==T$
9.    **Go to** Line 14;
10. **Else**
11.    Proceed to the next time slot and set $t=t+1$;
12.    **Go to** Line 4;
13. **End If**
14. Output the charging/discharging actions $\boldsymbol{P}^{bess} = [P_1^{bess}, \ldots, P_T^{bess}]$.

**End**

---

## 8.4.2 Application example

In this application example, we consider a 7-ho power outage period. The power demand profile of a building over the outage period is shown in Table 8.4. The data considered is modified from the Australian "Smart Grid, Smart City" customer trial dataset [8]. The power demand data are sampled with a 30-min time resolution. There are 14 time slots during the outage period. From the table, it can be calculated that the total energy demand of the building over the 7 h is 38.55kWh. The BESS is assumed to possess the configuration parameters of Table 8.5, and that, at the beginning of the outage period, its stored energy is 35kWh, that is, $E_0 = 35$.

**Table 8.4** Power demand of the building during the power outage period (unit: kW).

| TS1[a] | TS2 | TS3 | TS4 | TS5 | TS6 | TS7 |
|---|---|---|---|---|---|---|
| 2.6 | 4.3 | 2.5 | 12.4 | 7.8 | 3.6 | 3.2 |
| **TS8** | **TS9** | **TS10** | **TS11** | **TS12** | **TS13** | **TS14** |
| 5.4 | 11.0 | 4.9 | 3.4 | 4.4 | 6.6 | 5.0 |

[a]"TS" means "time slot."

**Table 8.5** Model of the BESS used in the application example.

| Property | Value |
|---|---|
| $P^{bess,rd}$ | 10 kW |
| $P^{bess,rc}$ | 10 kW |
| $\eta^c$ | 90% |
| $\eta^d$ | 84% |
| $E^{bess,r}$ | 40 kWh |
| $SOC^{lower}$ | 20% |
| $SOC^{upper}$ | 90% |

By applying Algorithm 8.1, the discharging power of the BESS in each time slot can be determined. Fig. 8.5 shows the profile of discharging power over the 7-h outage period together with the building's power demand profile. The numerical values of the discharging power are given in Table 8.6. The shaded area in the figure indicates the energy demand not satisfied by the BESS that can be calculated as Eqs. (8.6) and (8.7).

$$E' = \sum_{t=1}^{T} E'_t, \tag{8.6}$$

$$E'_t = \left(L_t - \left|P_t^{bess}\right|\right) \times \Delta t, \tag{8.7}$$

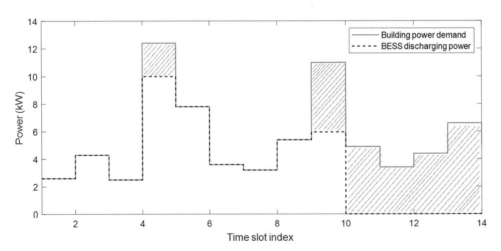

**Fig. 8.5** Profiles of the building's power demand and the BESS discharging power.

**Table 8.6** Discharging power of the BESS during the power outage period (unit: kW).

| TS1 | TS2 | TS3 | TS4 | TS5 | TS6 | TS7 |
|---|---|---|---|---|---|---|
| 2.6 | 4.3 | 2.5 | 10 | 7.8 | 3.6 | 3.2 |

| TS8 | TS9 | TS10 | TS11 | TS12 | TS13 | TS14 |
|---|---|---|---|---|---|---|
| 5.4 | 5.96 | 0 | 0 | 0 | 0 | 0 |

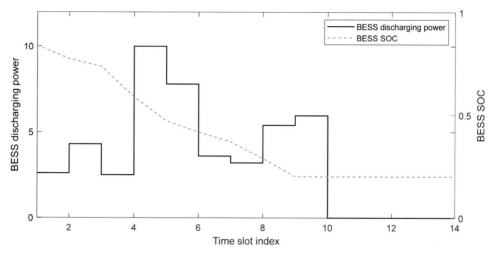

**Fig. 8.6** Profiles of the BESS's discharging power and SOC.

where $E_t'$ denotes the amount of energy demand of the building that is not served by the BESS during the time slot $t$ (in kWh); and $E'$ denotes the total amount of energy demand of the building that is not served by the BESS over the outage period (in kWh). By substituting the numbers in Tables 8.4 and 8.5, there is a total of 15.87 kWh energy demand of the building not served by the BESS, that is, the BESS serves $38.55 - 15.87 = 22.68$ kWh energy to the building demand over the power outage period. After time slot 9, the BESS's SOC reaches the lower limit and no discharging power is available. Therefore, the discharging power of the BESS is zero after time slot 9.

Based on the discharging power and the initial condition of the BESS (i.e., $E_0$), the SOC variation of the BESS can be calculated based on Eqs. (8.2) and (8.3). Fig. 8.6 plots the discharging power of the BESS together with the SOC values for each time slot.

## 8.5 BESS integration for time-varying energy tariff

### 8.5.1 Overview

In this section, we highlight the ability of BESSs to help improve buildings' energy performance in a grid-connected operation. For example, a BESS can assist an building occupant in dealing with time-varying energy tariffs to reduce the energy bill. A simple approach to exploit and achieve this involves the charging of the BESS from the grid in prespecified low-energy price hours and the discharging of the BESS to serve the building's energy demand in pre-specified high-energy price hours.

With consideration of the discharging actions of the BESS, the building's *net-load* profile $L^{net}$ over a period consisting of $T$ time slots can be calculated as

$$\boldsymbol{L}^{net} = \left[L_1^{net}, ..., L_t^{net}, ..., L_T^{net}\right], \tag{8.8}$$

$$L_t^{net} = L_t + P_t^{bess}, t = 1 : T, \tag{8.9}$$

where $L_t^{net}$ denotes the building's net load in time slot $t$ (in kW). The net load indicates the amount of power the building needs to be imported from the grid. Since a negative value of $P_t^{bess}$ indicates discharging, Eq. (8.9) shows that (i) when the BESS discharges power, the net-load of the building reduces and, consequently, the power taken from the grid reduces as well. If the discharged power of the BESS is larger than the building's power demand (i.e., $-1 \times P_t^{bess} > L_t$), the net load $L_t^{net}$ is negative, meaning that the power produced (i.e., discharged by the BESS) is larger than the building's demand; and (ii) when the BESS is charged, it acts as an energy load and it increases the building's net load.

The energy tariff rates over the $T$ time slots can be collected in the vector $\boldsymbol{\rho}$ as:

$$\boldsymbol{\rho} = [\rho_1, ..., \rho_t, ..., \rho_T], \tag{8.10}$$

where $\rho_t$ denotes the energy tariff rate in the time slot $t$ (in \$/kWh). Based on the adopted notation, a building's energy cost over $T$ time slots (denoted as $C$) is calculated as

$$C = \sum_t^T C_t, \tag{8.11}$$

$$C_t = \begin{cases} 0, \text{if } L_t^{net} \leq 0 \\ L_t^{net} \times \rho_t \times \Delta t, \text{otherwise} \end{cases} \tag{8.12}$$

where $C_t$ denotes the building's energy cost in time slot $t$ (in \$) and $\Delta t$ is the duration of a time slot (in hours). When $L_t^{net} \leq 0$, the building can be self-sustained in terms of energy needs, and the energy cost is nil. Otherwise, the net load is charged at the energy tariff rate.

## 8.5.2 Energy management strategy

As discussed above, the BESS can be used to absorb energy from the grid in low-energy price time slots and to discharge energy to serve the building in high-energy price time slots. The sets of time slots with low- and high-energy tariff rates are here referred to as $\boldsymbol{\varpi}^{lp}$ and $\boldsymbol{\varpi}^{hp}$, respectively. The BESS is scheduled to be charged and discharged in the time slots belonging to the two sets $\boldsymbol{\varpi}^{lp}$ and $\boldsymbol{\varpi}^{hp}$, respectively. For simplicity, in the following we consider the BESS does not charge nor discharge in other time slots.

Based on the building's power demand and the energy tariff, the energy management strategy for the BESS is described in Algorithm 8.2.

In each time slot, the BEMS forecasts the building's power demand (Line 4) and checks the amount of energy that can be discharged from the BESS (Line 5) and that can be charged into the BESS (Line 6) based on the BESS's current condition. In the case of a low-energy price time slot, the BEMS determines the charging power of the

**ALGORITHM 8.2 Energy management strategy for a BESS to account for time-varying energy tariff.**

**Start**
1. Input the BESS model ($P^{bess,\ rc}$, $P^{bess,\ rd}$, $E^{bess,\ r}$, $\eta^c$, $\eta^d$, $SOC^{upper}$, and $SOC^{lower}$);
2. Input the initial energy stored in the BESS (denoted as $E_0$) and the number of time slots $T$;
3. Start from the first time slot by setting the time index $t=1$;
4. Forecast the value of $L_t$;
5. Calculate the current dischargeable energy in the BESS as $E_t^{dcb}=E_{t-1}^{bess}-E^{bess,\ re}\times SOC^{lower}$;
6. Calculate the current chargeable energy in the BESS as $E_t^{cb}=E^{bess,\ re}\times SOC^{upper}-E_{t-1}^{bess}$;
7. **If** $t\in\varpi^{lp}$
8.      Set $P_t^{bess}=min\left(\frac{E_t^{cb}}{\eta^c\times\Delta t},P^{bess,rc}\right)$;
9. **Else If** $t\in\varpi^{hp}$
10.     Set $P_t^{bess}=-1\times min\left(\frac{E_t^{dcb}\times\eta^d}{\Delta t},L_t,P^{bess,rd}\right)$;
11. **Else**
12.     Set $P_t^{bess}=0$;
13. **End If**
14. **If** $t==T$
15.     **Go to** Line 20;
16. **Else**
17.     Proceed to the next time slot and set $t=t+1$;
18.     **Go to** Line 4;
19. **End If**
20. Output the charging/discharging actions $\boldsymbol{P}^{bess}=[P_1^{bess},\ldots,P_T^{bess}]$.
**End**

BESS as the smaller value between the two following two values (Lines 7 and 8): (i) The charging power that can fully charge the BESS in one time slot (i.e., $\frac{E_t^{cb}}{\eta^c\times\Delta t}$); and (ii) the BESS's charging power capacity. When the current time slot is a high-energy price time slot, the BEMS determines the discharging power of the BESS by taking the smallest value among the following three values (Lines 9 and 10): (i) The building's power demand, (ii) the BESS' discharging power capacity, and (iii) the discharging power that can draw out all the dischargeable energy in the BESS in one time slot (i.e., $\frac{E_t^{dcb}\times\eta^d}{\Delta t}$). The BEMS does not apply any charging or discharging actions to the BESS (Lines 11 and 12) when the current time slot is neither a low-energy price time slot nor a high-energy price time slot. Such an energy management logic repeats until the energy management horizon is reached (Lines 14–19).

### 8.5.3 Application example

We design an application example to illustrate the energy management strategy of Algorithm 8.2. We consider one-day energy management for a building charged by a time-of-use electricity tariff. The energy management starts at 0 am and the energy management interval is 30 min, that is, there are a total of 48 time slots over the energy management period. The building's power demand, given in Table 8.7, and the BESS's charging/discharging power are assumed to remain constant during each time slot. It is assumed that the building is charged by the time-of-use electricity tariff described in Table 8.8. Fig. 8.7 plots the building's power demand profile and the time–of-use tariff.

The BESS is assumed to possess the properties specified in Table 8.5. It is also assumed that, at the beginning of the energy management period, no dischargeable energy is stored in the BESS, that is, $E_0 = E^{bess,r} \times SOC^{lower} = 40 \times 20\% = 8$ kWh.

Based on the electricity tariff rates of Table 8.8, the low-electricity price period runs between 10 pm and 7 am and the high-electricity price period falls between 5 and 8 pm, that is, $\boldsymbol{\varpi}^{lp} = \{1,2,3,4,5,6,7,8,9,10,11,12,13,14,45,46,47,48\}$ and $\boldsymbol{\varpi}^{lp} = \{35,36,37, 38,39,40\}$ where, in the definition of the sets, the physical time is converted into time slot indices. By applying the energy management procedures of Algorithm 8.2, the charging/discharging power of the BESS is determined as shown in Fig. 8.8 together with the BESS's SOC profile when charging and discharging. The building's net load profile is also plotted in Fig. 8.8 for ease of reference. The energy cost of the building over the 24 h is $22.68. With the use of the BESS, the energy cost reduces to $10.86, with a cost saving of $11.82.

**Table 8.7** Building power demand used in the application example (unit: kW).

| TS1 | TS2 | TS3 | TS4 | TS5 | TS6 | TS7 | TS8 | TS9 | TS10 | TS11 | TS12 |
|-----|-----|-----|-----|-----|-----|-----|-----|-----|------|------|------|
| 8.2 | 4.8 | 4.2 | 2.2 | 1.3 | 1.2 | 1.2 | 1.4 | 1.3 | 1.2 | 1.2 | 1.4 |
| TS13 | TS14 | TS15 | TS16 | TS17 | TS18 | TS19 | TS20 | TS21 | TS22 | TS23 | TS24 |
| 1.3 | 4.5 | 4.5 | 3.6 | 2.1 | 0.8 | 0.2 | 0 | 0.6 | 0 | 0.3 | 1.2 |
| TS25 | TS26 | TS27 | TS28 | TS29 | TS30 | TS31 | TS32 | TS33 | TS34 | TS35 | TS36 |
| 2.2 | 0 | 0.4 | 0.2 | 0.2 | 0.6 | 1.3 | 2.0 | 1.3 | 1.9 | 6.5 | 5.7 |
| TS37 | TS38 | TS39 | TS40 | TS41 | TS42 | TS43 | TS44 | TS45 | TS46 | TS47 | TS48 |
| 9.8 | 10.4 | 7.0 | 1.6 | 5.2 | 12.9 | 9.3 | 3.5 | 5.8 | 5.6 | 4.9 | 5.2 |

**Table 8.8** Time-of-use electricity tariff used in the application example.

| Period | Time | Rate (in $/kWh) |
|--------|------|------------------|
| Peak hours | 5–8 pm | 0.76 |
| Secondary peak hours | 2–5 pm | 0.35 |
| Shoulder hours | 7 am–2 pm, 8–10 pm | 0.15 |
| Off-peak hours | 0–7 am, 10 pm–0 am | 0.08 |

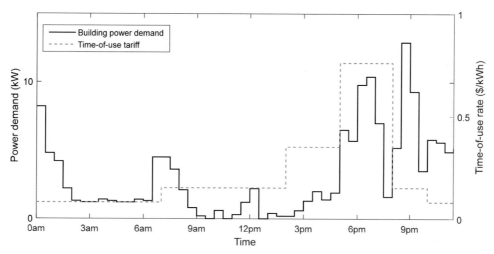

**Fig. 8.7** Profiles of the building's power demand and the time-of-use electricity tariff.

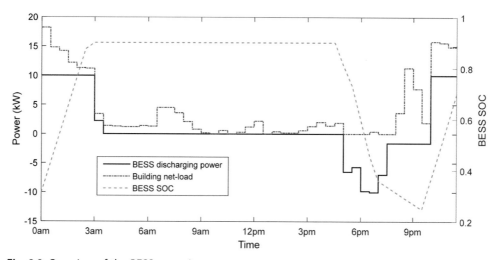

**Fig. 8.8** Overview of the BESS operation.

## 8.6 BESS integration with on-site renewable energy

### 8.6.1 Overview

With the widespread deployment of distributed RESs in buildings, it is becoming more common to use BESSs for absorbing the surplus renewable energy and to discharge it when there is no sufficient renewable energy. Such a scenario is illustrated in Fig. 8.9. Using BESSs to accommodate RESs can help a building to better utilize energy generated by its on-site RESs and to reduce the amount of energy imported from the grid. As a result, the building can operate efficiently as a self-sustained energy system.

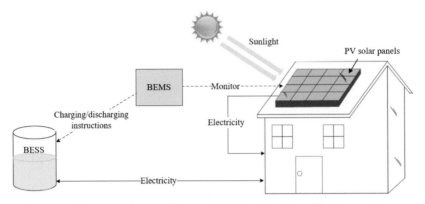

**Fig. 8.9** Illustration of the BEMS for a building-side BESS and rooftop PV solar panels.

Let us consider a building integrated with a BESS and a RES and denote the RES's power output over a period consisting of $T$ time slots as a $T$-dimensional vector $\boldsymbol{P}^{res}$:

$$\boldsymbol{P}^{res} = \left[ P_1^{res}, \ldots, P_t^{res}, \ldots, P_T^{res} \right], \tag{8.13}$$

where the element $\boldsymbol{P}_t^{res}$ denotes the power output of the RES in time slot $t$ (kW, $t=1:T$). With consideration of the charging/discharging of the BESS, the building's *net load* profile $\boldsymbol{L}^{net} = [L_1^{net}, \ldots, L_t^{net}, \ldots, L_T^{net}]$ can be calculated as

$$L_t^{net} = L_t + P_t^{bess} - P_t^{res}, t = 1 : T. \tag{8.14}$$

Since there could be on-site renewable power output ($P_t^{res}$) and the BESS could discharge power (represented by a negative value of $P_t^{bess}$), the value of $L_t^{net}$ can be either positive, nil, or negative. The sign of $L_t^{net}$ indicates the direction of the power flowing between the building and the grid: (i) A positive value means the building needs to import power from the grid to meet its own demand; (ii) a negative value means the building generates surplus power that is fed back to the grid; and (iii) a zero value means the building's power demand is exactly served by the energy it generates through the on-site RES and/or BESS.

### 8.6.2 Energy management strategy

The basic principle of using BESSs to accommodate buildings' on-site renewable energy is to determine the BESS's charging/discharging power based on the RES's power output, the BESS's operational condition and the building's power demand. Considering an energy management period consisting of $T$ time slots, the energy management procedures can be performed based on Algorithm 8.3.

---

**ALGORITHM 8.3 Energy management strategy for a BESS to accommodate renewable energy.**

**Start**
1. Input the BESS model ($P^{bess, rc}$, $P^{bess, rd}$, $E^{bess, r}$, $\eta^c$, $\eta^d$, $SOC^{upper}$, and $SOC^{lower}$);
2. Input the initial energy stored in the BESS (denoted as $E_0$) and the number of time slots $T$;
3. Input the RES model (e.g., the wind and PV solar power models in Chapter 2);
4. Start from the first time slot by setting the time index $t=1$;
5. Forecast the value of $L_t$;
6. Forecast the renewable power output in the time slot (denoted as $P_t^{res}$) based on the RES model;
7. Calculate the current dischargeable energy in the BESS as $E_t^{dcb} = E_{t-1}^{bess} - E^{bess, r} \times SOC^{lower}$;
8. Calculate the current chargeable energy in the BESS as $E_t^{cb} = E^{bess, r} \times SOC^{upper} - E_{t-1}^{bess}$;
9. **If** $P_t^{res} < L_t$
10.   Set the deficit power demand as $L_t^{def} = L_t - P_t^{res}$;
11.   Set $P_t^{bess} = -1 \times min\left(\frac{E_t^{dcb} \times \eta^d}{\Delta t}, L_t^{def}, P^{bess,rd}\right)$;
12. **Else**
13.   Set the surplus renewable power as $P_t^{sur} = P_t^{res} - L_t$;
14.   Set $P_t^{bess} = min\left(\frac{E_t^{cb}}{\eta^c \times \Delta t}, P^{bess,rc}, P_t^{sur}\right)$;
15. **End If**
16. **If** $t==T$
17.   **Go to** Line 22;
18. **Else**
19.   Proceed to the next time slot and set $t=t+1$;
20.   **Go to** Line 5;
21. **End If**
22. Output the charging/discharging actions $\boldsymbol{P}^{bess} = [P_1^{bess}, ..., P_T^{bess}]$.
**End**

---

The BEMS collected information on the BESS model and the RES model (Lines 1–3). It then determines the charging/discharging power of the BESS in each time slot. The BEMS also forecasts the building's power demand and the RES's power output (Lines 5 and 6). If the RES is a wind turbine or a PV solar panel, its power output can be estimated following the models introduced in Chapter 2. The BEMS can then calculate the dischargeable and chargeable energy of the BESS (Lines 7 and 8).

If the renewable power is smaller than the building's power demand (Line 9), the BESS needs to be discharged. The BEMS determines the discharging power of the BESS by taking the smallest value among the following three values (Lines 10 and 11): (i) The discharging power that can be drawn out from the BESS in one-time slot (i.e., $\frac{E_t^{dcb} \times \eta^d}{\Delta t}$);

(ii) the deficit power demand of the building that cannot be served by the RES; and (iii) the discharging power capacity of the BESS. When the renewable power is not smaller than the building's power demand, the BESS is charged by the surplus renewable power and the BEMS determines the charging power of the BESS by taking the smallest value among the following values: (i) The charging power that can fully charge the BESS in one-time slot (i.e., $\frac{E_t^{cb}}{\eta^c \times \Delta t}$); (ii) the surplus renewable power; and (ii) the BESS's charging power capacity.

### 8.6.3 Application example

We consider a building with a rooftop PV solar panel and a BESS. The solar panel has an area of $40 \text{ m}^2$ and an energy conversion efficiency factor of 0.3. The BESS properties are given in Table 8.5. At the beginning of the energy management period, it is assumed that no dischargeable energy is stored in the BESS, that is, $E_0 = E^{bess,r} \times SOC^{lower} = 40 \times 0.2 = 8$ kWh. In this example, we consider a one-day energy management period with an energy management interval of 15 min, that is, there are a total of 96 time slots. The PV solar power output model presented in Chapter 2 is used to calculate the power output of the PV solar panel.

Fig. 8.10 shows the profiles of the building's power demand and the solar power output produced by the PV solar panel in which the power demand and solar power output are assumed to remain constant during each 15-min time slot.

The charging/discharging power of the BESS is determined based on Algorithm 8.3 and the results are plotted in Fig. 8.11. The figure also shows the SOC profile of the BESS

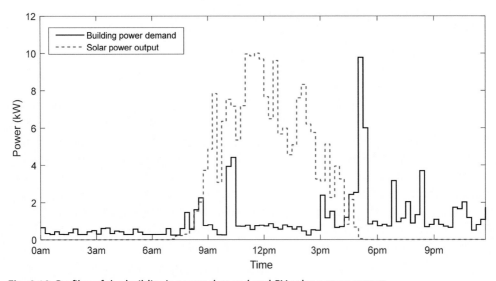

**Fig. 8.10** Profiles of the building's power demand and PV solar power output.

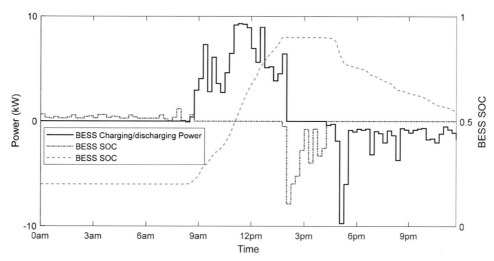

**Fig. 8.11** Results of the BESS operation.

and the building's net load under the charging/discharging power. It can be seen that, in some time slots, the part of solar power that cannot be charged into the BESS or that cannot be consumed by the building, is fed back to the grid (indicated by a negative value of the net-load). In other time slots, the building's power demand cannot be fully served by the BESS and the PV solar panel, and power needs to be imported from the grid to meet the demand (depicted by a positive value of the net load).

The total amounts of energy that needs to be imported from the grid (denoted as $E^{in}$) over the energy management period is calculated as follows:

$$E^{in} = \sum_{t=1}^{T} E_t^{in}, \tag{8.15}$$

$$E_t^{in} = \begin{cases} 0, if\ L_t^{net} < 0 \\ L_t^{net} \times \Delta t, otherwise \end{cases}. \tag{8.16}$$

The amount of energy that is fed back to the grid (denoted as $E^{out}$) over the energy management period is determined based on

$$E^{out} = \sum_{t=1}^{T} E_t^{out}, \tag{8.17}$$

$$E_t^{out} = \begin{cases} -L_t^{net} \times \Delta t, if\ L_t^{net} < 0 \\ 0, otherwise \end{cases} \tag{8.18}$$

where $E_t^{in}$ and $E_t^{out}$ denote the amounts of energy imported from the grid and fed back to the grid in time slot $t$, respectively (in kWh). In this application example, the values of $E_t^{in}$ and $E_t^{out}$ are 3.34 kWh and 8.40 kWh, respectively.

## 8.7 Optimization-based BESS energy management

### 8.7.1 Overview

In Sections 8.4–8.6, we have designed heuristic strategies to manage the charging/discharging power of the building-side BESSs. If forecasting for the renewable power output and the power demand of the building can be obtained over a finite future time horizon, the BESS energy management can also be performed by formulating a BESS charging/discharging power optimization model by using the optimization techniques introduced in Chapter 7. In this section, we demonstrate how optimization techniques can be applied to perform energy management for building-side BESSs.

We consider the design of a BEMS, which determines the charging and discharging power of a BESS aiming at minimizing the energy cost of the building over a period consisting of $T$ time slots. It is assumed the building is charged by a time-varying energy tariff. It is also assumed the building is equipped with a PV solar panel, which produces local energy supply support for the building.

### 8.7.2 Problem formulation

We start by defining the three elements of the optimization problem (that consist of the decision variables, the objective function, and the relevant constraints) to formulate the optimal BESS energy management problem. For ease of reference, the notation adopted in this problem is summarized in Table 8.9 (partly taken from the previous sections).

**Table 8.9** Summary of notation adopted for the BESS energy management optimization model.

| Notation | Explanation |
|---|---|
| $P^{bess,\ rc}$ | Charging power capacity of the BESS (in kW) |
| $P^{bess,\ rd}$ | Discharging power capacity of the BESS (in kW) |
| $\eta^c$ | Charging efficiency of the BESS (in %) |
| $\eta^d$ | Discharging efficiency of the BESS (in %) |
| $P_t^{bess}$ | Charging/discharging power of the BESS in time slot $t$ (in kW) |
| $E^{bess,\ r}$ | Energy capacity of the BESS (in kWh) |
| $SOC_t$ | SOC of the BESS in time slot $t$ (in %) |
| $E_t^{bess}$ | Energy stored in the BESS at $t$th time slot (in kWh) |
| $\Delta t$ | Duration of a time slot (in hours) |
| $SOC^{lower}$ | Lower allowable SOC limit of the BESS (in %) |
| $SOC^{upper}$ | Upper allowable SOC limit of the BESS (in %) |
| $L_t$ | Power demand of the building in time slot $t$ (in kW) |
| $\rho_t$ | Electricity tariff rate in time slot $t$ (in \$/kWh) |
| $P_t^{res}$ | Renewable power output in time slot $t$ (in kW) |

### 8.7.2.1 Decision variables

As the purpose of the optimization is to identify the charging/discharging power of the BESS at each time slot, we need to introduce $T$ decision variables, that is, a variable depicting the charging/discharging power at each time slot, and we collect the sought $T$ decision variables in the following vector:

$$\boldsymbol{P}^{bess} = \left[ P_1^{bess}, P_2^{bess}, ..., P_T^{bess} \right]. \tag{8.19}$$

### 8.7.2.2 Objective function

The objective function of the BESS energy management problem aims at minimizing the energy cost for the building over the $T$ time slots and this can be expressed as follows:

$$min\, F = \sum_{t=1}^{T} C_t, \tag{8.20}$$

where $C_t$ denotes the energy cost of the building in time slot $t$ (in \$) and it is calculated as

$$C_t = \begin{cases} 0 \; if \; L_t^{net} \leq 0 \\ L_t^{net} \times \Delta t \times r_t \; otherwise \end{cases}, \tag{8.21}$$

where the building's net load ($L_t^{net}$) is determined with Eq. (8.14).

### 8.7.2.3 Constraints

The optimization problem is subjected to several constraints. The values of the decision variables (i.e., the charging/discharging power of the BESS) are restricted by the BESS's power capacity.

$$-P^{bess,rd} \leq P_t^{bess} \leq P^{bess,rc}, t = 1 : T \tag{8.22}$$

In addition to the boundary constraints of the decision variables, the SOC of the BESS must remain within the allowable SOC range

$$SOC^{lower} \leq SOC_t \leq SOC^{upper}. \tag{8.23}$$

As a demonstration of how evolutionary algorithms can be used to solve energy management problems, we use the Differential Evolution (DE) algorithm (see Chapter 7) to solve the optimization model represented by Eqs. (8.19)–(8.23).

## 8.7.3 Setting up the solution with the DE algorithm

Once the optimization problem is defined (i.e., once the decision variables, the objective function and the relevant constraints are identified), it is possible to detail the solution process to be performed with the DE algorithm. Each individual of the population considered in the DE algorithm is described by the vector of decision variables introduced in

Eq. (8.19) that can vary within the range set in Eq. (8.22). The entire population consists of $N$ individuals and can be represented by the following matrix:

$$
\begin{bmatrix} \boldsymbol{P}_1^{bess} \\ \cdot \\ \cdot \\ \cdot \\ \boldsymbol{P}_N^{bess} \end{bmatrix} = \begin{bmatrix} P_{1,1}^{bess} & \cdots & P_{1,T}^{bess} \\ \cdot & & \cdot \\ \cdot & & \cdot \\ \cdot & & \cdot \\ P_{N,1}^{bess} & \cdots & P_{N,T}^{bess} \end{bmatrix},
\tag{8.24}
$$

where the subscript $i$ in $\boldsymbol{P}_i^{bess} = [P_{i,1}^{bess}, ..., P_{i,t}^{bess}, ..., P_{i,T}^{bess}]$ represents the index of an individual in the population.

The solution procedure to be adopted in the optimal BESS energy management model based on the DE algorithm is described in Algorithm 8.4. The output of the algorithm $\boldsymbol{P}^{bess,*}$ denotes the optimal charging/discharging schedule of the BESS generated by the algorithm.

In Algorithm 8.4, the BEMS starts by collecting information on the relevant parameters, models and inputs (Lines 1–6). It then uses the DE algorithm to perform the BESS charging/discharging optimization. The DE generates a population of individuals and records the individual with the minimum objective function value (Lines 8–14). As explained in Chapter 7, the *minByFitness*($\Theta$, *n*) function returns the first *n* individuals with minimum objective function values (i.e., fitness values) from the inputted population $\Theta$. In each generation of DE, the BEMS performs the subroutine *SubRoutine_GenerateTrialIndividual* to generate a trial individual for each individual of the population (Line 17). This subroutine is equivalent to Lines 9–12 of Algorithm 7.4 in which a trial individual is generated with the procedure of the DE algorithm. The *fitness*(·) function in Algorithm 9.4 calculates the objective function value of the inputted individual based on Eqs. (8.20) and (8.21).

It is important to note that the objective function needs to be calculated only for individuals who satisfy all specified constraints. In the above formulated BESS energy management problem, there are two constraints, that is, Eqs. (8.22) and (8.23). Because of this, the compliance of each individual with the adopted two constraints needs to be checked and, if required, adjusted to satisfy the constraints before evaluating its objective function value (Lines 10 and 18 in Algorithm 8.4).

In the solution adopted here, this task is performed by a separate function *constraintHandle*(•) that is described between Lines 10 and 18 in Algorithm 8.4. This function considers an individual $\boldsymbol{P}_i^{bess}$ and checks if it satisfies the constraints of Eqs. (8.22) and (8.23). If these conditions are not satisfied, the function adjusts $\boldsymbol{P}_i^{bess}$ to enable it to comply with the constraint. The function then outputs an individual that satisfies the considered constraint. The details of this function are presented in Algorithm 8.5.

---

**ALGORITHM 8.4 Building-side BESS operation optimization performed with the DE algorithm.**

**Start**
1. Input $T$ and $\Delta t$;
2. Input the BESS model ($P^{bess,\,rc}$, $P^{bess,\,rd}$, $E^{bess,\,r}$, $\eta^c$, $\eta^d$, $SOC^{upper}$, and $SOC^{lower}$);
3. Input the initial energy stored in the BESS (denoted as $E_0$);
4. Input the time-varying electricity rates $\rho = [\rho_1, ..., \rho_T]$;
5. Input the forecasted renewable power $P^{res} = [P_1^{res}, ..., P_T^{res}]$ using the models in Chapter 2;
6. Input the forecasted building power demand $L = [L_1, ..., L_T]$;
7. Set DE's control parameters: $N$, $G$, $Cr$, and $F$;
8. **For** $i = 1:N$
9.    Initialize $P_i^{bess}$ by randomly setting $P_{i,\,t}^{bess}$ ($t = 1:T$) in the range $[-P^{bess,\,rd}, P^{bess,\,rc}]$;
10.    Set $P_i^{bess} = constraintHandle(P_i^{bess})$;
11. **End For**
12. Set the population as $\Theta = \{P_1^{bess}, ..., P_N^{bess}\}$;
13. Set the generation index $g = 1$;
14. Set $P^{bess,\,*} = minByFitness(\Theta, 1)$;
15. **While** ($g <= G$)
16.   **For** $i = 1:N$
17.    Generate the trial individual $P_i^{trial}$ by performing *SubRoutine_GenerateTrialIndividual*;
18.    $P_i^{trial} = constraintHandle(P_i^{trial}, E_0, P^{bess,\,rc}, P^{bess,\,rd}, E^{bess,\,r}, \eta^c, \eta^d, SOC^{upper}, SOC^{lower})$;
19.    Set score1 $= fitness(P_i^{trial})$;
20.    Set score2 $= fitness(P_i^{bess})$;
21.    **If** (score1 < score2)
22.     Set $P_i^{bess} = P_i^{trial}$;
23.    **End If**
24.    **If** $fitness(P_i^{bess}) < fitness(P^{bess,\,*})$
25.     Set $P^{bess,\,*} = P_i^{bess}$;
26.    **End If**
27.   **End For**
28. **End While**
29. Output $P^{bess,\,*}$.

**End**

---

Algorithm 8.5 adjusts the inputted individual $P_i^{bess,in}$ generated by DE, which represents the charging and discharging power of the BESS over $T$ time slots, to satisfy the constraints of Eqs. (8.22) and (8.23). It checks the charging/discharging power of the BESS in each time slot and adjusts it when necessary to ensure the charging/discharging power does not exceed the charging/discharging power capacity and the BESS's SOC

---

**ALGORITHM 8.5 Pseudocodes of the function of constraint handling logics for the BESS energy management application.**

**Function** $P_i^{bess,\,out} = constraintHandle(P_i^{bess,\,in}, E_0, P^{bess,\,rc}, P^{bess,\,rd}, E^{bess,\,r}, \eta^c, \eta^d, SOC^{upper}, SOC^{lower})$

1. Set $P_i^{bess,\,out} = P_i^{bess,\,in}$;
2. **For** $t=1{:}T$
3.   **If** $P_{i,\,t}^{bess,\,out} < 0$
4.     Set the dischargeable energy as $E_t^{dcb} = E_{t-1}^{bess} - E^{bess,\,r} \times SOC^{lower}$;
5.     Set $P_{i,t}^{bess,out} = -1 \times \min\left(\frac{E_t^{dcb} \times \eta^d}{\Delta t}, \left|P_{i,t}^{bess,out}\right|, P^{bess,rd}\right)$;
6.     Set $E_t = E_{t-1} + \frac{P_{i,t}^{bess,out}}{\eta^d} \times \Delta t$;
7.   **Else**
8.     Set the chargeable energy as $E_t^{cb} = E^{bess,\,r} \times SOC^{upper} - E_{t-1}^{bess}$;
9.     Set $P_{i,t}^{bess} = \min\left(\frac{E_t^{cb}}{\eta^c \times \Delta t}, P^{bess,rc}, P_t^{sur}\right)$;
10.    Set $E_t = E_{t-1} + P_{i,\,t}^{bess,\,out} \times \eta^c \times \Delta t$;
11.  **End If**
12. **End For**
13. **Return** $P_i^{bess,\,out}$.
**End**

---

does not exceed the allowable range. If the charging/discharging power violates of one or two of the constraints, then the charging/discharging power is assigned the boundary value required by the BESS. During discharging, this is achieved by taking the minimum value among the following three values (Line 5): (i) The discharging power that can draw out all the dischargeable energy in the BESS in one time slot (i.e., $\frac{E_t^{dcb} \times \eta^d}{\Delta t}$); (ii) the discharging power generate by DE; and (iii) the discharging power capacity of the BESS. During charging, this is achieved by taking the minimum value among the following three values (Line 9): (i) The charging power that can charge all the chargeable energy into the BESS in one-time slot (i.e., $\frac{E_t^{cb}}{\eta^c \times \Delta t}$); (ii) the charging power generate by DE; and (iii) the charging power capacity of the BESS.

## 8.7.4 Application example

We use an application example to illustrate the optimization-based BESS energy management. The application scenario consists of a building with a dedicated BEMS and equipped with rooftop PV solar panels and a BESS. For illustrative purposes, we consider two BESSs with different capacities. The model parameters of the two BESSs are shown in Table 8.10. At the beginning of the simulation, it is assumed that the BESSs

**Table 8.10** Properties of the BESSs.

| Variable name | BESS 1 | BESS 2 |
|---|---|---|
| $P^{bess,\ rc}$ | 6 kW | 4 kW |
| $P^{bess,\ rd}$ | 6 kW | 4 kW |
| $E^{bess,\ r}$ | 12 kWh | 6 kWh |
| $SOC^{lower}$ | 0.1 | 0.1 |
| $SOC^{upper}$ | 0.8 | 0.8 |
| $\eta^c$ | 92% | 92% |
| $\eta^d$ | 88% | 88% |

possess initial SOC values (denoted as $SOC_0$) equal to 50%, that is, $SOC_0 = 0.5$ for both BESSs.

For the adopted PV solar panels, we assume that their area is 30 m$^2$ and that their energy conversion factor is constant and equal to 30%, that is, $\eta = 0.3$. The 24-h profiles of the building's power load and solar radiation density are summarized in Table 8.11.

**Table 8.11** Forecast data of the building's power load, solar radiation density, and solar power output.

| Time slot | Building power load (W) | Solar radiation (W/m$^2$) | Solar power output (W) |
|---|---|---|---|
| 0–1 am | 1191 | 0 | 0 |
| 1–2 am | 645 | 0 | 0 |
| 2–3 am | 193 | 0 | 0 |
| 3–4 am | 176 | 0 | 0 |
| 4–5 am | 176 | 0 | 0 |
| 5–6 am | 183 | 0 | 0 |
| 6–7 am | 163 | 0 | 0 |
| 7–8 am | 1538 | 9.2 | 55 |
| 8–9 am | 2455 | 98.6 | 592 |
| 9–10 am | 959 | 276.7 | 1660 |
| 10–11 am | 248 | 340.4 | 2043 |
| 11 am–12 pm | 224 | 494.1 | 2964 |
| 12–1 pm | 230 | 368.1 | 2209 |
| 1–2 pm | 203 | 290.8 | 1745 |
| 2–3 pm | 266 | 294.4 | 1766 |
| 3–4 pm | 518 | 181.2 | 1087 |
| 4–5 pm | 238 | 86.3 | 518 |
| 5–6 pm | 432 | 0.6 | 3 |
| 6–7 pm | 540 | 0 | 0 |
| 7–8 pm | 1818 | 0 | 0 |
| 8–9 pm | 2655 | 0 | 0 |
| 9–10 pm | 2224 | 0 | 0 |
| 10–11 pm | 2624 | 0 | 0 |
| 11 pm–0 am | 2520 | 0 | 0 |

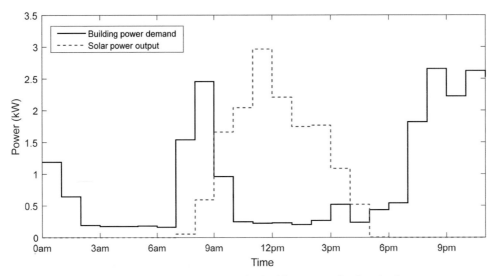

**Fig. 8.12** Profiles of the forecast data related to the building power load and solar power output.

The tabulated values have been provided based on a time discretization in which each time slot has a duration of 1 h, that is, $\Delta t = 1$, and, therefore, we have 24-time slots in total, that is, $T = 24$. It is assumed the power demand and solar radiation do not change during a time slot. By applying the PV solar power model in Chapter 2, it is possible to determine the solar power output profile that is collected in the right column of Table 8.11. For clarity, the building's power demand and the solar power profiles are plotted in Fig. 8.12. It can be seen the solar panel generates excessive solar energy during noon hours, therefore exceeding the building's demand.

It is assumed that the building is charged with the time-of-use electricity tariff summarized in Table 8.8. The parameters of the DE algorithm are set as follows: $N = 100$, $G = 200$, $F = 0.9$, and $Cr = 0.1$.

By running the BEMS based on the procedures described in Algorithm 8.4, the BESS is charged/discharged to accommodate the PV solar source and to feed the energy demand of the building. Fig. 8.13 shows the final charging/discharging power profile and the corresponding SOC variation profile for the three different cases considered: (i) Application scenario without the use of the BESS—in this case, the energy required by the building demand not satisfied by the solar power production is supplied by the grid; (ii) application scenario in which the larger BESS is monitored and controlled by the BEMS—the larger BESS is referred to as BESS 1; and (iii) application scenario in which the smaller BESS, denoted as BESS 2, is monitored and controlled by the BEMS. Fig. 8.14 shows the building's net load profiles with the two BESSs of different sizes as well as the case without the use of the BESS.

The plotted curves in Fig. 8.13 enable us to identify some expected operation patterns that occur in the presence of the BESSs. In the morning times when there is sufficient

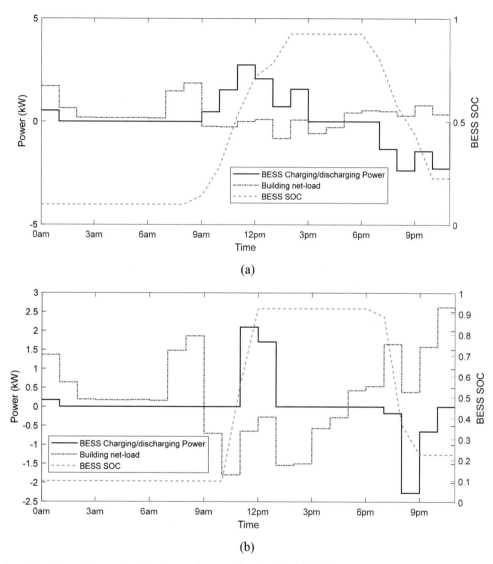

**Fig. 8.13** Operation results for the two BESSs: (A) BESS 1, (B) BESS 2.

solar power, the BESS is charged. For the smaller capacity BESS (i.e., BESS 2), the BESS is charged to a high SOC that approaches the allowable upper limit of 80%. In the evening times (after around 7 pm) when there is no solar power and the energy demand of the building is high, the BESS is discharged to serve the building's energy demand. In the scenarios evaluated with the use of BESS 2, the maximum power that can be charged into or drawn from BESS 2 is less than that of BESS 1, as expected by the smaller storage capacity of BESS 2.

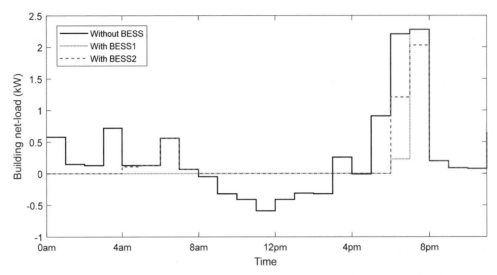

**Fig. 8.14** Net load profile of the building without the use of a BESS and with BESS of different sizes.

**Table 8.12** Economic evaluation for the building without the use of a BESS and with a BESS of different sizes.

|  | Without BESS | With BESS 1 | With BESS 2 |
|---|---|---|---|
| Total energy imported from the grid | 18.89 kWh | 12.3 kWh | 16.0 kWh |
| Energy cost | $2.41 | $1.36 | $1.96 |

Figs. 8.13 and 8.14 show that, by increasing the size of the BESS, it is possible to reduce the amount of power consumption that the building needs to import from the grid and this is reflected in a negative or nil value for the building's net load in some of the time slots. The selection of the size of the BESS needs to be based on a careful evaluation of its initial cost as well as its operational costs.

The economic evaluation of the BESS in building energy management operations is summarized in Table 8.12. By using the BESSs to accommodate the on-site solar energy, the amount of energy that the building needs to import from the grid is reduced. In this application scenario, the building's energy cost is approximately halved with the use of a BESS when considering the case without any BESS as a reference. As expected, the larger BESS (i.e., BESS 1) leads to greater savings when compared to the performance of the smaller one (i.e., BESS 2).

## 8.8 Introduction to vehicle-to-building/home integration

Conventional EVs use a unidirectional charger. by using a converter built either into the EV or the charger itself, the alternating current (AC) electricity sourced from the grid or solar panels can be converted to direct current (DC) electricity. To implement the V2B/V2H function, bidirectional chargers are needed, which can take DC electricity from the EV battery and convert it into AC electricity that can be used in a building. Currently, several EV models are possessing V2B/V2H functions.

While the energy management strategies introduced in the early parts of this chapter can be applied to manage the battery of the EV when it is plugged into a building, there are several differences between the V2B/V2H integration and the use of stationary BESSs to power the building. These differences should be taken into account when making energy management strategies for V2B/V2H integration. When compared with stationary BESSs, the EV battery cannot be used to power the building all the time. The time availability of an EV to perform V2B/V2H service depends on the life schedule and use patterns of the EV owner. Since EVs are primarily used as a vehicle, the lifetime depreciation of the EV battery in performing the V2B/V2H service needs to be considered. For example, the frequent discharging of the EV battery would shorten its lifetime and this could not be acceptable for an EV owner. The EV owner would impose specific requirements on the energy stored in the EV battery. For example, he/she could allow the EV battery to be used to power the home during the night and could require that at least 60% of battery energy capacity be stored by 8 am when he/she departs the house. Such requirements need to be considered by the BEMS when it manages the charging/discharging actions of an EV battery.

## References

[1] M.A. Hannan, S.B. Wali, P.J. Ker, M.S. Abd Rahman, M. Mansor, V.K. Ramachandaramurthy, K.M. Muttaqi, T.M.I. Mahlia, Z.Y. Dong, Battery energy-storage system: a review of technologies, optimization objectives, constraints, approaches, and outstanding issues, J. Energy Storage 42 (2021) 103023.

[2] S.M. Mousavi, F. Faraji, A. Majazi, K. Al-Haddad, A comprehensive review of flywheel energy storage system technology, Renew. Sustain. Energy Rev. 67 (2017) 477–490.

[3] B.K. Sakia, S.M. Benoy, M. Bora, J. Tamuly, M. Pandey, D. Bhattacharya, A brief review on supercapacitor energy storage devices and utilization of natural carbon resources as their electrode materials, Fuel 282 (2020) 118796.

[4] C.N. Truong, M. Naumann, R.C. Karl, M. Muller, A. Jossen, H.C. Hesse, Economics of residential photovoltaic battery systems in Germany: the case of Tesla's powerwall, Batteries 2 (2) (2016).

[5] Mike-fiesta, This file is made available and licensed under the Creative Commons Attribution-Share Alike 4.0 International license. Attribution: Mike-fiesta, 2021. https://commons.wikimedia.org/wiki/File:Lead-acid_batteries_components_2.jpg.

[6] Claus Ableiter, This file is made available and licensed under the Creative Commons Attribution-Share Alike 4.0 International license. Attribution: Claus Ableiter, 2008. https://commons.wikimedia.org/wiki/File:Lithium-Ionen-Accumulator.jpg.

[7] Iron Edison, This file is made available and licensed under the Creative Commons Attribution-Share Alike 3.0 International license. Attribution: Iron Edison, 2011. https://commons.wikimedia.org/wiki/File:Nickel_Iron_Battery_-_Depth_of_Discharge_life.jpg.

[8] "Smart Grid, Smart City" Customer Trial Data, 2015. [Online]. Available at: https://data.gov.au/dataset/ds-dga-4e21dea3-9b87-4610-94c7-15a8a77907ef/details, Accessed 22 June 2023.

# CHAPTER 9

# Energy management of flexible electric appliances

## Contents

## 9.1 Introduction

With the advances in information and communication technologies, an increasing number of appliance products have been incorporating wireless communication functions, therefore enabling them to receive command signals from Building Energy Management Systems (BEMSs) that control their operations. Depending on the occupant's lifestyle and requirements, the operation time or power of some appliances can be adjusted. These appliances are here denoted as flexible electric appliances (as we assume to deal with electric appliances) and their operational flexibility can improve the building's energy efficiency and enhance its demand response capability.

Regardless of types, functions, and manufacturing models, flexible electric appliances can be generally subdivided into two categories: (i) Time Shiftable Appliances (TSAs), referring to as the appliances whose operation time can be shifted, but whose operating power cannot be changed; and (ii) Power Adjustable Appliances (PAAs), referring to as the appliances whose operation time is fixed (determined by the building occupant) but their operating power can be adjusted. In this chapter, we provide a description of TSAs and PAAs, and present possible model representations that enable them to be included in building energy management implementations. Different types of energy management strategies are presented for the identification of possible scheduling schemes for the appliances. Simple energy management strategies are designed for TSAs and PAAs to demonstrate how to exploit their operational flexibilities to reduce buildings' energy costs subjected to time-varying energy tariffs. Particular attention is provided to formulate optimization-based energy management strategies for TSAs and PAAs and to describe the details of the solution process implemented with the differential evolution (DE) algorithm (introduced in Chapter 7). The optimization problems are expressed by minimizing the building's energy cost or by minimizing the amount of energy the building needs to import from the grid (considering the penetration of on-site renewable energy sources). The satisfaction of the building occupant with the flexible appliances is considered by incorporating multiple operational constraints in the optimization models. Application examples are performed to highlight some of the features of the numerical techniques.

More sophisticated optimization strategies have been presented in the literature for the scheduling of the appliance operations and their details can be found in the specialized literature, for example, [1–5]. The complexity of the proposed algorithms and application problems is kept to a minimum for illustrative purposes, for example, by considering only a small number of appliances in the application examples. The presented algorithms and solution strategies can be applied to more complex and larger energy management problems.

## 9.2 Modeling time shiftable appliances
### 9.2.1 Model of a time shiftable appliance
Time shiftable appliances are denoted as TSAs. The power consumption of these appliances is fixed,[a] and their operation time is flexible as it can be shifted within a certain time

---

[a] In the operation of an appliance, its actual power consumption is not a fixed value as its value fluctuates with a certain range. From an energy management prospective, it is usually acceptable to consider the operating of a time shiftable appliance to remain fixed and equal to the rated power specified in the product specification.

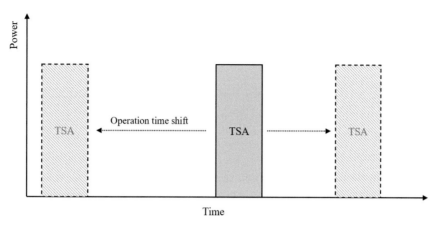

**Fig. 9.1** Depiction of the operation time flexibility of TSAs.

range [6], as depicted in Fig. 9.1. Examples of TSAs include washing machines, dish washers, and coffee machines. The operation time of these appliances is driven by the occupants' behavior, and they can be shifted to a certain extent, depending on the preferences and requirements of the building occupants.

We now consider three key features of a TSA that need to be included in a numerical representation that could be used in an optimization process. We consider that the time domain of energy management is composed of multiple time slots, and each time slot covers a certain time duration. The key features of a TSA are defined below:

- *Task duration*: It represents the time duration that is required by a TSA to complete its task. For example, a washing machine may take 50 min to complete a clothes washing task. The task duration of a TSA is denoted as $T^d$ and is measured as the number of time slots required to complete the task.
- *Operating power*: A TSA is considered to consume a certain amount of power when running and we denote it as $P^{op}$ (in kW).
- *Allowable operation time range*: Each TSA is expected to run and complete its task within the time range specified by the preferences of the building occupant. This time range represents the allowable operation time range and it depends on various factors, including the lifestyle preferences of the occupants. For example, an occupant may want to run the rice cooker to cook the rice between [4 pm, and 6 pm] to ensure that the rice is ready for dinner. The start and end of the allowable operation time range for a TSA are depicted by the variables $t^{tsa,s}$ and $t^{tsa,e}$ (expressed in terms of time slot indices). Obviously, $t^{tsa,e} - t^{tsa,s}$ should be greater or equal to $T^d$ to ensure that the appliance has sufficient time to complete its task.

Based on the above features, a TSA can be modeled as a vector consisting of four tuples:

$$\text{TSA} = \left[ T^d, P^{op}, t^{tsa,s}, t^{tsa,e} \right] \tag{9.1}$$

## 9.2.2 Worked example

Let us consider a washing machine as a TSA. The operating power of the washing machine is 1.2 kW, and it takes 90 min to complete the washing task. Let us consider a 10-h energy management period between 12 pm and 10 pm. The period consists of 60 time slots and each time slot lasts for 10 min. The building occupant allows the washing machine to run between 3 pm and 8 pm. Determine the 4-tuple vector depicting the key characteristics of the washing machine.

**Solution**

From the description, it can be seen that the operating power of the washing machine is $P^{op} = 1.2$ kW. Since it takes 90 min to complete the task and each time slot lasts for 10 min, the washing machine's task duration expressed in time slots is $T^d = \frac{90}{10} = 9$.

The energy management period runs between 12 pm and 10 pm. In the adopted discretization, time slot 1 depicts the period 12–12:10 pm; time slot 2 refers to 12:10–12:20 pm; and so on. The time slot indices of 3 pm and 7:50 pm are 19 and 48, respectively (note the time slot 48 covers the period of 7:50 pm to 8 pm). Therefore, the 4-tuple vector depicting the key characteristics of the washing machine (denoted as WM) can be expressed as:

$$\text{WM} = \left[ T^d, P^{op}, t^{tsa,s}, t^{tsa,e} \right] = [9, 1.2, 19, 48]$$

## 9.3 Modeling power adjustable appliances

### 9.3.1 Model of a power adjustable appliance

Power adjustable appliances are denoted in the following as PAAs. This category includes appliances whose operation time is set by the building occupant and cannot be shifted, while their power consumption can be adjusted within a certain range [7]. This is illustrated in Fig. 9.2. For example, an artificial lighting system's power consumption could be adjusted based on the level of light present in a particular building space. Three key features are introduced in the following to account for the presence of a PAA in an energy management process.

- *Operation time slots*: A PAA is associated with a period when its operating power can be adjusted. The start and end time slots of the time period are denoted as $t^{paa,s}$ and $t^{tsa,e}$, respectively. In each time slot during the period of $[t^{paa,s}, t^{paa,e}]$, the PAA's power consumption can be adjusted.
- *Power adjustment range*: A PAA is associated with a building occupant-specified power consumption range, defined as $[P^{min}, P^{max}]$. $P^{min}$ (in kW) represents the minimum power that the PAA needs to consume in a time slot; and $P^{max}$ (in kW) represents the maximum power that the PAA can consume in a time slot. In each time slot during the period $[t^{paa,s}, t^{paa,e}]$, the power consumption of the PAA can be adjusted within the range of $[P^{min}, P^{max}]$.

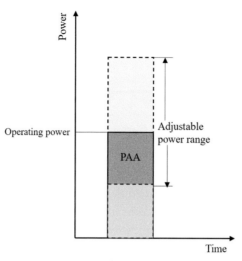

**Fig. 9.2** Graphical representation of the adjustable operating power range of PAAs.

- *Desirable operating power:* For a PAA, it is assumed that the occupant specifies desirable operating power values for each operation time slot, here denoted as $P^{dsr}$ (in kW). In practice, the occupant usually specifies his/her desired level or intensity of the function of the PAA and the BEMS then maps it to the power level.
- *Power adjustment ramping rates:* For some PAAs, over-intensive variation of operating power in neighboring time slots can cause discomfort to building occupants and/or induce a negative effect on the health of an appliance. For example, adjusting the power consumption of a light can lead to a variation of brightness, and a sudden intensive change of brightness could lead to visual discomfort to an occupant. Based on this consideration, the operating power adjustment of a PAA is subjected to a ramp up rate ($\gamma^{up}$) and a ramp down rate ($\gamma^{down}$). Both rates are expressed in kW, and they restrict the power adjustment degree between two neighboring power adjustable time slots within the interval $[t^{paa,s}, t^{paa,e}]$:

$$\begin{cases} P^{paa}_{t+1} - P^{paa}_t \leq \gamma^{up}, if\ P^{paa}_{t+1} > P^{paa}_t \\ P^{paa}_t - P^{paa}_{t+1} \leq \gamma^{down}, if\ P^{paa}_{t+1} < P^{paa}_t \end{cases}, \forall t = t^{paa,s} : t^{paa,e} - 1 \qquad (9.2)$$

where $P^{paa}_t$ (in kW, $t = t^{paa,s} : t^{paa,e}$) denotes the power consumption of the PAA in time slot $t$ (in kW). Eq. (9.2) expressed the fact that, in two neighboring time slots within $[t^{paa,s}, t^{paa,e}]$, the change of an appliance's operating power is restricted by the ramping rates.

Based on the above properties, a PAA can be modeled as a 7-tuple vector:

$$PAA = \left[t^{paa,s}, t^{paa,e}, P^{min}, P^{max}, P^{dsr}, \gamma^{up}, \gamma^{down}\right] \qquad (9.3)$$

It is considered that for a PAA, no matter how its operating power is adjusted, its total energy consumption (denoted as $E^{paa}$) should not be lower than the PAA's energy consumption when it operates at the desirable operating power, here denoted as $E^{dsr}$ and calculated as:

$$E^{paa} \geq E^{dsr} = \sum_{t=t^{paa,s}}^{t^{paa,e}} P^{dsr} \times \Delta t \qquad (9.4)$$

### 9.3.2 Worked example

Let us consider an artificial lighting system as a PAA. The energy management period is set to be 5 h, from 6 to 11 pm. It is assumed that 20 time slots occur over the energy management period, and each time slot covers 15 min. The lighting system operates from 8 pm to 10 pm. The desirable operating power of the light is 800 W, and the operating power is allowed to be adjusted in the range of [600 W, 1200 W]. The ramp up and down rates of the power consumption adjustment for the light is set to be 200 W and 150 W, respectively.

Based on the above description, determine the 7-tuple vector describing the lighting system.

**Solution**

(1) Since the energy management period is between 6 and 11 pm and each energy management time slot covers 15 min, the first time slot occurs between 6 and 6:15 pm. The operation time of the lighting system (i.e., 8–10 pm) corresponds to the time slot range of 9–16: $t^{paa,s} = 9$ and $t^{paa,e} = 16$.
(2) The power adjustment range of the light is $[P^{min}, P^{max}]$, where $P^{min} = 600\text{W} = 0.6\,\text{kW}$ and $P^{max} = 1200\,\text{W} = 1.2\,\text{kW}$.
(3) The desirable operating power of the light is $P^{dsr} = 800\text{W} = 0.8\,\text{kW}$.
(4) The power adjustment ramping rates of the light are: $\gamma^{up} = 200\text{W} = 0.2$ kW; $\gamma^{down} = 150\text{W} = 0.15\,\text{kW}$.

The above values can be collected in the 7-tuple vector describing the PAA as follows:

$$\text{Light} = \left[ t^{paa,s}, t^{paa,e}, P^{min}, P^{max}, P^{dsr}, \gamma^{up}, \gamma^{down} \right] = [9,16,0.6,1.2,0.8,0.2,0.15]$$

## 9.4 A simple energy management scheme for TSAs

### 9.4.1 Energy management strategy

Based on the appliance model presented in Section 9.2, a BEMS can be designed to manage the operations of TSAs and this is usually performed with the aim of reducing the overall building's energy cost while ensuring that the TSAs complete their tasks within the occupant-specified allowable time ranges.

For the monitoring and control of TSAs, a BEMS determines a schedule of the operation times for each TSA that are mapped within a discretized time domain (assumed to be formed by $T$ time slots, each with a duration of $\Delta t$ that is usually expressed in hours). Denote the time-varying electricity tariff rates over the $T$ time slots as $\boldsymbol{\rho} = [\rho_1, ..., \rho_t, ..., \rho_T]$, where $\rho_t$ represents the electricity price over the time slot $t$ (in \$/kWh, $t = 1:T$). In the following, we present a simple heuristic algorithm in which a BEMS manages $K$ TSAs whose pseudocode is described in Algorithm 9.1. With this approach, the BEMS considers the allowable operation time range of each TSA and schedules the appliances to operate in the low-price period while ensuring the tasks are finished within the allowable operation time range.

The subscript $k$ ($k = 1:K$) in the notation of Algorithm 9.1 is used to indicate the model parameters of the $k$th TSA. In the algorithm, the BEMS starts by defining the input model parameters of the $K$ TSAs and the available electricity rates (Lines 1 and 2). For each TSA, the BEMS checks if the appliance can start to operate at the beginning of the time period that has the lowest electricity rate within the appliance's allowable operation time range (denoted as $t_k^*$) (Line 4). If there is sufficient time for the appliance to execute its task, then the appliance will be scheduled to start operating at that time slot (Lines 5 and 6). If this arrangement is not possible, the algorithm would then anticipate the start of the operation by reducing the value for $t_k^*$ by one time slot and continue to do so until a feasible set of time slots is allocated (Lines 7–9).

---

**ALGORITHM 9.1  A simple energy management scheme for TSAs.**

**Start**
1.  Input the models of the $K$ TSAs;
2.  Input the electricity tariff rates $\boldsymbol{\rho} = [\rho_1, ..., \rho_t, ..., \rho_T]$;
3.  **For** $k = 1:K$
4.      Get the earliest time slot of the cheapest time period within $[t_k^{tsa,s}, t_k^{tsa,e}]$, denoted as $t_k^*$;
5.      **If** $t_k^* + T_k^d - 1 \leq t_k^{tsa,e}$
6.          Output the $k$th TSA's schedule as $[t_k^*, t_k^* + T_k^d - 1]$;
7.      **Else**
8.          Set $t_k^* = t_k^* - 1$;
9.          **Go to** Line 5.
10. **End If**
11. **End For**
**End**

## 9.4.2 Application example

In this section, we consider an application example to deploy the procedure for the BEMS based on Algorithm 9.1. In this example, we consider six controllable NTC-TSAs whose key parameters, previously introduced in Section 9.2 (i.e., operation power, task durations, and allowable operation time ranges), are summarized in Table 9.1. The building is assumed to be charged by a four-segment time-of-use tariff that is specified in Table 9.2. In the solution process, we set $T = 288$ and $\Delta t = 1/12$ (hours).

In this example, we consider two scenarios for the scheduling of the appliances:

**(1)** *Scenario 1*: The BEMS ensures that each appliance starts operating at the beginning of its allowable operation time range and keeps running until the task is completed; and

**(2)** *Scenario 2*: The BEMS schedules the appliances based on Algorithm 9.1.

The power consumption of the scheduled appliances is illustrated in Figs. 9.3 and 9.4 for scenarios 1 and 2, respectively. For ease of reference, the time-of-use rates are also provided on the right vertical axis.

During scenario 1, several appliances, that is, washing machine, dish washer, and clothes dryer, operate during peak hours (with the highest time-of-use rate). With the use of Algorithm 9.1 their operations are shifted from peak hours to shoulder hours while the operations of the remaining appliances remain unchanged. From an economical viewpoint, the overall energy costs for scenarios 1 and 2 are $3.50 and $1.49, respectively, therefore highlighting the potential benefit of exploiting (even simple) algorithms in the scheduling of appliances monitored and controlled by a BEMS.

**Table 9.1** Configurations of the TSAs.

|  | Rated power | Task duration | Allowable operation time range |
| --- | --- | --- | --- |
| Washing machine | 0.8 kW | 70 min (14 time slots) | [4 pm, 10 pm] |
| Clothes dryer | 1.4 kW | 60 min (12 time slots) | [5 pm, 10 pm] |
| Oven | 1.2 kW | 40 min (8 time slots) | [10 am, 4:30 pm] |
| Pool pump | 0.9 kW | 120 min (24 time slots) | [10 am, 8:30 pm] |
| Rice cooker | 0.4 kW | 50 min (10 time slots) | [8 am, 12 pm] |
| Dish washer | 0.7 kW | 70 min (14 time slots) | [6 pm, 12 am] |

**Table 9.2** Time-of-use electricity tariff.

| Time | Rate ($/kWh) |
| --- | --- |
| Peak: 3–8 pm | 0.86 |
| Secondary peak: 12–3 pm | 0.52 |
| Shoulder: 7 am–12 pm, 8–10 pm | 0.27 |
| Off-peak: 10 pm–7 am | 0.08 |

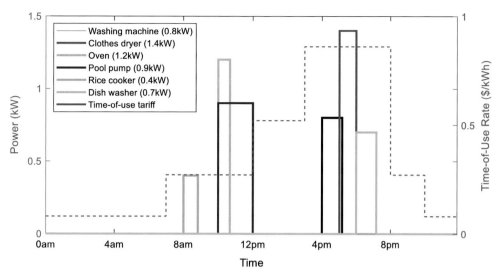

**Fig. 9.3** Schedule of the operations of the TSAs for scenario 1 (i.e., each appliance starts operating at the beginning of its allowable operation time range).

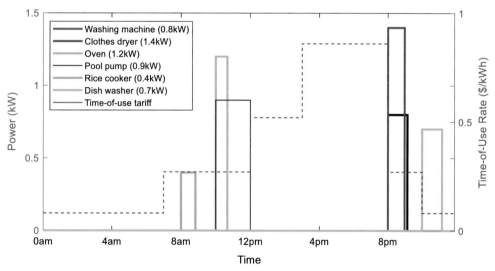

**Fig. 9.4** Schedule of the operations of the TSAs for scenario 2 (i.e., schedule determined with Algorithm 9.1).

## 9.5 A simple energy management schemes for PAAs

### 9.5.1 Energy management strategy

In this section, we design a simple BEMS that performs energy management for a group of PAAs with the aim of reducing the building's energy cost subjected to a time-of-use tariff while ensuring the operational requirements of the PAAs. In the following, we present a simple heuristic algorithm in which a BEMS manages $M$ PAAs whose pseudo-code is described in Algorithm 9.2.

The subscript $m$ $(m=1{:}M)$ in the notation of Algorithms 9.2 is used to indicate the model parameters of the $m$th PAA. With this approach, the BEMS always tries to schedule the PAAs to consume power energy in the time slots with low energy prices.

---

**ALGORITHM 9.2 A simple energy management scheme for PAAs.**

**Start**
1. Input models of the $M$ PAAs;
2. Input $T$ and $\Delta t$;
3. Input the electricity tariff rates $\rho = [\rho_1, \ldots, \rho_t, \ldots, \rho_T]$;
4. **For** $m = 1{:}M$
5.     Initialise a set of processed time slots $\mathbf{\Phi}_m$ to be an empty set, i.e. $\mathbf{\Phi}_m = \{\}$.
6.     Initialise a set of unprocessed time slots $\mathbf{\Theta}_m = [t^{paa,s}, t^{paa,s}+1, \ldots, t^{paa,e}]$;
7.     Initialise the $(t^{paa,e} - t^{paa,s}+1)$-dimensional operating power vector as $\boldsymbol{P}_m^{op} = [0, \ldots, 0]$;
8.     **If** $\mathbf{\Theta}_m$ is empty
9.         **Terminate**;
10.     **Else**
11.         Get the time slot with the cheapest electricity tariff rate from $\mathbf{\Theta}_m$, denoted as $t^*$;
12.         Set $P_{m,t^*}{}^{op} = maxConsumablePower(\mathbf{\Phi}_m, t_m^{paa,s}, t_m^{paa,e}, P_m^{min}, P_m^{max}, P_m^{dsr}, \boldsymbol{P}_m^{op}, \Delta t)$;
13.         **If** time slot $t^*-1$ is in $\mathbf{\Phi}_m$
14.            **If** $P_{m,-1}^{op} < P_m^{op}$
15.                **If** $(P_{m,t^*}{}^{op} - P_{m,t^*-1}{}^{op}) > \gamma^{up})$
16.                    Set $P_{m,t^*}{}^{op} = P_{m,t^*-1}{}^{op} + \gamma^{up}$;
17.                **End If**
18.            **End If**
19.         **End If**
20.         **If** time slot $t^*+1$ is in $\mathbf{\Phi}_m$
21.            **If** $P_{m,t^*+1}{}^{op} < P_{m,t^*}{}^{op}$
22.                **If** $(P_{m,t^*}{}^{op} - P_{m,t^*+1}{}^{op}) > \gamma^{down})$
23.                    Set $P_{m,t^*}{}^{op} = P_{m,t^*+1}{}^{op} + \gamma^{down}$;
24.                **End If**
25.            **End If**
26.         **End If**
27.         Remove $t^*$ from $\mathbf{\Theta}_m$;
28.         Set $\mathbf{\Phi}_m = \mathbf{\Phi}_m \cup t^*$;
29.         **Go to** Line 8.
30. **End For**
**End**

The BEMS accepts the PAA models, electricity tariff and time slot information as inputs (Lines 1–3 in Algorithm 9.2). Then, it performs energy management tasks for each PAA (Lines 5–29).

For each PAA (indexed by the subscript $m$), the energy management algorithm determines its power consumption in each of its operation time slots. It first initializes two sets: (i) a set $\Phi_m$ storing the time slots that have been processed (i.e., in which the PAA's power consumption has been scheduled) – the set is initialized to be empty (Line 5); and (ii) a set $\Theta_m$ storing the time slots that have not been processed (i.e. in which the PAA's power consumption has not been scheduled) – this set is initialized to contain all the time slots between $t_m^{paa,s}$ and $t_m^{paa,e}$ (Line 6). The BEMS then repeatedly finds the time slot with the cheapest energy price from the unprocessed time slots in $\Theta_m$ (denoted as $t^*$, Line 11) and invokes Algorithm 9.3 to determine the maximum power it can consume in that time slot (denoted as $P_{m,t^*}^{op}$, Line 12). If there are neighboring time slots of $t^*$ that have already been processed, then $P_{m,t^*}^{op}$ is further adjusted and subjected to the ramping constraints based on the power consumptions of the PAA in the neighboring time slots (Lines 13–26). After the processing, the power consumption of the PAA in time slot $t^*$ is determined, and $t^*$ is removed from $\Theta_m$ and is added into $\Phi_m$ (Lines 27 and 28). This process is iteratively performed until all the time slots have been processed.

Algorithm 9.3 presents the procedures of determining the maximum power a PAA can consume in a time slot given the sets of processed and unprocessed time slots (i.e., $\Phi$ and $\Theta$). The algorithm firstly calculates the sum of two parts of energy (Line 1): (i) the amount of energy the PAA has been scheduled to consume – this is calculated as $\sum_{t \in \Phi} P_t^{op} \times \Delta t$; and (ii) the minimum amount of energy the PAA is expected to consume in the unprocessed time slots – this is calculated as $P^{max} \times |\Theta| \times \Delta t$. The algorithm then calculates the maximum amount of energy the PAA can consume in the considered time slot without accounting for the upper limit of power consumption $P^{max}$ (Lines 2 and 3). This is calculated as $E^{dsr} - E'$. Based on this, the maximum power is determined as the smaller value between: (i) the power needed for PAA to consume the energy calculated as $E^{dsr} - E'$; and (ii) $P^{max}$ (Line 4). In Line 4, the function min($\cdot$) returns the minimum value among the inputted numbers.

---

**ALGORITHM 9.3 Function of calculating a PAA's maximum consumable power in a time slot.**

**Function** $P^{max} = maxConsumablePower(\Phi, \Theta, t^{paa,s}, t^{paa,e}, P^{min}, P^{max}, P^{dsr}, \mathbf{P}^{op}, \Delta t)$
1. Set $E' = \sum_{t \in \Phi} P_t^{op} \times \Delta t + P^{max} \times |\Theta| \times \Delta t$;
2. Set $E^{dsr} = P^{dsr} \times \Delta t \times (t^{paa,e} - t^{paa,s} + 1)$;
3. Set $E'' = E^{dsr} - E'$;
4. **Return** $P^{max} = \min\left(\frac{E''}{\Delta t}, P^{max}\right)$.
**End**

**Table 9.3** Electricity tariff rates.

| Time period | Rate ($/kWh) |
|---|---|
| 9 am–12 pm | 0.20 |
| 12–6 pm | 0.35 |
| 6–8 pm | 0.63 |
| 8–9 pm | 0.45 |

**Table 9.4** Models of the PAAs.

| | $t^{paa,s}$ | $t^{paa,e}$ | $p^{min}$ | $p^{max}$ | $p^{dsr}$ | $\gamma^{up}$ | $\gamma^{down}$ |
|---|---|---|---|---|---|---|---|
| Light 1 | 3 | 9 | 0.1 | 0.3 | 0.2 | 0.04 | 0.04 |
| Light 2 | 17 | 24 | 0.15 | 0.45 | 0.28 | 0.08 | 0.05 |
| Pool pump | 5 | 8 | 0.5 | 1.2 | 0.8 | 0.2 | 0.2 |
| Amplifier | 19 | 24 | 0.1 | 0.4 | 0.3 | 0.04 | 0.08 |

## 9.5.2 Application example

In this section, we set up an application example to demonstrate the energy management strategy in Algorithm 9.2. We consider an energy management period consisting of 12 h, from 9 am to 9 pm. Each time slot lasts for 30 min. Based on this input we have: $T = 24$ and $\Delta t = 0.5$.

The electricity tariff rates over the 24-time slot period are shown in Table 9.3. We consider four PAAs, whose models are given in Table 9.4.

By applying Algorithm 9.2 to the PAA models, the operating power consumption of the PAAs can be determined. Figs. 9.5 and 9.6 visualize the PAAs' operation under two cases:

**(1)** *Scenario 1:* Without energy management, that is, each PAA operates at the desirable power in its operation time slots; and

**(2)** *Scenario 2:* With the energy management strategy in Algorithm 9.2.

It can be seen that with the energy management strategy, the PAAs' power consumption in high electricity price hours (i.e., 6–8 pm) is reduced. As a result, the energy cost of running the PAAs is reduced from $1.71 to $1.34. It can also be seen from Fig. 9.5 that the power ramping up and down constraints are satisfied for the four PAAs.

## 9.6 Energy management for TSAs: An optimization-based approach

### 9.6.1 Problem formulation

In the previous sections, we have highlighted the potential economic benefit of defining the scheduling of appliances based on considerations of the varying energy costs that

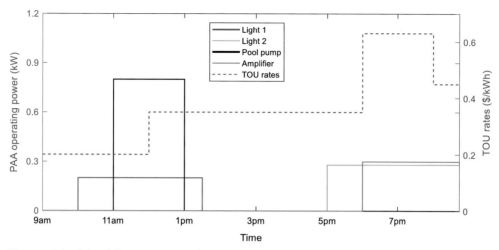

**Fig. 9.5** Schedule of the operations of the PAAs for scenario 1 (i.e., each appliance consumes the desirable power in its operation time slots).

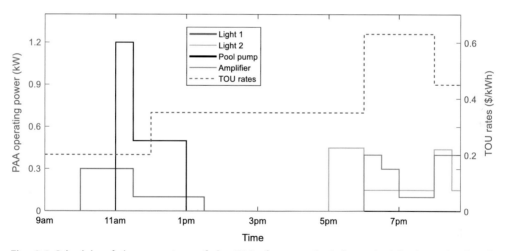

**Fig. 9.6** Schedule of the operations of the PAAs for scenario 2 (i.e., schedule determined with Algorithm 9.2).

occur during the day. The optimization techniques presented in Chapter 7 can be applied to determine the optimal operation plans for flexible appliances. In this section, we formulate an optimization scheduling problem for TSAs that is solved by means of an evolutionary algorithm. In this initial problem formulation, we identify the decision variables, the objective function, and the relevant constraints.

For convenience purposes, the notation used in this section is presented in Table 9.5.

**Table 9.5** Summary of the notation adopted for the TSA energy management optimization model.

| Variable Name | Definition |
| --- | --- |
| $K$ | Number of TSAs |
| $x_k$ | Time slot when the $k$th TSA starts to operate |
| $T$ | Number of time slots over the energy management period |
| $\Delta t$ | Duration of a time slot (in hours) |
| $P_k^{op}$ | Operating power of the $k$th TSA (in kW) |
| $T_k^d$ | Task duration of the $k$th TSA (in hours) |
| $t_k^{tsa,s}$ | Starting time slot of the $k$th TSA's allowable operation time range |
| $t_k^{tsa,e}$ | End time slot of the $k$th TSA's allowable operation time range |

### 9.6.1.1 Decision variables

Let us assume that there are $K$ TSAs whose optimal operation time is to be optimized. The decision variables, here denoted as $x_k$ (with $k = 1{:}K$), represent the index of the time slot when the $k$th TSA starts to operate. We have here assumed that the time domain, that is, the scheduling horizon, has been subdivided into $T$ time slots. The $K$ decision variables that depict the starting time slot for the $K$ TSAs are collected in the vector $\boldsymbol{x}$ as

$$\boldsymbol{x} = [x_1, x_2, \ldots, x_K] \tag{9.5}$$

### 9.6.1.2 Objective function

It aims at minimizing the energy cost of the TSAs during the $T$ time slots and this can be expressed as:

$$\text{Minimize } F = \sum_{t=1}^{T} C_t \tag{9.6}$$

$$C_t = \sum_{k=1}^{K} (P_{k,t} \, \Delta t \, \rho_t) \tag{9.7}$$

in which $C_t$ denotes the energy cost of the TSAs in time slot $t$ (in \$); $P_{k,t}$ depicts the power consumption of the $n$th TSA (measured in kW) in the time slot $t$. The value of $P_{k,t}$ depends on the decision variable $x_k$ as follows:

$$P_{k,t} = \begin{cases} P_k^{op} & \text{if } x_k \leq t \leq x_k + T_k^d - 1 \\ 0 & \text{otherwise} \end{cases} \tag{9.8}$$

Eq. (9.8) implies that during the operation period of the $k$th TSA, that is, during $[x_k, x_k + T_k^d - 1]$, the power consumption of the appliance is equal to its rated power, while in the remaining time slots, its power consumption is nil.

### 9.6.1.3 Constraints

The constraints are required to ensure that the appliances operate in the allowable operation time ranges and that the operations are completed within these ranges. For this reason, $x_k$ is bounded as follows:

$$t_k^{tsa,e} \leq x_k \leq t_k^{tsa,e} - T_k^d + 1, k = 1 : K \qquad (9.9)$$

## 9.6.2 Solution process based on the DE algorithm

The optimization problem introduced in Eqs. (9.5)–(9.9) is solved in this section using the DE algorithm. Each decision variable (collected in the vector defined in Eq. (7.9) in Chapter 7) is treated as an individual of the population considered by the DE algorithm and the whole population of the $N$ individuals is collected in the following matrix:

$$\begin{bmatrix} \boldsymbol{x}_1 \\ \cdot \\ \cdot \\ \cdot \\ \boldsymbol{x}_N \end{bmatrix} = \begin{bmatrix} x_{1,1} & \cdot & \cdot & \cdot & x_{1,K} \\ \cdot & & & & \cdot \\ \cdot & & & & \cdot \\ \cdot & & & & \cdot \\ x_{N,1} & \cdot & \cdot & \cdot & x_{N,K} \end{bmatrix} \qquad (9.10)$$

The pseudocode describing the use of the DE algorithm for the identification of the optimal scheduling of the TSA is described in Algorithm 9.4.

In Algorithm 9.4, the *minByFitness*($\boldsymbol{\Theta}$, *n*) function returns the first *n* individuals with minimum objective function values from the inputted population $\boldsymbol{\Theta}$. The subroutine *SubRoutine_GenerateTrialIndividual* is equivalent to Lines 9–12 of Algorithm 7.4 in Chapter 7. It generates a trial individual for an individual based on DE's heuristic rules. The *obj*($\boldsymbol{x}$,...) function takes an individual and other parameter (i.e., the energy tariff rates tariff rate vector $\boldsymbol{\rho}$ and TSA models, which are represented by "..." in the function notation) as inputs; it calculates the objective function value of the inputted individual following Eqs. (9.6)–(9.8).

For the BEMS described in Algorithm 9.4, it accepts the input data of the time slot information, electricity tariff rates, TSA models, and the control parameters of DE (Lines 1–4). It then initializes a population of candidate TSA operation plans (Lines 5–7) and records the best solution in the population (Line 10). After that, the BEMS performs the evolution process of the DE to iteratively generate variations of the individuals, to evaluate the objective function value of the individuals and their variants (i.e., the energy cost of different candidate TSA operation plans), and to update the best solution it finds (Lines 11–23). After a finite number of evolution generations, the BEMS outputs the best TSA operation plan it has found (Line 24).

## 9.6.3 Application example

In this Section, we reconsider the application example previously considered in Section 9.4.2, with the same TSAs' details and time-of-use tariff data specified in

---

**ALGORITHM 9.4 Pseudocode for optimal scheduling of TSAs based on the DE algorithm.**

**Start**

1. Input $T$ and $\Delta t$;
2. Input electricity tariff rates $\rho = [\rho_1, \ldots, \rho_T]$;
3. Input TSA models ($P_k^{op}$, $T_k^d$, $t_k^{tsa,s}$, $t_k^{tsa,e}$, $k = 1{:}K$);
4. Set the parameters of DE: $N$, $G$, $Cr$, and $F$;
5. **For** $i = 1{:}N$
6.    Initialize $\boldsymbol{x}_i$ by randomly setting $x_k$ between $[t_k^{tsa,s}, t_k^{tsa,e} - T_k^d + 1]$;
7. **End For**
8. Set the population as $\Theta = \{\boldsymbol{x}_1, \ldots, \boldsymbol{x}_N\}$;
9. Set the generation index $g = 1$;
10. Set the historically recorded best individual $\boldsymbol{x}^* = minByFitness(\Theta, 1)$;
11. **While** $(g <= G)$
12.    **For** $n = 1{:}N$
13.       Generate the trial individual $\boldsymbol{x}_n^{trial}$ by performing *SubRoutine_GenerateTrialIndividual*;
14.       Set score1 $= obj(\boldsymbol{x}_n^{trial}, \ldots)$;
15.       Set score2 $= obj(\boldsymbol{x}_n, \ldots)$;
16.       **If** (score1 < score2)
17.          Set $\boldsymbol{x}_n = \boldsymbol{x}_n^{trial}$;
18.       **End If**
19.       **If** $obj(\boldsymbol{x}_n) < obj(\boldsymbol{x}^*)$
20.          Set $\boldsymbol{x}^* = \boldsymbol{x}_n$;
21.       **End If**
22.    **End For**
23. **End While**
24. Output $\boldsymbol{x}^*$.

**End**

---

Tables 9.1 and 9.2. In the solution process, we employ Algorithm 9.4 and consider the following parameters for the DE algorithm: $N = 40$, $G = 30$, $F = 0.8$, and $Cr = 0.2$.

The calculated optimized schedule for the six TSAs is presented in Fig. 9.7 and it highlights how the BEMS avoids running appliances during the peak and secondary peak electricity price hours while still ensuring that all TSAs complete their tasks within their allowable operation time ranges.

The energy cost associated to the operation schedule of Fig. 9.7 is evaluated to be $0.87. This value is lower than the costs related to the schedules identified in scenarios 1 and 2 in Section 9.4.2 as reported in Table 9.6. These comparisons highlight the advantages of adopting optimized strategies in the management and control of TSAs.

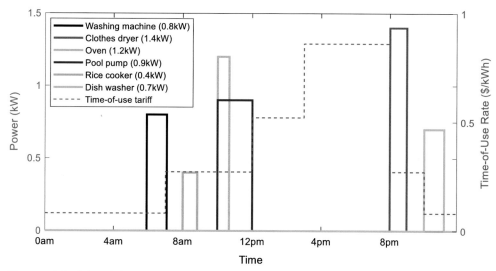

**Fig. 9.7** Schedule of the operations of the TSAs obtained with the DE algorithm presented in Algorithm 9.4.

**Table 9.6** Comparison of the expected energy cost obtained from scenarios 1 and 2 in Section 9.4.2 and based on the use of Algorithm 9.4.

|  | With the scheduling approach of scenario 1 in Section 9.4.2 | With the scheduling approach of scenario 2 in Section 9.4.2 (simple algorithm) | With optimization-based scheduling based on Algorithm 9.4 |
|---|---|---|---|
| Energy cost | $3.50 | $1.49 | $0.87 |

## 9.7 Energy management scheme for both TSAs and PAAs with renewable energy penetration: An optimization-based approach

### 9.7.1 Problem formulation

In Section 9.6, we have presented an optimization-based energy management framework for TSAs. In this section, we extend the energy management framework to include both TSAs and PAAs, and consider the penetration of building-side renewable energy sources in the energy management process. For convenience, the notation used in this section is defined in Table 9.7, in which some variables have been taken from previous sections and Table 9.5.

### 9.7.2 Decision variables

Let us consider that there are $K$ TSAs and $M$ PAAs described by the models in Sections 9.2 and 9.3. The operation time of the TSAs and operating power consumption of the PAAs are to be optimized. The decision variables of the flexible appliance scheduling

**Table 9.7** Summary of the notation adopted for the energy management optimization model for both TSAs and PAAs.

| Variable name | Definition |
|---|---|
| $K$ | Number of TSAs |
| $M$ | Number of PAAs |
| $T$ | Number of time slots over the energy management period |
| $\Delta t$ | Duration of a time slot (in hours) |
| $\mathbf{\Phi}^{tsa}$ | List of TSA models |
| $\mathbf{\Phi}^{paa}$ | List of PAA models |
| $\Phi_k^{tsa}$ | Model of the $k$th TSA |
| $\Phi_m^{paa}$ | Model of the $m$th PAA |
| $P_k^{op}$ | Operating power of the $k$th TSA (in kW) |
| $T_k^d$ | Task duration of the $k$th TSA (in hours) |
| $t_k^{tsa,s}$ | Start time slot of the $k$th TSA's allowable operation time range |
| $t_k^{tsa,e}$ | End time slot of the $k$th TSA's allowable operation time range |
| $t_m^{paa,s}$ | Start time slot of the $m$th PAA's operation time range |
| $t_m^{paa,e}$ | End time slot of the $m$th PAA's operation time range |
| $P_m^{min}$ | Minimum power consumption of the $m$th PAA (in kW) |
| $P_m^{max}$ | Maximum power consumption of the $m$th PAA (in kW) |
| $P_m^{dsr}$ | Desirable power consumption of the $m$th PAA (in kW) |
| $\gamma_m^{up}$ | Power ramp up limit of the $m$th PAA (in kW) |
| $\gamma_m^{down}$ | Power ramp down limit of the $m$th PAA (in kW) |
| $P_t^{res}$ | Power output of the renewable energy source in time slot $t$ (in kW) |
| $P_{k,t}^{tsa}$ | Power consumption of the $k$th TSA in time slot $t$ (in kW) |
| $P_{m,t}^{paa}$ | Power consumption of the $m$th PAA in time slot $t$ (in kW) |
| $P_t^{fa}$ | Total power consumption of the flexible appliances in time slot $t$ (in kW) |
| $x_i$ | $i$th decision variable in the decision variable vector |

problem include two parts: (i) the start time slot of each TSA; and (ii) the operating power of each PAA in each of its operation time slots. The decision variables can be represented as a vector $x$:

$$x = \left[ x_1, x_2, \ldots, x_K, x_{K+1}, \ldots, x_{K+T_1^{op}}, \ldots, x_{K+\Sigma_{m=1}^{M-1} T_m^{op}+1}, \ldots, x_{K+\Sigma_{i=m}^{M} T_m^{op}} \right] \qquad (9.11)$$

$$T_m^{op} = t_m^{paa,e} - t_m^{paa,s} + 1, \forall m = 1 : M \qquad (9.12)$$

There are a total of $K + \sum_{m=1}^{M} T_m^{op}$ decision variables in $x$, where $T_m^{op}$ represents the number of operation time slots of the $m$th PAA. Based on this, $x$ has dimension of $K + \sum_{m=1}^{M} T_m^{op}$. The decision variables are sequentially arranged in $x$. The first $K$ variables of this vector indicate the start time slot of the $K$ TSAs, that is, $x_k$ ($k = 1{:}K$) denotes the starting time slot of the $k$th TSA.

After the first $K$ variables, the subsequent variables represent the consecutive $T_m^{op}$ ($m = 1{:}M$) items that indicate the power consumption of a PAA in its operation time slots. These variables are grouped by the number of operation time slots of each PAA. For example: the $(K+1)$th to $(K+T_1^{op})$th variables denote the power consumption of the first PAA (in kW), in which $x_{j+K}$ $(j=1: T_1^{op})$ denotes the power consumption of the PAA in its $j$th operation time slot. The subsequent $(K+T_1^{op}+1)$th to $(K+T_1^{op}+T_2^{op})$th variables denote the power consumption of the second PAA, in which $x_{j+K+T_1^{op}}$ $(j=1: T_2^{op})$ denotes the power consumption of the PAA in its $j$th operation time slot.

### 9.7.3 Objective function

As discussed in Chapter 5, building energy management can be performed by considering different objectives. In Section 9.6, we have presented a TSAs' scheduling that minimizes the building's energy cost over a finite energy management period. In this section, we consider the energy management objective as minimizing the amount of energy that the building imports from the grid. In particular, the BEMS schedules TSAs and PAAs to maximize the building's local energy supply capability in terms of meeting its energy demand by using its on-site renewable energy output.

Based on this, the objective function of the appliance scheduling can be expressed as:

$$\text{Minimize } F = \sum_{t=1}^{T} \widetilde{E}_t \tag{9.13}$$

$$\widetilde{E}_t = \begin{cases} \left(P_t^{fa} - P_t^{res}\right)\Delta t, \text{if } P_t^{fa} > P_t^{res} \\ 0, otherwise \end{cases} \tag{9.14}$$

$$P_t^{fa} = \sum_{k=1}^{K} P_{k,t}^{tsa} + \sum_{m=1}^{M} P_{m,t}^{paa} \tag{9.15}$$

$$P_{k,t}^{tsa} = \begin{cases} P_k^{op}, \text{if } x_k \le t \le x_k + T_k^d - 1 \\ 0, otherwise \end{cases}, \forall k = 1:K \tag{9.16}$$

$$P_{m,t}^{paa} = \begin{cases} 0, \text{if } t < t_m^{paa,s} \text{ or } t > t_m^{paa,e} \\ x_{K+\sum_{m=1}^{M-1} T_m^{op}+(t-t_m^{paa,s}+1)}, otherwise \end{cases} \tag{9.17}$$

The objective function calculates the sum of the energy the building needs to import from the grid in each time slot to meet the energy demand of the flexible appliances (Eq. (9.13)). In each time slot, the energy the building imports from the grid is calculated with Eq. (9.14) by comparing the total power consumption of the flexible appliances and the power output of the building's on-site renewable power output.

In each time slot, the power consumption of each TSA and PAA depends on the corresponding decision variables and are expressed as Eqs. (9.16) and (9.17), respectively. Eq. (9.16) represents that, for a TSA and during its task execution time determined from $x$, its power consumption equals the appliance's operating power, while in other time slots, its power consumption is zero. Eq. (9.17) highlights how, for a PAA and during its operation time period, its power consumption values are determined from $\boldsymbol{x}$ (while in other time slots, its power consumption is nil).

### 9.7.4 Constraints

The appliance scheduling model is subjected to the constraints listed in the following:

**(1)** Operation time constraint of TSAs – each TSA's operation must remain within its allowable operation time range:

$$t_k^{tsa,s} \le x_k \le t_k^{tsa,e} - T_k^d + 1, \forall k = 1 : K \tag{9.18}$$

**(2)** (2) Power consumption constraint of PAAs – each PAA's operating power must fall within the adjustable range:

$$P_m^{min} \le x_{\sum_{j=1}^{M-1} T_j^{op}+z} \le P_m^{max}, \forall m = 1 : M, z = 1 : t_m^{paa,e} - t_m^{paa,s} + 1 \tag{9.19}$$

**(3)** Power ramping up and down constraint of PAAs – for each PAA, its operating power variation in two consecutive time slots needs to satisfy:

$$x_{\sum_{j=1}^{M-1} T_j^{op}+z+1} - x_{\sum_{j=1}^{M-1} T_j^{op}+z} \le \gamma^{up}, if\ x_{\sum_{j=1}^{M-1} T_j^{op}+z+1} > x_{\sum_{j=1}^{M-1} T_j^{op}+z},$$
$$\forall m = 1 : M, z = 1 : t_m^{paa,e} - t_m^{paa,s} + 1 \tag{9.20}$$

$$x_{\sum_{j=1}^{M-1} T_j^{op}+z} - x_{\sum_{j=1}^{M-1} T_j^{op}+z+1} \le \gamma^{down}, if\ x_{\sum_{j=1}^{M-1} T_j^{op}+z+1} < x_{\sum_{j=1}^{M-1} T_j^{op}+z},$$
$$\forall m = 1 : M, z = 1 : t_m^{paa,e} - t_m^{paa,s} + 1 \tag{9.21}$$

**(4)** Desired energy consumption constraint of PAAs:

$$\sum_{t=1}^{t_m^{paa,e}-t_m^{paa,s}+1} x_{\sum_{j=1}^{M-1} T_j^{op}+t} \times \Delta t \ge \sum_{t=1}^{t_m^{paa,e}-t_m^{paa,s}+1} P_m^{dsr} \times \Delta t, \forall m = 1 : M \tag{9.22}$$

### 9.7.5 Solution process based on the DE algorithm

The optimization problem defined above is solved in this section using the DE algorithm introduced in Chapter 7. The whole population of the $N$ individuals is collected in the following matrix:

$$\begin{bmatrix} \boldsymbol{x}_1 \\ . \\ . \\ . \\ \boldsymbol{x}_N \end{bmatrix} = \begin{bmatrix} x_{1,1} & . & . & . & x_{1,K+\sum_{m=1}^M T_m^{op}} \\ & . & & & . \\ & . & & & . \\ & . & & & . \\ x_{N,1} & . & . & . & x_{N,K+\sum_{m=1}^M T_m^{op}} \end{bmatrix} \tag{9.23}$$

The pseudocode describing the use of the DE algorithm for the identification of the optimal scheduling of TSAs and PAAs is described in Algorithm 9.5.

---

**ALGORITHM 9.5 Pseudocode for optimal scheduling of TSAs and PAAs based on the DE algorithm.**

**Start**
1. Input $T$ and $\Delta t$;
2. Input electricity tariff rates $\rho = [\rho_1, \ldots, \rho_T]$;
3. Input the RES model (e.g., the wind and PV solar power models in Chapter 2);
4. Input the forecasted renewable resource values (e.g. wind speed and solar radiation);
5. Forecast the renewable power (denoted as $\boldsymbol{P}^{res} = [P_1^{res}, \ldots, P_T^{res}]$) based on the RES model;
6. Input TAA models $\boldsymbol{\Phi}^{tsa} = [\Phi_1^{tsa}, \ldots, \Phi_K^{tsa}]$;
7. Input TAA models $\boldsymbol{\Phi}^{paa} = [\Phi_1^{paa}, \ldots, \Phi_M^{tsa}]$;
8. Set the parameters of DE: $N$, $G$, $Cr$, and $F$;
9. For $n=1{:}N$
10.    Initialize $\boldsymbol{x}_n$ as $\boldsymbol{x}_n = inititaliseIndividual(\boldsymbol{\Phi}^{tsa}, \boldsymbol{\Phi}^{paa})$;
11.    $\boldsymbol{x}_n = constraintHandle(\boldsymbol{x}_n, \boldsymbol{\Phi}^{tsa}, \boldsymbol{\Phi}^{paa})$;
12. **End For**
13. Set the population as $\boldsymbol{\Theta} = \{\boldsymbol{x}_1, \ldots, \boldsymbol{x}_N\}$;
14. Set the generation index $g=1$;
15. Set the historically recorded best individual $\boldsymbol{x}^* = minByFitness(\Theta, 1)$;
16. **While** $(g <= G)$
17.    **For** $n=1{:}N$
18.      Generate the trial individual $\boldsymbol{x}_n^{trial}$ by performing *SubRoutine_GenerateTrial Individual*;
19.      $\boldsymbol{x}_n^{trial} = constraintHandle(\boldsymbol{x}_n^{trial}, \boldsymbol{\Phi}^{tsa}, \boldsymbol{\Phi}^{paa})$;
20.      Set score1 $= fitness\ (\boldsymbol{x}_n^{trial})$;
21.      Set score2 $= fitness\ (\boldsymbol{x}_n)$;
22.      **If** (score1 < score2)
23.        $\boldsymbol{x}_n = \boldsymbol{x}_n^{trial}$;
24.      **End If**
25.      **If** $fitness(\boldsymbol{x}_n) < fitness(\boldsymbol{x}^*)$
26.        $\boldsymbol{x}^* = \boldsymbol{x}_n$;
27.      **End If**
28.    **End For**
29. **End While**
30. Output $\boldsymbol{x}^*$.
**End**

---

In Algorithm 9.5, the BEMS firstly accepts the input data (Lines 1–8). The renewable power output can be generated from the models introduced in Chapter 2 and each item $P_t^{res}$ ($t=1{:}T$) collected in $\boldsymbol{P}^{res}$ (Line 5) denotes the power output of the building's on-site renewable energy source in time slot $t$. For each TSA, its model representation $\Phi_k^{tsa}$

(Line 6, $k=1{:}K$) consists of the properties $T_k^d$, $P_k^{op}$, $t_k^{tsa,s}$, and $t_k^{tsa,e}$. For each PAA, its model $\Phi_m^{paa}$ (Line 7, $m=1{:}M$) consists of the properties $t_m^{paa,s}$, $t_m^{paa,e}$, $P_m^{min}$, $P_m^{max}$, $P_m^{dsr}$, $\gamma_m^{up}$, and $\gamma_m^{down}$.

Once the input values are specified, the BEMS initializes a population of individuals, where each individual represents a candidate operation plan of the TSAs and PAAs. For each individual, its elements (i.e., the decision variables) are randomly initialized within their value ranges by invoking the function $initaliseIndividual(\cdot)$. The logic of the function is shown in Algorithm 9.6.

In Algorithm 9.6, the function $rdn(a, b)$ returns a random real number in the range of two real numbers $[a, b]$; the function $rdnInt(a, b)$ returns a random integer within $[a, b]$. For each individual, there are a total of $K + \sum_{m=1}^{M}\left(t_m^{paa,e} - t_m^{paa,s} + 1\right)$ decision variables that need to be initialized. The first $K$ decision variables are initialized based on the $K$ TSAs' allowable operation time range (Lines 1–3 in Algorithm 9.6). After that, consecutive $(t_m^{paa,e} - t_m^{paa,s} + 1, m=1{:}M)$ decision variables are initialized based on a PAA's adjustable power range (Lines 4–9 in Algorithm 9.6).

In the algorithm of scheduling TSAs and PAAs (i.e., Algorithm 9.5), after initializing each individual, the initialized individual is adjusted by the $constraintHandle(\cdot)$ function to ensure that the individual is feasible (Line 11). The objective function values of the individuals are calculated based on Eqs. (9.13)–(9.17) and the individual with the minimum objective function value is recorded (Lines 15). After that, the individuals are iteratively updated based on DE's evolution process (Lines 16–19). In each generation, a trial

---

**ALGORITHM 9.6 Pseudocodes of the function of initializing an individual in the scheduling of TSAs and PAAs.**

**Function** $x = initialiseIndividual(\Phi^{tsa}, \Phi^{paa})$
1. **For** $k=1{:}K$
2.    Set $x_k=rdnInt(t_k^{tsa,s}, t_k^{tsa,e} - T_k^d + 1)$;
3. **End For**
4. **For** $m=1{:}M$
5.    For $z=1{:}t_m^{paa,e} - t_m^{paa,s} + 1$
6.      Set $p = \sum_{j=1}^{m-1}\left(t_j^{paa,e} - t_j^{paa,s} + 1\right) + z$;
7.      Set $x_p=rdn(P_m^{min}, P_m^{max})$;
8.    **End For**
9. **End For**
10. **Return** $x=[x_1, \ldots, x_{K+\sum_{m=1}^{M}(t_m^{paa,e}-t_m^{paa,s}+1)}]$.

**End**

individual is generated for each individual in the population by invoking the subroutine *SubRoutine_GenerateTrialIndividual*, which corresponds to Lines 9–12 of Algorithm 7.4 in Chapter 7. The trial individual is then adjusted by the *constraintHandle*(·) function. Selection is made between each individual and its trial individual based on their objective function values and the best individual ($\mathbf{x}^*$). After a pre-specified number of generations, the best individual is outputted (Lines 30).

The *constraintHandle*(·) function contains the logic of checking and adjusting an inputted individual to make sure that it satisfies the constraints (9.18)–(9.22), described in Algorithm 9.7. Firstly, the start operation time slot of each TSA is checked (Lines 2–8). If the value exceeds the allowable operation time range, it is then adjusted to be the boundary of the range. For each PAA, its power consumption in each of its operation time slots is checked (Lines 9–24). If the power consumption value generated by the individual is smaller than the minimum power, it has to be consumed at that time slot (calculated by the *minConsumablePower*(·) function of Algorithm 9.8), and the power consumption value is adjusted to the minimum power (Lines 11–14). If the power consumption value is larger than the maximum power it can consume, it is adjusted to be the maximum power (Lines 15–17). Afterward, the power consumption value is checked and adjusted (if necessary) subjected to the power ramping constraints (Lines 18–22).

Algorithm 9.8 shows the procedures of calculating the minimum power a PAA has to consume in a particular time slot. The algorithm calculates the amount of energy that has been scheduled to be consumed in the previous operation time slots of the PAA (Line 1). It then calculates how much energy is left for the PAA to consume based on its desirable energy consumption amount $E_m^{dsr}$ (Lines 2 and 3). The minimum power the PAA has to consume in the time slot is then determined as the smaller value between the following two values: (i) the power needed for achieving $E_m^{dsr}$ based on the previously scheduled energy and based on the situation that the PAA only consumes the maximum power in all the operation time slots after the given one - this is denoted as $\frac{E'''}{\Delta t}$ in Line 5; and (ii) the lower boundary value of the PAA's power adjustment range (i.e. $P_m^{min}$).

## 9.7.6 Application example

We provide an application example to demonstrate the energy management strategy presented in this section. Let us consider 5 TSAs and 3 PAAs, whose models are shown in Tables 9.8 and 9.9, respectively. We assume a 12-h energy management period from 8 am to 8 pm and the duration of each time slot is set to be 10 min (leading to a total of 72 time slots). The building is considered to be equipped with a Photovoltaic (PV) solar panel. The area and energy conversion efficiency factor of the solar panel are $30\,m^2$ and 25%, respectively. Fig. 9.8 shows the solar radiation density received by the solar panel. The

---

**ALGORITHM 9.7 Pseudocodes of the function of constraint handling logics for scheduling of TSAs and PAAs.**

**Function** $x^{new}$=constraintHandle($x$, $\Phi^{tsa}$, $\Phi^{paa}$)
1.  Set $x^{new}=x$;
2.  **For** $k=1:K$
3.      **If** $x_k < t_k^{tsa,s}$
4.        Set $x_k = t_k^{tsa,s}$;
5.      **Else If** $x_k > t_k^{tsa,e} - T_k^d + 1$
6.        Set $x_k = t_k^{tsa,e} - T_k^d + 1$;
7.      **End If**
8.  **End For**
9.  **For** $m=1:M$
10.    **For** $z=1:t_m^{paa,e} - t_m^{paa,s} + 1$
11.      Set $p = \sum_{j=1}^{m-1}\left(t_j^{paa,e} - t_j^{paa,s} + 1\right) + z$;
12.      Set $P^* = minConsumablePower(x^{new}, \Phi^{paa}, m, z, \Delta t)$;
13.      **If** $x_p < P^*$
14.        Set $x_p = P^*$;
15.      **Else If** $x_p > P_m^{max}$
16.        Set $x_p = P_m^{max}$;
17.      **End If**
18.      **If** $z>1$ **And** $x_p - x_{p-1} > \gamma^{up}$
19.        Set $x_p = x_{p-1} + \gamma^{up}$;
20.      **Else If** $z>1$ **And** $x_{p-1} - x_p > \gamma^{down}$
21.        Set $x_p = x_{p-1} - \gamma^{down}$;
22.      **End If**
23.    **End For**
24. **End For**
25. **Return** $x^{new}$;
**End**

---

control parameters of the DE are specified as follows: $N=200$, $G=100$, $F=0.6$, and $Cr=0.2$.

Figs. 9.9 and 9.10 show the optimized operation schedules of the flexible appliances without and with the implementation of energy management, respectively. In the case without energy management, each TSA operates starting from the allowable operation time range until it completes its task, and each PAA operates at the desirable power consumption. The solar power output values over the 12h are also plotted in the figures, which are calculated based on the solar radiation density following the PV solar power model presented in Chapter 2.

---

**ALGORITHM 9.8 Function of calculating a PAA's minimum consumable power in a time slot.**

**Function** $P^* = minConsumablePower(\mathbf{x}, \Phi^{paa}, m, z, \Delta t)$
1. Set the energy amount that has been scheduled as $E' = \sum_{o=1}^{z-1} x_{\sum_{j=1}^{m-1}(t_j^{paa,e} - t_j^{paa,s} + 1) + o} \times \Delta t$;
2. Set $E_m^{dsr} = P_m^{dsr} \times \Delta t \times (t_m^{paa,e} - t_m^{paa,s} + 1)$;
3. Set the energy amount that has not been scheduled as $E'' = E^{dsr} - E'$;
4. Set $E''' = E'' - \sum_{o=z+1}^{t_m^{paa,e} - t_m^{paa,s} + 1} P_m^{max} \times \Delta t$
5. **Return** $P^* = maximum\left(\frac{E'''}{\Delta t}, P_m^{min}\right)$.
**End**

---

**Table 9.8** Models of the TSAs.

| | $P^{op}$ | $T^d$ | $[t^{tsa,s}, t^{tsa,e}]$ |
|---|---|---|---|
| TSA1 | 1.4 kW | 90 min (9 time slots) | [16, 54] ([10 am, 5 pm])[a] |
| TSA2 | 0.9 kW | 50 min (5 time slots) | [7, 39] ([9 am, 2:30 pm]) |
| TSA3 | 0.7 kW | 70 min (7 time slots) | [1, 48] ([8 am, 4 pm]) |
| TSA4 | 1.0 kW | 60 min (6 time slots) | [19, 42] ([11 am, 3 pm]) |
| TSA5 | 0.5 kW | 100 min (10 time slots) | [31, 72] ([1 pm, 8 pm]) |

[a]The time range in the bracket in the fourth column indicates the physical time range covered by $[t^{tsa,s}, t^{tsa,e}]$.

**Table 9.9** Models of the PAAs.

| | $[t^{paa,s}, t^{paa,e}]$ | $P^{min}$ | $P^{max}$ | $P^{dsr}$ | $\gamma^{up}$ | $\gamma^{down}$ |
|---|---|---|---|---|---|---|
| PAA1 | [19, 32] ([11 am, 1:20 pm])[a] | 0.2 kW | 0.6 kW | 0.4 kW | 0.06 kW | 0.06 kW |
| PAA2 | [61, 72] ([6 pm, 8 pm]) | 0.4 kW | 0.9 kW | 0.5 kW | 0.1 kW | 0.07 kW |
| PAA3 | [40, 46] ([2:30 pm, 3:40 pm]) | 0.8 kW | 1.8 kW | 1.3 kW | 0.2 kW | 0.2 kW |

[a]The time range in the bracket in the second column indicates the physical time range covered by $[t^{paa,s}, t^{paa,e}]$.

It can be seen from the figures that by applying the energy management scheme of Algorithm 9.5, the appliances are scheduled to operate in the time slots with sufficient solar power as much as possible, to maximize the utilization of solar energy and reduce the amount of energy import from the grid. Without energy management, there is a total of 7.25 kWh energy that needs to be imported from the grid, while this value reduces to 4.08 kWh when using energy management. Fig. 9.11 shows the net-load profiles

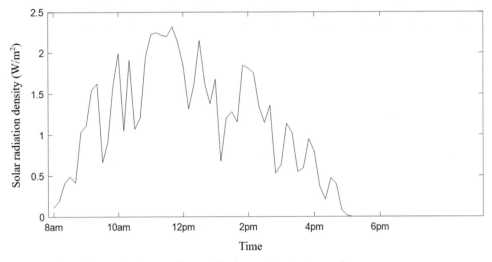

**Fig. 9.8** Solar radiation density profile used in the application example.

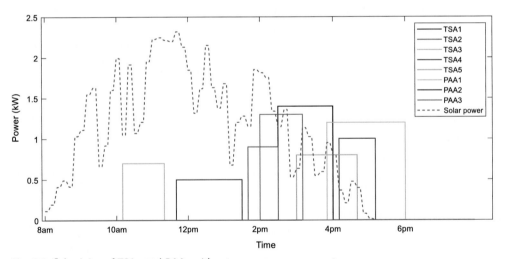

**Fig. 9.9** Schedules of TSAs and PAAs without energy management.

calculated from the flexible appliance schedules and solar power output with and without energy management. It can be seen with energy management, that both the amounts of energy load that needs to be served by the grid and the solar energy that is fed back to the grid are smaller than those evaluated without the use of the energy management, therefore indicating a better utilization of the on-site solar energy.

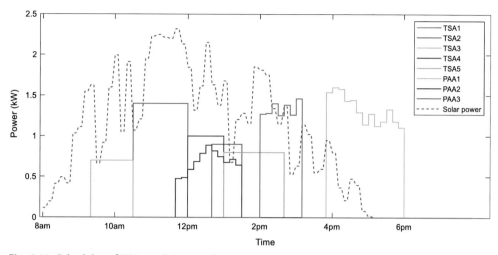

**Fig. 9.10** Schedules of TSAs and PAAs with energy management.

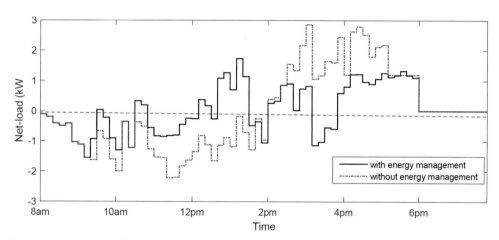

**Fig. 9.11** Net-load profiles with and without energy management.

## References

[1] M. Rasteger, M. Fotuhi-Firuzabad, F. Aminifar, Load commitment in a smart home, Appl. Energy 96 (2012) 45–54.
[2] F. Luo, W. Kong, G. Ranzi, Z.Y. Dong, Optimal home energy management with demand charge tariff and appliance operational dependencies, IEEE Trans. Smart Grid 11 (2020) 4–14.
[3] M.A. Pedrasa, T. Spooner, I. MacGill, Coordinated scheduling of residential distributed energy resources to optimize smart home energy services, IEEE Trans. Smart Grid 1 (2) (2010) 134–143.
[4] F. Luo, G. Ranzi, G. Liang, Z.Y. Dong, Stochastic residential energy resource scheduling by multi-objective natural aggregation algorithm, IEEE PES Gen. Meet. (2017).

[5] Z. Zhao, F. Luo, Y. Zhang, G. Ranzi, and S. Su, "Integrated household appliance scheduling with modelling of occupant satisfaction and appliance heat gain," Front. Energy Res., early access. https://doi.org/10.3389/fenrg.2021.724189.

[6] Z. Zhao, W. Lee, Y. Shin, K. Song, An optimal power scheduling method for demand response in home energy management system, IEEE Trans. Smart Grid 4 (3) (2013) 1391–1400.

[7] G. Zhao, L. Li, J. Zhang, and K.B. Letaief, "Residential demand response with power adjustable and unadjustable appliances in smart grid," in Proceedings of 2013 IEEE International Conference on Communications Workshops (ICC), Jun. 2013.

# CHAPTER 10

# Energy management of HVAC systems

## Contents

## 10.1 Introduction

Heating, Ventilation and Air Conditioning (HVAC) systems play a vital role in the energy consumption of modern buildings. It is estimated that the HVAC in a typical office building accounts for approximately 40% of the building's total energy consumption [1]. They also account for up to 50% of a commercial building's energy consumption and cause high peak power consumption [2]. Therefore properly managing the energy consumption of HVACs can improve the building's energy performance in terms of reducing its energy and peak power consumption and reducing the energy amount the building needs to be important from the grid (i.e., reducing the building's net load).

This chapter intends to provide an introduction to energy management techniques for HVAC systems. After introducing the key concepts related to HVACs, a thermal model suitable for simple modeling applications are presented to couple the thermal transient behavior of a building volume to the energy cost and HVAC operational management that can be deployed with a Building Energy Management System (BEMS). For this chapter, the building volume is considered based on the simplest possible scenario that consists of a building room to enable the key aspects of the model to be presented. More sophisticated thermal models could be considered for the building and coupled with the HVAC energy management strategies introduced for BEMSs. Nevertheless, the implementations of such refined models are considered outside the scope of this book. Based on the thermal model, simple energy management strategies are presented to manage the operation of HVACs while preserving the indoor thermal comfort of the building occupant. These include strategies for pre-cooling and pre-heating the room to reduce the energy cost and a strategy for controlling the HVAC's operation to increase the utilization level of the building's locally produced renewable energy. Application examples are provided throughout the chapters to highlight the particularities of the algorithms and to present some qualitative results.

## 10.2 Introduction to HVAC systems

An HVAC is a system installed inside a building to ensure good quality, proper airflow, and main comfort for the building occupant. The main function of an HVAC system is to circulate air indoors and exhaust it out of the building to maintain the air quality and indoor temperature in a satisfactory condition. The main components of an HVAC system include a filter, compressor, coils, condenser, blower, exhaust outlets, ducts, and other electrical components.

Different types of HVACs are available. For example, residential applications could consist of a ductless split type or a window type. Commercial HVAC systems can include a single split system or packaged terminal air conditioner (PTAC).

From an energy management perspective, HVACs can be grouped within thermostatically controlled loads (TCLs). TCLs refer to energy loads whose operation is driven by the thermal energy they produce and by the temperature changes they induce in the working environment [3–5]. Other TCLs include heaters, refrigerators, and water heaters. The operations of TCLs are usually driven by the need to establish particular environmental conditions to support operational requirements of equipment and building services. For HVACs, this usually means maintaining specific indoor temperature levels to satisfy the thermal comfort requirements of building occupants. This chapter will focus on the energy management of HVACs, while the energy management techniques for other kinds of TCLs can be found in dedicated references [6–9].

## 10.3 Thermal model for the modeling of indoor conditions

### 10.3.1 Overview

The thermal transient response of a building enables to description of the temperature variations that take place over time while accounting for specific boundary conditions, for example, temperature values, depicting expected indoor and outdoor conditions. The model representation of a building can varies in the level of sophistication introduced depending on the modeling requirements and objectives. In the literature, several advanced models have been developed over the last decades that can capture the thermal transient response of buildings, and whose results have been validated over numerous experimental measurements collected under different climatic conditions.

The model presented in this chapter assumes that the building is formed by a room only so that the complexity of the modeling can be kept to a minimum. This follows the approach used also in other chapters in which the model description and application examples are limited in size and complexity to enable the variables to be easily tabulated and plotted while focusing on the underlying optimization features of the modeling. The thermal model introduced in this chapter is presented in [10] and the temperature variation in a building, here taken without any loss of generality to represent a building room, is governed by the following expression:

$$\frac{du^{in}}{dt} = \frac{Q^{gain}}{E^{air}} \tag{10.1}$$

where $u^{in}$ represents the temperature in the interior environment of the room (in °C or K). $\frac{du^{in}}{dt}$ represents the temperature variation rate in the room; $Q^{gain}$ represents the rate (or density) of heat the room gains from other sources (in Watt); and $E^{air}$ represents the amount of energy needed for heating air by one degree (in Wh) that can be estimated as:

$$E^{air} = C^{air} \times \rho^{air} \times V \tag{10.2}$$

where $C^{air}$ and $\rho^{air}$ represent the heat capacity of air (in Wh/kg°C or Wh/kg·K) and the air density (in kg/m$^3$), respectively; and $V$ is the room's volume (in m$^3$). Given a specific building room, the values of $C^{air}$, $\rho^{air}$, and $V$ can be considered as constant parameters, therefore producing a constant value for $E^{air}$. Based on this, the indoor temperature variation rate is determined by the heat gain of the room. Heat gain can be positive or negative: a positive value indicates heat is brought into the room while a negative value indicates heat loss, that is, heat is taken out of the room. Eq. (10.1) can describe both the increase and decrease of the indoor temperature.

In the thermal model presented in this chapter, the heat gain of the room ($Q^{gain}$) is considered to include the following items:

- *Conduction heat gain* ($Q^{con}$)—This consists of the rate of heat transferred through the opaque envelope of the room (typically walls) by thermal conduction;
- *Ventilation heat gain* ($Q^{ven}$)—This denotes the rate of heat being transferred into the room (heat gain) or out of the room (heat loss) using ventilation;
- *Internal heat gain* ($Q^{int}$)—Refers to the rate of heat generated from internal heat sources located in the room. For example, internal heat sources can include appliances and human bodies;
- *Solar radiation heat gain* ($Q^{solar}$)—This depicts the rate of heat being transferred due to solar radiation through transparent components of the room's envelope (e.g., windows); and
- *HVAC heating/cooling gain* ($Q^{hvac}$)—This refers to the rate of heating or cooling energy induced by the HVAC in the room.

All five heat gain items are measured in Watt. Among these, the values of $Q^{int}$ and $Q^{solar}$ are always positive. The values of $Q^{con}$, $Q^{ven}$, and $Q^{hvac}$ can be positive or negative. Eq. (10.1) can then be re-written as:

$$\frac{du^{in}}{dt} = \frac{Q^{gain}}{E^{air}} = \frac{Q^{con} + Q^{ven} + Q^{hvac} + Q^{int} + Q^{solar}}{C^{air} \times \rho^{air} \times V} \tag{10.3}$$

In the following, we present possible estimation methods for the above heat gain components together with some worked examples.

## 10.3.2 Conduction heat gain

Thermal conduction refers to the process in which heat diffuses through a material from a hotter location to a colder point. The conduction heat gain depicts the rate of heat that occurs through the opaque room envelope (as we are assuming the building to consist of one room only) and it depends on the temperatures present in the interior and exterior environments of the room as well as the thermal property of the materials forming the components of the room's envelope. This thermal response is well depicted by the U–value, also denoted as thermal transmittance, and it provides an estimate of the insulating ability of the room's envelope. The U–value is determined as the ratio between the heat transfer rate that occurs through an envelope component (over the envelope's area) and the temperature difference across the component.

By considering a room's envelope, we can calculate the conduction heat gain as follows:

$$Q^{con} = U \times A^{suf} \times \left( u^{ex} - u^{in} \right) \tag{10.4}$$

where $A^{suf}$ is the surface area of the room's envelope (in m²); $u^{ex}$ and $u^{in}$ are the temperatures in the exterior and interior environments of the room, respectively, (in °C or K). $U$ is the U-value of the room's envelope walls (in W/m² K or W/m² °C).

If $u^{ex} > u^{in}$, $Q^{con}$ is positive, heat is transferred from the exterior to the interior of the component, and the room gains heat. If $u^{ex} < u^{in}$, heat is moving in the opposite direction the room loses heat to the exterior environment by thermal conduction in the envelope walls.

### 10.3.3 Ventilation heat gain

Ventilation heat gain is the heat gain due to air flowing through the air leakages and openings of the room. The estimation of the ventilation heat gain is based on the Air Changes per Hour (ACH, also known as air change rate). ACH represents the number of times that the total air volume in a given room or space is completely removed and replaced with air supply and/or circulated air in an hour. For example, if the air in a room can be completely replaced three times in four hours, then $ACH = \frac{3}{4} = 0.75$.

Based on ACH, the room's ventilation heat gain can be estimated as:

$$Q^{ven} = C^{air} \times \rho^{air} \times V \times \left(u^{ex} - u^{in}\right) \times ACH \tag{10.5}$$

The value for $Q^{ven}$ is positive when $u^{ex} > u^{in}$ and this indicates that the heat moves with airflow from the exterior environment into the room. $Q^{ven}$ is negative when $u^{ex} < u^{in}$, and, in this case, the heat moves with the airflow out of the room.

### 10.3.4 Worked example on conduction and ventilation heat gains

Let us consider a single-room building with dimensions shown in Fig. 10.1. The U-value of the room's envelope is 0.83 W/m² °C. The indoor and outdoor temperatures are 14°C and 8°C, respectively. It takes six hours to completely replace the air in the building three times. The heat capacity and density of the air are 0.3 Wh/kg°C and 1.3 kg/m³, respectively.

Determine the conduction heat gain and ventilation heat gains of the single-room building.

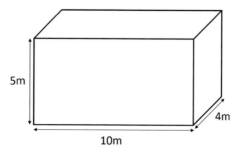

**Fig. 10.1** Dimensions of the assumed single-room building.

**Solution**

Based on the problem description, $ACH = \frac{3}{6} = 0.5$. The volume of the building is $V = 5 \times 10 \times 4 = 200 \text{m}^3$. The surface area of the building's envelope, which includes six walls, is:

$$A^{suf} = (10 \times 4 \times 2) + (10 \times 5 \times 2) + (4 \times 5 \times 2) = 80 + 100 + 40 = 220 \text{ m}^2$$

The conduction heat gain can then be obtained as:

$$Q^{con} = U \times A^{suf} \times \left(u^{ex} - u^{in}\right) = 0.83 \times 220 \times (14 - 8) = 1{,}095.6 \text{ W}$$

and the ventilation heat gain is:

$$Q^{ven} = C^{air} \times \rho^{air} \times V \times \left(u^{ex} - u^{in}\right) \times ACH = 0.3 \times 1.3 \times 200 \times (14 - 8) \times 0.5$$
$$= 234.0 \text{ W}$$

### 10.3.5 Internal heat gain

Heat sources could be present in a building, and these influence the indoor thermal conditions. For example, internal heat can be generated from human breath and appliances' operation. Internal heat gain of a room can be simply modeled by relating a coefficient with the room's floor area:

$$Q^{int} = \varepsilon \times A^{floor} \tag{10.6}$$

where $\varepsilon$ is the internal heat gain coefficient (in $\text{W/m}^2$); and $A^{floor}$ is the floor area of the room (in $\text{m}^2$).

### 10.3.6 Solar radiation heat gain

Solar radiation through fenestration components (e.g., windows and glass doors) in a building's envelope is an important factor influencing the indoor climate. The solar radiation heat gain through a fenestration component in a room's envelope is estimated as:

$$Q^{solar} = I \times A^{fen} \times SHGC \tag{10.7}$$

where $I$ is the solar radiation density ($\text{W/m}^2$); $A^{fen}$ is the area of the fenestration component (in $\text{m}^2$), and SHGC (in %) is the solar heat gain coefficient of the fenestration component and it is usually provided by the manufacturer. SHGC represents the fraction of solar radiation that is admitted through a fenestration component. A small value for SGHC denotes the fact that a small amount of solar radiation can be admitted through the fenestration component. In this chapter, we consider $I$ to be known, while the methods for estimating its value can be found in dedicated references [11, 12].

### 10.3.7 Worked example

Let us reconsider the single-room building in the worked example (Section 10.3.4) and we introduce a window in the front wall of the room. The dimensions are shown in Fig. 10.2. We assume the internal heat gain coefficient of the room to be $3.5\,\mathrm{W/m^2}$, the SHGC of the window to equal 30%, and the solar radiation density received by the window to be $420\,\mathrm{W/m^2}$.

Determine the internal heat gain and solar radiation heat gain of the room.

**Solution**

Based on the problem description, the floor area of the room is $A^{floor} = 10 \times 4 = 40\,\mathrm{m^2}$ and the area of the fenestration component (in this case, the window) is $A^{fen} = 2 \times 3 = 6\,\mathrm{m^2}$. Therefore the internal heat gain is determined as:

$$Q^{int} = \varepsilon \times A^{floor} = 3.5 \times 40 = 140.0\ \mathrm{W}$$

and the solar radiation heat gain is evaluated as:

$$Q^{solar} = I \times A^{fen} \times SHGC = 420 \times 6 \times 0.3 = 756.0\ \mathrm{W}$$

### 10.3.8 HVAC heating/cooling gain

The heating/cooling gain that occurs during the operations of an HVAC can be estimated as:

$$Q^{hvac} = m \times P^{op} \times COP \tag{10.8}$$

where $m$ indicates the operating mode of the HVAC: 1-heating, $-1$-cooling; $P^{op}$ is the HVAC's operating power (in W). When the HVAC operates in heating mode ($m=1$), $Q^{hvac}$ is positive as it depicts a heating gain. When the HVAC operates in cooling mode ($m=-1$), $Q^{hvac}$ is negative to reflect the cooling gain.

COP in Eq. (10.8) is defined as the coefficient of performance of the HVAC and it describes how effective an HVAC is at transferring heat vs. the amount of electrical power it consumes. For air-sourced HVACs, the COP is related to the difference

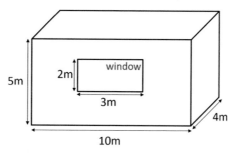

**Fig. 10.2** Dimensions of the assumed single-room building with a window.

between the temperatures in the exterior and outdoor environments of the room and can be estimated through the following fitting expression:

$$COP = -\theta \times |u^{ex} - u^{in}| + \delta \tag{10.9}$$

where $\theta$ and $\delta$ are fitting parameters that depict the HVAC's characteristics.

## 10.3.9 Worked example on HVAC heating/cooling gain

Let us consider a room conditioned by an HVAC that operates in a heating mode. The operating power of the HVAC is 800 W. The temperatures in the interior and exterior environments of the room are 14°C and 8°C, respectively. The fitting parameters of the HVAC's COP are $\theta = 0.04$ and $\delta = 3.91$.

Determine the heating gain of the room from the HVAC.

**Solution**
The HVAC's COP is calculated as:

$$COP = -\theta \times |u^{ex} - u^{in}| + \delta = -0.04 \times |14 - 8| + 3.91 = 3.67$$

and the HVAC heating gain is obtained from:

$$Q^{hvac} = m \times P^{hvac} \times COP \times s = 1 \times 800 \times 3.67 \times 1 = 2,936.0 \text{ W}$$

## 10.3.10 Summary of the thermal model

Table 10.1 Summarizes the notations previously introduced for different aspects of the presented thermal model.

The governing equation of the model can be obtained by substituting Eqs. (10.4)–(10.9) into Eq. (10.3) as follows [10]:

$$\frac{du^{in}}{dt} = \frac{Q^{con} + Q^{ven} + Q^{hvac} + Q^{int} + Q^{solar}}{C^{air} \times \rho^{air} \times V}$$

$$= \frac{U \times A^{suf} \times (u^{ex} - u^{in}) + C^{air} \times \rho^{air} \times V \times (u^{ex} - u^{in}) \times ACH + \varepsilon \times A^{floor}}{C^{air} \times \rho^{air} \times V}$$

$$+ \frac{I \times A^{fen} \times SHGC + m \times P^{hvac} \times (-\theta \times |u^{ex} - u^{in}| + \delta) \times s}{C^{air} \times \rho^{air} \times V}$$

$$\tag{10.10}$$

Model (10.10) can be solved numerically by introducing a discretization to account for the thermal variation that occurs in the model domain. By introducing a time discretization based on $\Delta t$ time intervals (in hours) and by assuming all parameters to remain constant during $\Delta t$, the numerical solution of the thermal model can be written as:

$$\frac{\Delta u^{in}}{\Delta t} = \frac{Q^{con} + Q^{ven} + Q^{hvac} + Q^{int} + Q^{solar}}{C^{air} \times \rho^{air} \times V} \tag{10.11}$$

**Table 10.1** Thermal model variables.

| Variable name | Value |
|---|---|
| $Q^{gain}$ | Total heat gain of the room (in W) |
| $E^{air}$ | The energy needed for heating air by one degree (in Wh) |
| $Q^{con}$ | Conduction heat gain (in W) |
| $Q^{ven}$ | Ventilation heat gain (in W) |
| $Q^{int}$ | Internal heat gain (in W) |
| $Q^{solar}$ | Solar radiation heat gain (in W) |
| $Q^{hvac}$ | HVAC heating/cooling gain (in W) |
| $C^{air}$ | Air heat capacity of air (in Wh/kg°C or Wh/kg K) |
| $\rho^{air}$ | Air density (in kg/m$^3$) |
| $V$ | Volume of the room (in m$^3$) |
| $U$ | U-value of the room's envelope (in W/m$^2$ K or W/m$^2$ °C) |
| $A^{suf}$ | The surface area of the room's envelope (in m$^2$) |
| $u^{ex}$ | Temperature in the exterior environment of the room (in °C or K) |
| $u^{in}$ | Temperature in the indoor environment of the room (in °C or K) |
| $ACH$ | Air exchanges per hour |
| $\varepsilon$ | Internal heat gain coefficient (in W/m$^2$) |
| $A^{floor}$ | Floor area of the room (in m$^2$) |
| $I$ | Solar radiation density (W/m$^2$) |
| $A^{fen}$ | Area of the fenestration component in the room's envelope (in m$^2$) |
| $SHGC$ | Solar heat gain coefficient of the fenestration component (%) |
| $m$ | The operation mode of the HVAC (1-heating, 0-cooling) |
| $P^{hvac}$ | Operating power of the HVAC (W) |
| $COP$ | Coefficient of performance of the HVAC |
| $\theta, \delta$ | Fitting parameters of COP |

that can be re-arranged as follows to determine the variation of the indoor temperature over $\Delta t$ (denoted as $\Delta u^{in}$):

$$\Delta u^{in} = \left( \frac{Q^{con} + Q^{ven} + Q^{hvac} + Q^{int} + Q^{solar}}{C^{air} \times \rho^{air} \times V} \right) \times \Delta t$$

$$= \left[ \frac{U \times A^{suf} \times (u^{ex} - u^{in}) + C^{air} \times \rho^{air} \times V \times (u^{ex} - u^{in}) \times ACH + \varepsilon \times A^{floor}}{C^{air} \times \rho^{air} \times V} + I \times A^{fen} \times SHGC + m \times P^{hvac} \times (-\theta \times |u^{ex} - u^{in}| + \delta) \times s \right] \times \Delta t$$

(10.12)

In the following worked example, we illustrate the ease of use of the proposed numerical solution of the thermal model.

## 10.3.11 Worked example on thermal model

Let us reconsider the single-room building with the dimensions shown in Fig. 10.2. We now assume that the room is conditioned by means of an HVAC operating in a heating mode. Values of the other parameters related to the thermal model are given in Table 10.2.

Determine the temperature in the interior environment ($u^{in}$) after operating the HVAC for 15 min considering the following time discretization $\Delta t$:

**(1)** Case 1: $\Delta t = 15$ min,
**(2)** Case 2: $\Delta t = 5$ min

**Solution**

From the building's dimensions shown in Fig. 10.2, it can be calculated that $A^{floor} = 40\text{m}^2$, $A^{fen} = 6\text{m}^2$, $V = 200\text{m}^3$, and $A^{suf} = 220$ m$^2$

**(1)** For Case 1, $\Delta t = 0.25$ (hours). The variation of the room's temperature is determined by a one-step calculation. By substituting all parameters into Eq. (10.12), the variation of the room's temperature can be calculated to be $\Delta u = 0.82°$C. Therefore after 15 min, the room's temperature is $u^{in} + \Delta u = 14 + 0.82 = 14.82°$C.

**(2)** For Case 2, $\Delta t = \frac{1}{12}$ (hours). The variation of the room's temperature is determined by 3-step calculations:

   **(i)** Step 1: Substitute all parameters into Eq. (10.12) to calculate the room's temperature variation over the first 5 min (i.e., the first $\Delta t$). Denote the temperature variation as $\Delta u_1 = 0.27°$C. The room's temperature after the first 5 min (denoted as $u_1^{in}$) is $u_1^{in} = u^{in} + \Delta u_1 = 14 + 0.27 = 14.27°$C.

   **(ii)** Step 2: Input $u_1^{in}$ as $u^{in}$ into Eq. (10.12) to calculate the temperature variation over the second 5 min as $\Delta t_2 = 0.12°$C. The room's temperature after the first 5 min is $u_2^{in} = u_1^{in} + \Delta u_2 = 14.27 + 0.12 = 14.39°$C

**Table 10.2** Settings of thermal model parameters used in the application example.

| Name | Value |
|---|---|
| $C^{air}$ | 0.3 Wh/kg°C |
| $\rho^{air}$ | 1.3 kg/m$^3$ |
| $U$ | 0.83 W/m$^2$ °C |
| $u^{ex}$ | 8°C |
| $u^{in}$ | 14°C |
| $ACH$ | 0.5 |
| $\varepsilon$ | 3.5 W/m$^2$ |
| $I$ | 120 W/m$^2$ |
| $SHGC$ | 30% |
| $P^{hvac}$ | 800 W |
| $\theta$ | 0.4 |
| $\delta$ | 3.91 |

**(iii)** Step 2: Input $u_2^{in}$ as $u^{in}$ into Eq. (10.12) to calculate the temperature variation over the third 5 min as $\Delta u_3 = 0.05°C$. The room's temperature after the third 5 min is $u_3^{in} = u_2^{in} + \Delta u_3 = 14.39 + 0.05 = 14.44°C$. After 15 min, the room's temperature becomes 14.44°C.

Comparing the two cases, it can be seen that by setting the discretization interval $\Delta t$ to different values, different results can be obtained for estimating the room's temperature after the same period. Smaller discretization can lead to more refined results at the expense of more computations to be carried out. It is common to run simulations for decreasing values of $\Delta t$ until the differences in results, obtained with subsequent discretization are acceptable.

## 10.4 Model representation for the operation of an HVAC
### 10.4.1 Model describing the operations of an HVAC

We denote as $u^{set}$ the set temperature of the HVAC (in °C) that is specified by the building occupant. This temperature represents the value that maximizes the occupant's subjective thermal comfort in the room. The HVAC's operation task is to keep the room's temperature within the comfort temperature band depicted by $[u^{set} - \Delta u, u^{set} + \Delta u]$, where $\Delta u$ (in °C) is the temperature width of the comfort band. For example, if $u^{set} = 20°C$ and $\Delta u = 0.5°C$, then the comfort temperature band of the room is $20 \pm 0.5 = [19.5, 20.5]°C$.

The HVAC system is considered to have two states: an operation state and a standby state. We use $s$ to denote an HVAC's state: $s = 0$ and 1 represent the HVAC is in a standby and operating state, respectively. The power consumption of an HVAC under the two states can be expressed as:

$$P^{hvac} = \begin{cases} P^{op}, & \text{if } s = 1 \\ P^{std}, & \text{if } s = 0 \end{cases} \tag{10.13}$$

where $P^{std}$ represents the power consumption of the HVAC in standby mode (in W). The value of $P^{std}$ is small (usually several watts), and in energy management strategy designs it can be considered to be negligible and taken as zero, that is, $P^{std} = 0$. When the HVAC is in the operation state, it generates heating/cooling gain for the room based on the model depicted in Eq. (10.8).

When the temperature of the exterior environment is higher than the room's temperature the HVAC can operate in a cooling mode. When the room's temperature is lower than the lower boundary of the comfort temperature band, the HVAC is turned to its standby state, and the room's temperature is expected to rise. When the room's temperature is higher than the upper boundary of the comfort temperature band, the HVAC is turned to its operational state, and the room's temperature is expected to decrease. Such an HVAC operation logic can be expressed as:

$$\text{Cooling mode: } s = \begin{cases} 1, & \text{if } u^{in} > u^{set} + \Delta u \\ 0, & \text{if } u^{in} < u^{set} - \Delta u \end{cases} \qquad (10.14)$$

When the temperature in the exterior environment is lower than the room's temperature, the HVAC operates in its heating mode. When the room's temperature is higher than the upper boundary of the comfort temperature band, the HVAC is turned to its standby state, and the room's temperature is expected to decrease. When the room's temperature is lower than the lower boundary of the comfort temperature band, the HVAC is turned to its operation state, and the room's temperature is expected to rise. This can be expressed as:

$$\text{Heating mode: } s = \begin{cases} 0, & \text{if } u^{in} \geq u^{set} + \Delta u \\ 1, & \text{if } u^{in} \leq u^{set} - \Delta u \end{cases} \qquad (10.15)$$

## 10.4.2 Cycling operation model for an HVAC

Eqs. (10.14) and (10.15) represent a cycling operation strategy for an HVAC that implies that the HVAC keeps the operation state to warm up or cool down the room until the room's temperature reaches the upper boundary (for heating mode) or lower boundary (for cooling mode) of the comfortable temperature band. At this point, the HVAC switches to the standby state, and the room's temperature decreases or increases until it reaches the lower boundary (for heating mode) or upper boundary (for cooling mode) of the comfortable temperature band. The HVAC then switches to the operation mode and the above process is repeated. Algorithms 10.1 and 10.2 show the HVAC cycling operation logic for the heating mode and cooling mode, respectively.

We now go through the logic included in Algorithms 10.1 and 10.2. The algorithms accept the relevant input parameters (Lines 1–4) and set the HVAC's operation mode (Line 5). The HVAC's initial state is set to be the operation state (Line 6). Then, the algorithm determines the HVAC's state in each time slot sequentially (Lines 7–18). In each time slot, the current solar radiation density and temperature in the exterior environment are inputted into the algorithms (Line 8). The values of these two parameters could vary in different time slots, and the use of the subscript $t$ aims to distinguish among values obtained in different time intervals.

The algorithms then determine the state of the HVAC in the current time slot (i.e., determine the value of $s_t$, Lines 9–15). It checks the room's temperature at the beginning of the current time slot, which is equal to the room's temperature at the end of the previous time slot (i.e., $u_{t-1}^{in}$). If the room's temperature is out of the comfortable temperature range, the HVAC's state is set to be the operation mode or the standby mode accordingly (Lines 9–12); otherwise, the HVAC's state remains in the same mode as its state in the previous time slot (Lines 13 and 14). At the end of the current time slot, the

---

**ALGORITHM 10.1 Cycling operation of an HVAC in a heating mode.**

**Start**

1. Input parameters $C^{air}$, $\rho^{air}$, $U$, $ACH$, $\varepsilon$, $SHGC$, $P^{op}$, $P^{std}$, $\theta$, $\delta$, $A^{surf}$, $A^{floor}$, $A^{fen}$, and $V$;
2. Input the room's initial temperature $u_0^{in}$;
3. Input $u^{set}$ and $\Delta u$;
4. Input the duration of a HVAC control time slot $\Delta t$;
5. Set HVAC operation mode $m = 1$;
6. Initialise the HVAC's initial state as $s_0 = 1$;
7. Set the current time slot index as $t = 1$;
8. Measure the current solar radiation density $I_t$ and temperature in the exterior environment $u_t^{ex}$;
9. **If** $u_{t-1}^{in} > u^{set} + \Delta u$
10.    Set $s_t = 0$;
11. **Else If** $u_{t-1}^{in} < u^{set} - \Delta u$
12.    Set $s_t = 1$;
13. **Else**
14.    Set $s_t = s_{t-1}$;
15. **End If**
16. Use $I_t$, $u_t^{ex}$, $s_t$, $u_{t-1}^{in}$ and other parameters to update the room's temperature at the end of the current time slot (i.e., $u_t^{in}$) based on Eq. (10.12);
17. Proceed to the next time and set $t = t + 1$;
18. **Go to** Line 8.

**End**

---

room's temperature is updated based on the determined HVAC state (Line 16). Such cycling operation logic is applied to every time slot (Lines 17 and 18).

### 10.4.3 Application example on an HVAC cycling operation

The proposed application example aims to demonstrate the use of the HVAC's cycling operating model introduced in the previous section. We start by assuming that the HVAC operates in a heating mode (i.e., $m = 1$). The model parameters adopted in the calculations are specified in Table 10.3. The initial temperature in the room is set as $u_0^{in} = 17°C$ and the initial state of the HVAC is the operation state (i.e., $s_0 = 1$). The solar radiation density and the temperature in the exterior environment of the room are considered to be constant and equal to $40\,W/m^2$ and $7°C$, respectively (i.e., $I_t = 40$ and $u_t^{ex} = 7$).

Algorithm 10.1 is applied to the control of the HVAC that aims to keep the indoor temperature within the occupant-specified comfort range over 120 min, with the

---

**ALGORITHM 10.2 Cycling operation of an HVAC in a cooling mode.**

**Start**

1. Input parameters $C^{air}$, $\rho^{air}$, $U$, $ACH$, $\varepsilon$, $SHGC$, $P^{op}$, $P^{std}$, $\theta$, $\delta$, $A^{surf}$, $A^{floor}$, $A^{fen}$, and $V$;
2. Input the room's initial temperature $u_0^{in}$;
3. Input $u^{set}$ and $\Delta u$;
4. Input the duration of a HVAC control time slot $\Delta t$;
5. Set HVAC operation mode $m = -1$;
6. Initialise the HVAC's initial state as $s_0 = 1$;
7. Set the current time slot index as $t = 1$;
8. Measure the current solar radiation density $I_t$ and temperature in the exterior environment $u_t^{ex}$;
9. **If** $u_{t-1}^{in} < u^{set} - \Delta u$
10.    Set $s_t = 0$;
11. **Else If** $u_{t-1}^{in} > u^{set} + \Delta u$
12.    Set $s_t = 1$;
13. **Else**
14.    Set $s_t = s_{t-1}$;
15. **End If**
16. Use $I_t$, $u_t^{ex}$, $s_t$, $u_{t-1}^{in}$ and other parameters to update the room's temperature at the end of the current time slot (i.e., $u_t^{in}$) based on Eq. (10.12);
17. Proceed to the next time and set $t = t + 1$;
18. **Go to** Line 8.

**End**

---

duration of each HVAC control interval set as one minute (i.e., $\Delta t = 1/60$ h). Representative results are presented in Fig. 10.3 to highlight how the HVAC is periodically switched between operation and standby states to maintain the indoor temperature within the comfort band (i.e., 18–20°C). The initial room temperature and the initial HVAC state are also plotted in the figure at the time slot index of zero.

It can be seen from Fig. 10.3 that the room's temperature is maintained in the comfortable temperature band (18–20°C) most of the time, while during a few time instances, the room's temperature is temporarily slightly out of the band. This is a consequence of the adopted energy management logic (in Algorithms 10.1 and 10.2) because the BEMS checks the room's temperature and decides the HVAC's state at the beginning of each control time slot. If the room's temperature varies out of the comfortable temperature band during a time slot, it will be identified at the beginning of the subsequent time slot, and the HVAC's state will be adjusted accordingly to regulate the room's temperature back into the comfortable temperature band. When the HVAC is controlled more

**Table 10.3** Settings of thermal model parameters for the application example of the HVAC cycling operation.

| Variable name | Value |
|---|---|
| $C^{air}$ | $0.3\,\mathrm{Wh/kg\,^{\circ}C}$ |
| $\rho^{air}$ | $1.3\,\mathrm{kg/m^3}$ |
| $U$ | $0.4\,\mathrm{W/m^2\,^{\circ}C}$ |
| $ACH$ | $0.1$ |
| $\varepsilon$ | $3.5\,\mathrm{W/m^2}$ |
| $SHGC$ | $33\%$ |
| $P^{hvac}$ | $800\,\mathrm{W}$ |
| $\theta$ | $0.2$ |
| $\delta$ | $5.0$ |
| $V$ | $200\,\mathrm{m^3}$ |
| $A^{floor}$ | $40\,\mathrm{m^2}$ |
| $A^{surf}$ | $220\,\mathrm{m^2}$ |
| $A^{fen}$ | $6\,\mathrm{m^2}$ |
| $u^{set}$ | $19^{\circ}\mathrm{C}$ |
| $\Delta u$ | $1^{\circ}\mathrm{C}$ |

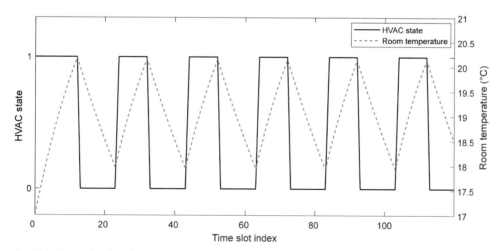

**Fig. 10.3** Example of cycling operation of an HVAC in a heating model.

frequently (i.e., the duration of an HVAC control time slot is shorter), the room's temperature regulation will be more responsive at the cost of more computationally intensive requirements. The setting up of the control frequency can be regarded as a trade-off between the indoor climate regulation responsiveness and the required computational cost.

Fig. 10.4 shows the variation of the COP of the HVAC operating in a heating mode (calculated with Eq. (10.9)), together with the difference of the temperatures in the interior and exterior environments of the room. It can be seen that when the temperature difference decreases, the COP increases, therefore indicating that the HVAC works more efficiently. When the temperature difference increases, the COP decreases because the HVAC's working performance decreases.

We now consider the case where the HVAC operates in a cooling mode. The parameter settings in Table 10.3 are used. The solar radiation density and the temperature in the exterior environment of the room are considered to be constant and equal to $40\,W/m^2$ and 32°C, respectively, (i.e., $I_t = 40$ and $u_t^{ex} = 32$). The initial temperature in the room is set to be $u_0^{in} = 17°C$ and the initial state of the HVAC is specified as the operation state (i.e., $s_0 = 1$).

Algorithm 10.2 is applied to the control of the HVAC that aims to keep the indoor temperature within the occupant-specified comfort range [18°C, 20°C]. Fig. 10.5 shows the operation result of the HVAC as well as the variation of the room's temperature. Similarly, to Fig. 10.3, the HVAC is periodically switched between the two states to regulate the room's temperature. Fig. 10.6 shows the variation of the COP of the HVAC operating in a cooling mode, together with the difference in the temperatures in the interior and exterior environments of the room.

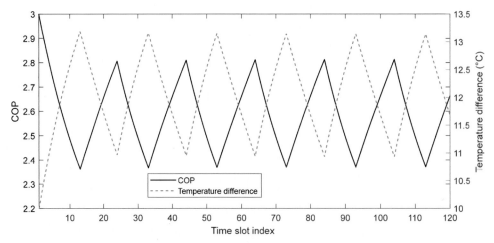

**Fig. 10.4** Variations of the HVAC's COP and the differences of the temperatures in the interior and exterior environments of the room when the HVAC works in a heating mode.

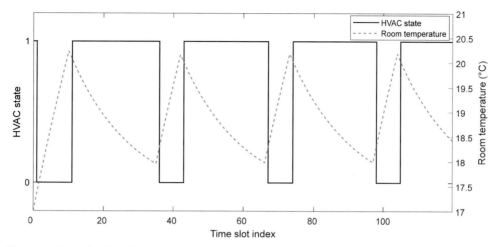

**Fig. 10.5** Example of cycling operations of an HVAC in a cooling model.

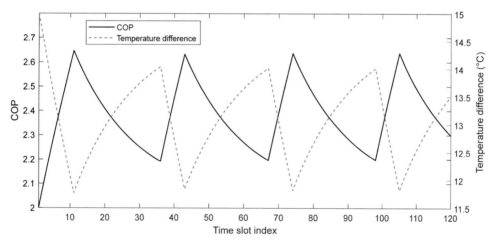

**Fig. 10.6** Variations of the HVAC's COP and the difference of the temperatures in the interior and exterior environments of the room when the HVAC works in a cooling mode.

## 10.5 Evaluation of the HVAC's energy consumption and energy cost

### 10.5.1 Calculation method

The HVAC's operation state series $s$ can be expressed as

$$s = [s_1, ..., s_T] \tag{10.16}$$

where $T$ is the number of time slots with the HVAC operation and $s_t \in \{0, 1\}$ $(t = 1:T)$ represents the state of the HVAC in time slot $t$. The total energy consumption of the HVAC over the $T$ time slots (denoted as $E^{hvac}$) is calculated as:

$$E^{hvac} = \sum_{t=1}^{T} P_t^{hvac} \Delta t \tag{10.17}$$

$$P_t^{hvac} = \begin{cases} P^{op}, & \text{if } s_t = 1 \\ P^{std}, & \text{if } s_t = 0 \end{cases} \tag{10.18}$$

where $P_t^{hvac}$ represents the power consumption of the HVAC in time slot $t$ (in kW) and $\Delta t$ represents the duration of a time slot (in hours). Given a time-varying electricity tariff $\rho$:

$$\rho = [\rho_1, ..., \rho_T] \tag{10.19}$$

where $\rho_t$ represents the electricity price in time slot $t$ (in \$/kWh, $t=1:T$), the energy cost of the HVAC over the $T$ time slots (denoted as $C^{hvac}$) can be determined as:

$$C^{hvac} = \sum_{t=1}^{T} P_t^{hvac} \Delta t \rho_t \tag{10.20}$$

## 10.5.2 Worked example

Consider an HVAC with an operating power of 900 W and a standby power of 10 W. The state trajectory of the HVAC over five-time slots is depicted by [1,0,0,1,1]. Each time slot lasts for ten minutes. Let us assume that the electricity price rates are equal to 0.50 \$/kWh for the first three time slots and \$0.26 \$/kWh for the last two time slots.

Determine the energy consumption and cost of the HVAC over the five-time slots.

### Solution

Since each time slot lasts for ten minutes, $\Delta t = \frac{1}{6}$ (hours).

The total energy consumption of the HVAC over the five time slots is:

$$E^{hvac} = \sum_{t=1}^{T} P_t^{hvac} \Delta t = (900 + 10 + 10 + 900 + 900) \times \frac{1}{6} = 453.3 \text{ Wh}$$

and the energy cost of the HVAC becomes:

$$C^{hvac} = \sum_{t=1}^{T} P_t^{hvac} \Delta t \rho_t$$

$$= \frac{0.9 \times 0.5}{6} + \frac{0.02 \times 0.5}{6} + \frac{0.02 \times 0.5}{6} + \frac{0.9 \times 0.26}{6} + \frac{0.9 \times 0.26}{6}$$

$$= 0.075 + 0.002 + 0.002 + 0.039 + 0.039 = 0.16\$$$

### 10.5.3  Introduction to an HVAC energy management

Based on the previous HVAC cycling operation model (Section 10.4), the energy management of HVACs usually aims to establish strategies that can "break" the cycling operation process. For example, a strategy could require switching the state of an HVAC that is cooling the room from the operation mode to the standby state even if the room's temperature does not reach the lower boundary of the comfort temperature band. Usually, the purpose of performing energy management for an HVAC is to reduce the energy consumption of the HVAC in specific periods to reduce the building's energy cost or to reduce the building's peak power consumption. When defining HVAC energy management strategies, the preservation of the building occupant's thermal comfort remains a fundamental requirement. In the rest of this chapter, we introduce some basic energy management strategies for HVACs.

## 10.6  Energy management of HVACs with varying set-temperatures

### 10.6.1  Overview

The energy consumption of an HVAC is affected when the set temperature changes. For an HVAC operating in a cooling mode, an increase in the set temperature $u^{set}$ reduces the operation time of the HVAC leads to a decrease in the HVAC's energy consumption, while its operation time increases when the set temperature is decreased (because the HVAC needs to spend more time in operation to achieve the lower temperature value) and it's the energy consumption increases. Similar trends can be observed when considering HVACs operating in a heating mode.

The approach of changing the set temperature to reduce the HVAC's power consumption requires a sacrifice in the occupant's thermal comfort. For example, on a hot summer day, by increasing the set temperature of the room from 21°C to 23°C the building occupant is expected to feel hotter. Therefore such an action relies on the willingness of the occupant to sacrifice thermal comfort in exchange for other benefits, for example, energy cost saving.

### 10.6.2  Energy management strategy

Let us consider a fixed value of the temperature width $\Delta u$ of the comfort band and two set-temperature values $u^{set,2}$ and $u^{set,1}$, where $u^{set,2} > u^{set,1}$. When $u^{set,1}$ is applied, the comfort temperature band is $[u^{set,1} - \Delta u, u^{set,1} + \Delta u]$, while when $u^{set,2}$ is applied, the comfort temperature band becomes $[u^{set,2} - \Delta u, u^{set,2} + \Delta u]$. Let us assume that the energy management period consists of $T$ time slots and two pre-specified time slot sets $\mathbf{\Phi}^1$ and $\mathbf{\Phi}^2$, which contain the time slots when $u^{set,1}$ and $u^{set,2}$ are applied, respectively. Based on the adopted definitions: $\mathbf{\Phi}^1 \cup \mathbf{\Phi}^2 = \{1, 2, \dots T\}$ and $\mathbf{\Phi}^1 \cap \mathbf{\Phi}^2 = \varnothing$. For example, when the

HVAC operates in a cooling mode, $\mathbf{\Phi}^2$ can contain the time slots that occur during the high electricity price periods. Since $u^{set,2} > u^{set,1}$, this approach allows the room's temperature to remain at a higher value during the time slots in $\mathbf{\Phi}^2$ and, therefore limits the energy cost associated with the HVAC's operation. Similarly, when the HVAC operates in a heating mode, $\mathbf{\Phi}^1$ can be defined to collect the time slots associated with the high electricity price periods to reduce the HVAC's operation time.

The cycling operation model of the HVAC (presented in Section 10.4) is applied in the following to implement an energy management strategy. For the time slots when $u^{set,1}$ is applied, the cycling operation is subjected to the temperature band of $[u^{set,1} - \Delta u, u^{set,1} + \Delta u]$; for the time slots when $u^{set,2}$ is applied, the cycling operation is subjected to the temperature band of $[u^{set,2} - \Delta u, u^{set,2} + \Delta u]$. Without any loss of generality, we consider the HVAC operates in a cooling mode and the energy management logic is expressed in Algorithm 10.3.

In Algorithm 10.3, the BEMS first accepts the input parameters and sets the room's initial operation condition (Lines 1–6). In each time slot, it measures the solar radiation density and the temperature in the exterior environment of the room (Line 8). It then checks if the set temperature $u^{set,1}$ or $u^{set,2}$ needs to be applied in the current time slot. When $u^{set,1}$ is applied, the BEMS determines the HVAC's state based on the room's temperature at the beginning of the time slot and on the comfortable temperature band $[u^{set,1} - \Delta u, u^{set,1} + \Delta u]$ (Lines 9–16); otherwise, the BEMS determines the HVAC's state based on the room's temperature at the beginning of the time slot and on the comfortable temperature band $[u^{set,2} - \Delta u, u^{set,2} + \Delta u]$ (Lines 17–25). At the end of the time slot, the room's temperature is updated (Line 26).

### 10.6.3 Application example

In this example, we consider a single-room building with thermal model parameters summarized in Table 10.4. We consider a 6-h energy management period, from 9 a.m. to 3 p.m. The control interval is set to be one minute (i.e., $\Delta t = \frac{1}{60}$ (hours)) and, therefore there are a total of 360 time slots. The solar radiation intensity, the temperature in the exterior environment of the room, and the energy price are assumed to be different in each hour as shown in Table 10.5.

$u^{set,1}$ and $u^{set,2}$ are set to be 19°C and 21°C, respectively, and $\Delta u$ is set to be 1°C. $u^{set,1}$ is applied during the low-energy price period that occurs between 9–11 a.m. and 2–3 p.m., and $u^{set,2}$ is applied over the high-energy price intervals between 11 a.m. and 2 p.m. The two sets can then be written as $\mathbf{\Phi}^1 = \{1, \ldots, 120, 301, \ldots, 360\}$ and $\mathbf{\Phi}^2 = \{121, \ldots, 300\}$.

Algorithm 10.3 is then applied with the above input parameters, and the HVAC's operation results over the 6-h energy management period are shown in Fig. 10.7. For comparative purposes, Fig. 10.8 shows the HVAC's cycling operation result with the

**ALGORITHM 10.3 Cycling operation of an HVAC in a cooling mode and with varying set temperatures.**

**Start**

1. Input parameters $C^{air}$, $\rho^{air}$, $U$, $ACH$, $\varepsilon$, $SHGC$, $P^{op}$, $P^{std}$, $\theta$, $\delta$, $A^{surf}$, $A^{floor}$, $A^{fen}$, and $V$;
2. Input $u^{set,1}$, $u^{set,2}$, $\Phi^1$, $\Phi^2$, and $\Delta u$;
3. Input the room's initial temperature $u_0^{in}$;
4. Input $T$ and $\Delta t$;
5. Set HVAC operation mode $m = -1$;
6. Initialise the HVAC's initial state as $s_0 = 1$;
7. Set the current time slot index as $t = 1$;
8. Input the current solar radiation density $I_t$ and temperature in the exterior environment $u_t^{ex}$;
9. **If** $t \in \Phi^1$
10.    **If** $u_{t-1}^{in} < u^{set,\,1} - \Delta u$
11.       Set $s_t = 0$;
12.    **Else If** $u_{t-1}^{in} > u^{set,\,1} + \Delta u$
13.       Set $s_t = 1$;
14.    **Else**
15.       Set $s_t = s_{t-1}$;
16.    **End If**
17. **Else**
18.    **If** $u_{t-1}^{in} < u^{set,\,2} - \Delta u$
19.       Set $s_t = 0$;
20.    **Else If** $u_{t-1}^{in} > u^{set,\,2} + \Delta u$
21.       Set $s_t = 1$;
22.    **Else**
23.       Set $s_t = s_{t-1}$;
24.    **End If**
25. **End If**
26. Use $I_t$, $u_t^{ex}$, $s_t$, $u_{t-1}^{in}$ and other parameters to update the room's temperature at the end of the current time slot (i.e., $u_t^{in}$) based on Eq. (10.12);
27. Proceed to the next time and set $t = t + 1$;
28. **If** $t > T$
29.    **Terminate**;
30. **Else**
31.    **Go to** Line 8.
32. **End If**

**End**

**Table 10.4** Settings of thermal model parameters used in the application example.

| Name | Value |
|---|---|
| $C^{air}$ | $0.3\,\mathrm{Wh/kg\,°C}$ |
| $\rho^{air}$ | $1.3\,\mathrm{kg/m^3}$ |
| $U$ | $0.4\,\mathrm{W/m^2\,°C}$ |
| $ACH$ | $0.05$ |
| $\varepsilon$ | $3.5\,\mathrm{W/m^2}$ |
| $SHGC$ | $33\%$ |
| $P^{hvac}$ | $800\,\mathrm{W}$ |
| $\theta$ | $0.2$ |
| $\delta$ | $4.0$ |
| $V$ | $240\,\mathrm{m^3}$ |
| $A^{floor}$ | $60\,\mathrm{m^2}$ |
| $A^{surf}$ | $300\,\mathrm{m^2}$ |
| $A^{fen}$ | $10\,\mathrm{m^2}$ |

**Table 10.5** Solar radiation intensity, temperature in the exterior environment, and energy price over the 6-h energy management period.

|  | 9–10 a.m. | 10–11 a.m. | 11 a.m.–12 p.m. | 12–1 p.m. | 1–2 p.m. | 2–3 p.m. |
|---|---|---|---|---|---|---|
| Solar radiation intensity $(\mathrm{W/m^2})$ | 40 | 48 | 60 | 70 | 64 | 52 |
| Temperature in the exterior environment $(°C)$ | 27 | 27.6 | 29 | 30.6 | 31.4 | 30.2 |
| Energy price $(\$/\mathrm{kWh})$ | 0.4 | 0.4 | 1.2 | 1.2 | 1.2 | 0.65 |

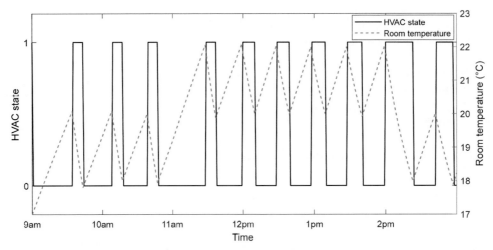

**Fig. 10.7** Variations of the HVAC's state and the room's temperature in a cooling mode with varying set temperatures.

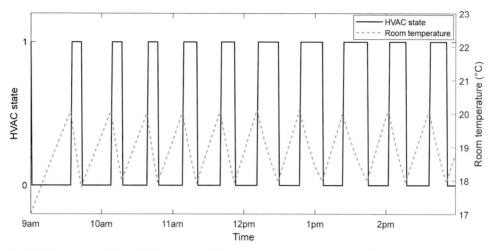

**Fig. 10.8** Variations of the HVAC's state and the room's temperature in a cooling mode with a fixed set temperature.

set temperature fixed at 19°C. It can be noted that, with the varying set-temperature energy management strategy, the room's temperature in the period 11 a.m.–2 p.m. is higher than during the other time intervals. As a result of this, the HVAC's operation time and energy consumption are reduced between 11 a.m. and 2 p.m., and this leads to a reduced energy cost in the high-energy price hours, as shown in Table 10.6. Such an energy cost reduction is achieved at the expense of indoor thermal comfort, that is, the increase of the room's temperature.

## 10.7 Pre-heating/cooling-based HVAC energy management

### 10.7.1 Overview

Pre-heating or pre-cooling is a commonly adopted energy management strategy performed with HVACs. For example, the HVAC can operate continuously over multiple time slots before a specific period to pre-heat or pre-cool a room. The pre-heating/cooling strategy is usually adopted for the following two scenarios:

– *Pre-heating/cooling for occupancy comfort enhancement*—The HVAC is operated to preheat or pre-cool a room before an occupancy period so that the room's temperature can fall into the comfortable temperature range at the beginning of the occupancy period when the building occupant enters the room;

– *Pre-heating/cooling for energy cost saving*—The HVAC is operated to pre-heat or pre-cool the room before a high electricity price period. This approach can reduce the HVAC's operation in the high electricity price period and lead to energy cost reductions.

**Table 10.6** Energy consumption and cost of the HVAC in the two cases.

| | | 9–10 a.m. | 10–11 a.m. | 11 a.m.–12 p.m. | 12–1 p.m. | 1–2 p.m. | 2–3 p.m. | Total |
|---|---|---|---|---|---|---|---|---|
| Operation time (min) | Case of varying set-temperature | 9 | 18 | 10 | 24 | 23 | 38 | 122 |
| | Case of fixed set-temperature | 9 | 18 | 24 | 31 | 27 | 30 | 139 |
| Energy consumption (Wh) | Case of varying set-temperature | 120 | 240 | 133.3 | 320 | 306.7 | 506.7 | 1626.7 |
| | Case of fixed set-temperature | 120 | 240 | 320 | 413.3 | 360 | 400 | 1853.3 |
| Energy cost ($) | Case of varying set-temperature | 0.05 | 0.10 | 0.16 | 0.38 | 0.37 | 0.33 | $1.39 |
| | Case of fixed set-temperature | 0.05 | 0.10 | 0.38 | 0.50 | 0.43 | 0.26 | $1.72 |

## 10.7.2 Energy management strategy

Let us consider two time points $t^1$ and $t^2$. Before $t^1$, the HVAC is turned off. In the period of $[t^1, t^2]$, the HVAC operates to pre-heat/cool the room. After $t^2$, the HVAC works in a cycling operation mode presented in Section 10.4 to regulate the room's temperature within the comfortable temperature band. This is illustrated in Fig. 10.9. Algorithm 10.4 shows the pre-heating energy management procedure, and the pre-cooling energy management procedure can be determined following a similar strategy.

In Algorithm 10.4, the BEMS starts by setting up the relevant parameters and measures the room's temperature in a one-time slot before the pre-heating period (Lines 1–6). During the pre-heating period $[t^1, t^2]$, the BEMS sets the HVAC's state to be in an operation state to continuously heat the room (Lines 9 and 10). After the time point $t^2$, the BEMS sets the HVAC's state based on the pre-specified comfortable temperature range and the state of the HVAC in the previous time slot (Lines 11–18). The room's temperature is updated at the end of each time slot based on the adopted thermal model (Line 20).

## 10.7.3 Application example

We designed a simple application example to demonstrate how the pre-heating and pre-cooling energy management strategies can enhance occupancy comfort and save energy bills for the building occupant. We consider a four hours, and $\Delta t$ is set to be $\frac{1}{60}$ hours (i.e., one minute). Therefore there are a total of 240 time slots. The solar radiation intensity and the temperature in the exterior environment of the room are considered to be constant and equal to $10\,\text{W/m}^2$ and $4°\text{C}$, respectively, (i.e., $I_t = 40$ and $u_t^{ex} = 32$). The room's temperature at the beginning of the 60 time slots is considered to be $8°\text{C}$, that is, $t_0^{in} = 8$. Other parameter settings are shown in Table 10.7.

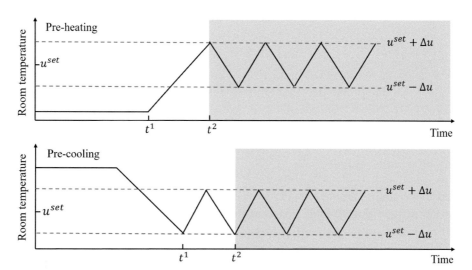

**Fig. 10.9** Illustration of pre-heating and pre-cooling energy management for HVACs.

---

**ALGORITHM 10.4 Pre-heating energy management of an HVAC.**

**Start**

1. Input parameters $C^{air}$, $\rho^{air}$, $U$, $ACH$, $\varepsilon$, $SHGC$, $P^{on}$, $P^{std}$, $\theta$, $\delta$, $A^{surf}$, $A^{floor}$, $A^{fen}$, and $V$;
2. Input $u^{set}$ and $\Delta u$;
3. Input $t^1$, $t^2$ and $\Delta t$;
4. Set HVAC operation mode $m=1$;
5. Set the HVAC's state before the pre-heating period as $s_{t^1-1}=0$;
6. Input the room's temperature before the pre-heating period as $u^{in}_{t^1-1}$;
7. Set the current time slot index as $t=t^1$;
8. Input the current solar radiation density $I_t$ and temperature in the exterior environment $u^{ex}_t$;
9. **If** $t^1 \leq t \leq t^2$
10.    Set the HVAC's state $s_t=1$;
11. **Else If** $t > t^2$
12.    **If** $u^{in}_{t-1} < u^{set} - \Delta u$
13.       Set $s_t=1$;
14.    **Else If** $u^{in}_{t-1} > u^{set} + \Delta u$
15.       Set $s_t=1$;
16.    **Else**
17.       Set $s_t=s_{t-1}$;
18.    **End If**
19. **End If**
20. Use $I_t$, $u^{ex}_t$, $s_t$, $u^{in}_{t-1}$ and other parameters to update the room's temperature at the end of the current time slot (i.e. $u^{in}_t$) based on Eq. (10.12);
21. Proceed to the next time and set $t=t+1$;
22. **Go to** Line 8.
23. **End If**

**End**

---

The HVAC is considered to operate in a heating mode. $t^1$ and $t^2$ are set to be 30 and 60, respectively, which represent the time slot indices. The value of $t^2$ indicates the start of the occupancy period. Between time slots 1 and 30, the HVAC is off and it then starts to operate to pre-heat the room in time slot 30 for 30 time slots (i.e., 30 min). After time slot 60, the HVAC turns into the cycling operation mode.

Figs. 10.10 and 10.11 show the room's temperature variation with and without the pre-heating strategy, respectively. For the case without pre-heating, the HVAC does not operate until time slot 20 and, after that, the HVAC performs the cycling operation. From the plotted results considering the pre-heating, the room's temperature is regulated within the comfortable temperature band [19°C, 21°C] at the start of the occupancy

**Table 10.7** Thermal model parameters used in the application example.

| Variable name | Value |
|---|---|
| $C^{air}$ | $0.3\,\text{Wh/kg}\,^{\circ}\text{C}$ |
| $\rho^{air}$ | $1.3\,\text{kg/m}^3$ |
| $U$ | $0.16\,\text{W/m}^2\,^{\circ}\text{C}$ |
| $ACH$ | $0.1$ |
| $\varepsilon$ | $2.0\,\text{W/m}^2$ |
| $SHGC$ | $33\%$ |
| $P^{hvac}$ | $1600\,\text{W}$ |
| $\theta$ | $0.2$ |
| $\delta$ | $4.0$ |
| $V$ | $240\,\text{m}^3$ |
| $A^{floor}$ | $60\,\text{m}^2$ |
| $A^{surf}$ | $300\,\text{m}^2$ |
| $A^{fen}$ | $20\,\text{m}^2$ |
| $u^{set}$ | $20^{\circ}\text{C}$ |
| $\Delta u$ | $1^{\circ}\text{C}$ |

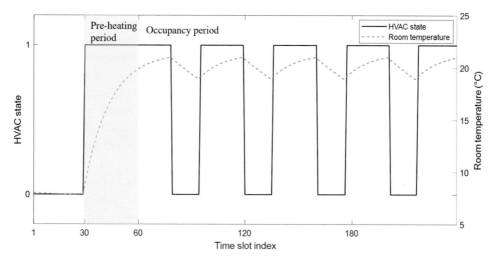

**Fig. 10.10** Variation of the HVAC's state and the room's temperature with a pre-heating strategy.

period. In contrast, when there is no pre-heating strategy applied, the HVAC starts to heat the room at the start of the occupancy period. At that time, the room's temperature is 8.6°C, and the HVAC takes 24 min to regulate the room's temperature into the comfortable temperature band. The pre-heating strategy can improve the comfort of the building occupant.

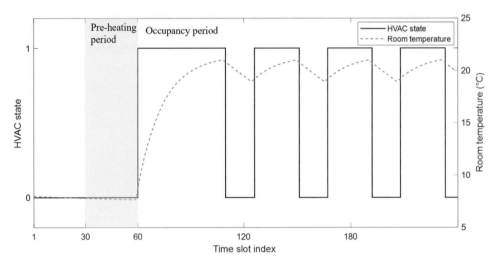

**Fig. 10.11** Variation of the HVAC's state and the room's temperature without a pre-heating strategy.

## 10.8 HVAC energy management for renewable energy accommodation

### 10.8.1 Overview

The operation of an HVAC can be managed to accommodate renewable energy sources locally deployed in a building. In such an application scenario, since renewable power output is intermittent and fluctuates, the HVAC's state needs to be controlled to operate more when there is renewable power available and to reduce its operation when only limited renewable power is available. This can increase the amount of the HVAC's energy consumption that is served by the renewable energy source and reduce the energy consumption that needs to be imported from the grid. This approach can improve the utilization of the building's on-site renewable energy sources. As a result, the local energy supply ability of the building can be enhanced, and the energy cost and net energy of the building can be reduced. In this process, the thermal comfort of the building occupant is a model requirement, and the HVAC's operation needs to regulate the room's temperature to comply with the pre-specified comfortable temperature band.

### 10.8.2 Energy management strategy

Let us consider a building integrated with a renewable energy source. Consider an energy management period consisting of $T$ time slots and the duration of each time slot to be $\Delta t$ (in hours). Let us depict the power output of the renewable energy source over the $T$ time slots to be $\boldsymbol{P}^{res} = [P_1^{res}, P_2^{res}, \ldots, P_T^{res}]$, where $P_t^{res}$ ($t=1:T$) represents the renewable power output in time slot $t$ (in W).

We now consider a simple energy management strategy to ensure that the HVAC's operation better adapts to the intermittent renewable power output: we set up two limit values for the renewable power output, denoted as $P^{res,low}$ and $P^{res,up}$ ($P^{res,low} < P^{res,up}$). When the renewable power output is less than $P^{res,low}$, the renewable power is insufficient, and the BEMS tries to set the HVAC's state to be in a standby mode to save energy. When the renewable power output is larger than $P^{res,up}$, the renewable power is sufficient and the BEMS tries to set the HVAC's state to be in operation mode. In other situations, the HVAC follows the cycling operation logic presented in Section 10.4.2. In this manner, the BEMS can increase the utilization level of renewable energy by providing energy to the HVAC operation. This energy management logic is shown in Algorithm 10.5 for the case when the HVAC operates in a cooling mode. The HVAC energy management procedure for a heating mode can be obtained similarly.

In Algorithm 10.5, the BEMS begins by specifying the relevant parameters and initializes the room's condition (Lines 1–7). In each time slot, the BEMS measures the solar radiation density and the temperature in the exterior environment. It also determines the value of the renewable power output at that time slot, and this can be done, for example, based on the renewable energy models presented in Chapter 2 (Lines 9 and 10). The BEMS then checks if the room's temperature is within the comfortable temperature band and it sets the HVAC's state to be on standby or operation state as required (Lines 11–14).

If the room's temperature is within the comfortable temperature band, the BEMS checks if the renewable power output in the current time slot is too low (i.e., $P_t^{res} < P^{res,low}$). If yes, then it sets the HVAC's state to be in the standby state (Lines 16 and 17). If the renewable power output level is high (i.e., $P_t^{res} > P^{res,up}$), the BEMS sets the HVAC's state to be in operation mode (Lines 18 and 19). If the renewable power output level is moderate (i.e., $P^{res,low} \le P_t^{res} \le P^{res,up}$), the BEMS requires the HVAC's state to remain unchanged and equal to its state in the previous time slot (Lines 20 and 21). At the end of the current time slot, the room's temperature is updated and the above energy management logic is applied to the subsequent time slot until the end of the energy management period is reached (Lines 24–30).

### 10.8.3 Application example

We use an application example to demonstrate the effectiveness of the energy management strategy in Algorithm 10.5. We consider a single-room building with thermal parameters summarized in Table 10.8. Based on the input data, the comfortable temperature band of the room is [18.5°C, 21.5°C]. The solar radiation density and the temperature in the exterior environment of the room are set to be constant and equal to $20\,W/m^2$ and 30°C, respectively. A 10-h energy management period is considered, and each time slot covers one minute (i.e., $\Delta t = \frac{1}{60}$) and the total number of time slots is $10 \times 60 = 600$. The building is assumed to be equipped with a wind turbine. The parameters of the wind turbine are given in Table 10.9. Fig. 10.12 shows the 10-h wind speed at the building's site as well as the wind power output that is calculated from the

## ALGORITHM 10.5 HVAC energy management procedures for renewable energy (cooling mode).

**Start**
1. Input parameters $C^{air}$, $\rho^{air}$, $U$, $ACH$, $\varepsilon$, $SHGC$, $P^{op}$, $P^{std}$, $\theta$, $\delta$, $A^{surf}$, $A^{floor}$, $A^{fen}$, and $V$;
2. Input $P^{res,low}$ and $P^{res,up}$;
3. Input the room's initial temperature $u_0^{in}$;
4. Input $u^{set}$ and $\Delta u$;
5. Input $T$ and $\Delta t$;
6. Set HVAC operation mode $m=-1$;
7. Initialise the HVAC's initial state as $s_0=1$;
8. Set the current time slot index as $t=1$;
9. Input the current solar radiation density $I_t$ and temperature in the exterior environment $u_t^{ex}$;
10. Determine the value of $P_t^{res}$;
11. **If** $u_{t-1}^{in}<u^{set}-\Delta u$
12.    Set $s_t=0$;
13. **Else If** $u_{t-1}^{in}>u^{set}+\Delta u$
14.    Set $s_t=1$;
15. **Else**
16.    **If** $P_t^{res}<P^{res,low}$
17.       Set $s_t=0$;
18.    **Else If** $P_t^{res}>P^{res,up}$
19.       Set $s_t=1$;
20.    **Else**
21.       Set $s_t=s_{t-1}$;
22.    **End If**
23. **End If**
24. Use $I_t$, $u_t^{ex}$, $s_t$, $u_{t-1}^{in}$ and other parameters to update the room's temperature at the end of the current time slot (i.e., $u_t^{in}$) based on Eq. (10.12);
25. Proceed to the next time slot and set $t=t+1$;
26. **If** $t>T$
27.    **Terminate**;
28. **Else**
29.    **Go to** Line 9.
30. **End If**
**End**

**Table 10.8** Thermal model parameters for the application example of the HVAC energy management strategy for renewable energy.

| Variable name | Value |
|---|---|
| $C^{air}$ | $0.3\,\mathrm{Wh/kg\,°C}$ |
| $\rho^{air}$ | $1.3\,\mathrm{kg/m^3}$ |
| $U$ | $0.2\,\mathrm{W/m^2\,°C}$ |
| $ACH$ | $0.1$ |
| $\varepsilon$ | $2.5\,\mathrm{W/m^2}$ |
| $SHGC$ | $20\%$ |
| $P^{hvac}$ | $1500\,\mathrm{W}$ |
| $\theta$ | $0.2$ |
| $\delta$ | $4.0$ |
| $V$ | $320\,\mathrm{m^3}$ |
| $A^{floor}$ | $80\,\mathrm{m^2}$ |
| $A^{surf}$ | $304\,\mathrm{m^2}$ |
| $A^{fen}$ | $20\,\mathrm{m^2}$ |
| $u^{set}$ | $20\,\mathrm{°C}$ |
| $\Delta u$ | $1.5\,\mathrm{°C}$ |

**Table 10.9** Wind turbine model used in the application example of the HVAC energy management for renewable energy.

| Blade radius | Cut-in wind speed | Rated wind speed | Cut-out wind speed |
|---|---|---|---|
| 3 m | 3 m/s | 8 m/s | 15 m/s |

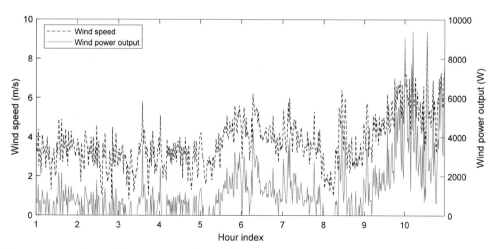

**Fig. 10.12** Wind speed and calculated wind power used in the application of the HVAC energy management for renewable energy.

wind speed data following the wind power output model presented in Chapter 2. $P^{res,low}$ and $P^{res,up}$ are set to be 100 W and 1000 W, respectively.

Figs. 10.13 and 10.14 show the variation of the HVAC's state and the room's temperature with and without the energy management strategy of Algorithm 10.5, respectively. For the latter case, the HVAC is considered to perform the cycling operation following Algorithm 10.2. The wind power output profile is also plotted in the figures.

It can be seen that, with the energy management strategy, the cycling operation of the HVAC is frequently "interrupted" by the BEMS, to better utilize the wind power. This increases the HVAC operation when there is sufficient wind power (e.g., between the 5th and 8th hours and between the 9th and 10th hours) and it reduces its operation when the wind power is limited (e.g., between the 1st and 5th hours). As a result, with the energy management strategy, 4323.3 Wh out of 5625.0 Wh energy is served by the wind turbine over 10 h, and 1301.7 Wh energy needs to be served by the grid. Without the implementation of the energy management strategy, 2746.8 out of 5100.0 Wh energy is served by the wind turbine and 2353.2 Wh energy needs to be imported from the grid. The result indicates that, in this application example and with the use of the energy management strategy, although the total energy consumption of the HVAC increases (because the HVAC is controlled to operate over a longer period to exploit the wind power), the amount of energy that the building needs to take from the grid to meet the HVAC's energy demand is significantly reduced when compared to the case that does not adopt an energy management strategy.

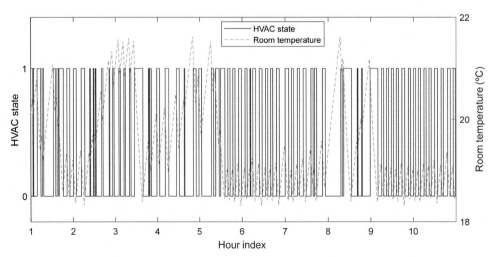

**Fig. 10.13** Variation of the HVAC's state and the room's temperature with the energy management strategy.

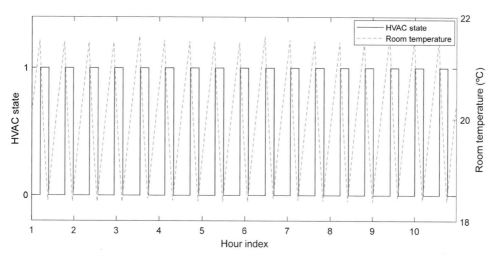

**Fig. 10.14** Variation of the HVAC's state and the room's temperature without the energy management strategy (i.e., with the cycling operation mode).

## References

[1] HVAC Factsheet—Energy Breakdown, Department of the Environment and Energy, Australian Government, Sep. 2013. [Online], Available: https://www.energy.gov.au/sites/default/files/hvac-factsheet-energy-breakdown.pdf. (Accessed 11 July 2023).

[2] HVAC [Online]. 2020. Available: https://www.energy.gov.au/business/equipment-and-technology-guides/hvac (Accessed 13 June 2023).

[3] N. Lu, S. Katipamula, Control strategies of thermostatically controlled appliances in a competitive electricity market, in: Proceedings of IEEE Power Engineering Society General Meeting, 2005, San Fancisco, CA, USA, June 2005.

[4] M. Vanouni, N. Lu, Improving the centralized control of thermostatically controlled appliances by obtaining the right information, IEEE Trans. Smart Grid 6 (2) (2015) 946–948.

[5] M. Tostado-Veliz, M. Bayat, A.A. Ghadimi, F. Jurado, Home energy management in off-grid dwellings: exploiting flexibility of thermostatically controlled appliances, J. Clean. Prod. 310 (2021).

[6] Y. Ding, Y. Song, H. Hui, C. Shao, Integration of Air Conditioning and Heating into Modern Power Systems: Enabling Demand Response and Energy Efficiency, Springer International Publishing, New York, USA, 2019.

[7] H. Wang, K. Meng, F. Luo, Z.Y. Dong, Demand response through smart home energy management using thermal inertia, in: Proceedings of 2013 Australia Universities Power Engineering Conferences, Hobart, Australia, Sep. 2013.

[8] G. Escriva-Escriva, I. Segura-Heras, M. Alcazar-Ortega, Application of an energy management and control system to assess the potential of different control strategies in HVAC systems, Energ. Buildings 42 (11) (2010) 2258–2267.

[9] H. Chaouch, C. Ceken, S. Ari, Energy management of HVAC systems in smart buildings by using fuzzy logic and M2M communication, J. Building Eng. 44 (2021) 102606.

[10] D. Ryder-Cook. (2009). Thermal Modelling of Buildings [Online]. Available: https://inference.org.uk/is/papers/DanThermalModellingBuildings.pdf (Accessed 11 July 2023).

[11] G.M. Masters, Renewable and Efficient Electric Power Systems, John Wiley & Sons, Hoboken, New Jersey, USA, 2013.

[12] Grid-Connected PV Systems: Design and Installation, eighth ed., Global Sustainable Energy Solutions Pty Ltd., 2022.

# CHAPTER 11

# Energy sharing among buildings

## Contents

## 11.1 Introduction

Historically, buildings have been operating as pure energy consumers with the energy being delivered through dedicated supply systems (e.g., gas networks, electrical power grids, and heat pipes) by energy providers (e.g., power companies and natural gas companies). With the deployment of distributed renewable energy sources, such as rooftop/envelope solar panels and wind turbines, buildings are transforming from being pure energy consumers to *prosumers* that are capable of behaving as producers and consumers at the same time. In this manner, buildings can satisfy their energy demand by a combination of energy supply received from the external macro energy grid system as well as generated on-site.

The surplus energy generated by the building energy resources, for example, in the case of electricity, is reinjected into the external power grid at zero or specified feed-in subsidy, i.e., the user receives payment through feeding electricity back into the external grid. The amount of the payment depends on the feed-in tariff rate. For example, the benchmark range of the feed-in tariff rate in New South Wales (Australia) during the financial year 2019/2020 has ranged between 8.5 and 10.4 Australian cents per kilowatt

*Building Energy Management Systems and Techniques*
https://doi.org/10.1016/B978-0-323-96107-3.00010-2

hour [1]. In some regions, there are no feed-in tariff schemes, meaning that the surplus power generated by the distributed renewable energy sources is fed back to the grid for free.

Besides feeding the surplus energy into the grid, the widespread deployment of distributed renewable energy sources in buildings provides the foundation for possible energy-sharing strategies among buildings. For example, let us consider the scenario in which we can sell the surplus energy generated by our rooftop Photovoltaic (PV) solar panels to our neighbors when they need energy. Alternatively, when we need more energy to cover our consumption, we could buy electricity from our neighbors at a price lower than the price at which we would buy the energy from our electricity retailer. Such an energy-sharing scenario between buildings can lead to multiple benefits: it can create an opportunity for the occupants and/or the building managers to access more affordable energy with reduced energy bills; it can increase the value of clean energy assets installed in buildings and foster local economics of building communities; and it can reduce buildings' dependencies on the external power grid. In this chapter, we focus on the following approaches that can facilitate energy sharing among buildings: (i) community-level energy management systems that aggregately manage the energy production and consumption of a group of buildings subjected to certain community-scale objectives that cannot be achieved by individual buildings; and (ii) local energy trading mechanisms that enable buildings to autonomously trade energy with each other.

In the first part of the chapter, we present an internal energy-sharing pricing scheme that can be deployed in communities formed by multiple buildings. Key aspects defining the trading process are described with a model representation that can be used to perform optimizations and numerical simulations. Particular attention is devoted to the role of buildings as prosumers in which they can act as consumers and producers of energy depending on the time period and their energy conditions. In the second part of the chapter, peer-to-peer energy trading strategies among buildings are introduced. In the final part of the chapter, the use of blockchain technology in the implementation of peer-to-peer energy trading is considered. Selected case studies in which this technology has been introduced in pilot projects are briefly presented. Application examples are included in different parts of the chapter to highlight the use of the proposed formulations and to gain insight into the possible benefits of the trading strategies.

## 11.2 An internal energy-sharing pricing scheme for buildings

### 11.2.1 Introduction

In this section, we consider a community that consists of multiple buildings, where each building is equipped with rooftop PV solar panels. Let us assume that we can set up an internal energy-sharing mechanism between these buildings. Without any loss of generality, we consider the buildings to be autonomous, i.e., they are managed by different building managers/owners. We also assume that the energy sharing among the buildings

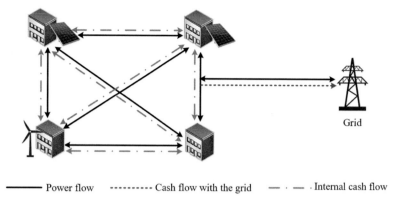

——— Power flow   ---------- Cash flow with the grid   — · — Internal cash flow

**Fig. 11.1** Schematic of a building community energy management scenario.

is not free and that it is controlled by financial drivers. To do so, we need to define a mechanism for the setting up of the energy rates. For such a sharing framework, an energy management system needs to be established to interact with all buildings as well as the grid, as shown in Fig. 11.1. In this scenario, we consider the following three types of electricity prices:

- *Electricity retail price from the grid* is denoted as $\rho^{grid}$ (in \$/kWh) and consists of the price with which buildings purchase the energy from the external grid;
- *Feed-in tariff* is referred to as $\rho^{fit}$ (in \$/kWh) and represents the price paid by the grid to the buildings for the amount of renewable power that is reversely fed into the grid; and
- *Internal energy-sharing price*, denoted as $\rho^{es}$ (in \$/kWh), is the price with which the buildings share among themselves the renewable energy generated on their premises.

From an economic viewpoint, it is expected that the following condition applies: $\rho^{fit}<\rho^{es}<\rho^{grid}$, i.e., the internal energy-sharing price $\rho^{es}$ is expected to be smaller than the price of purchasing energy from the grid but larger than the feed-in subsidy. In this way, the energy rates can encourage buildings to share energy among themselves. In each time slot, the renewable energy generated by the buildings is equally shared by all the buildings at the rate $\rho^{es}$, and only the surplus electricity that cannot be consumed by the whole community is purchased from the grid at the rate $\rho^{fit}$. The relevant notation introduced for the internal energy-sharing pricing scheme is defined in Table 11.1.

## 11.2.2 Formulation of internal energy-sharing pricing
### 11.2.2.1 Overview
In this section, we consider a community formed by $N$ buildings. The net-power load of the $n$th building during the $t$th time slot is calculated as:

$$L_{n,t}^{net} = L_{n,t} - P_{n,t}^{res},$$ (11.1)

**Table 11.1** Definition of variables relevant to the internal energy-sharing pricing scheme.

| Variable | Definition |
| --- | --- |
| $N$ | Number of buildings |
| $\Delta t$ | Duration of one time slot (in hours) |
| $L_{n,t}$ | Forecast power demand of the $n$th building during the $t$th time slot (in kW) |
| $P_{n,t}^{res}$ | Forecast power output from the $n$th building's renewable energy source during the $t$th time slot (in kW) |
| $L_{n,t}^{net}$ | Net-load of the $n$th building during the $t$th time slot (in kW) |
| $\widetilde{L}_{t}^{net}$ | Net-load of the community during the $t$th time slot (in kW) |
| $\rho^{es}$ | Internal energy-sharing price in the community (in \$/kWh) |
| $\rho^{fit}$ | Feed-in tariff (in \$/kWh) |
| $\rho^{grid}$ | Price of energy purchased from the grid (in \$/kWh) |

where the notation is defined in Table 11.1. A positive value of $L_{n,t}^{net}$ indicates that the building needs energy from external sources, and a negative value means that the building has surplus energy to share with other buildings or to feed back into the external grid. Based on the adopted notation, the net-power demand of the whole community during the $t$th time slot can be expressed as:

$$\widetilde{L}_{t}^{net} = \sum_{n=1}^{N} L_{n,t}^{net} = \sum_{i=1}^{N} L_{n,t} - \sum_{i=1}^{N} P_{n,t}^{res} \tag{11.2}$$

During the $t$th time slot, the buildings that need energy, i.e., the buildings with $L_{n,t}^{net} > 0$, are expected to purchase energy from other buildings that possess surplus renewable energy at the internal energy-sharing price. In these transactions, the buildings with surplus energy, i.e., the buildings with $L_{n,t}^{net} < 0$, are expected to receive money from the buildings that need energy.

If the renewable power generation of the whole building community cannot cover the community's energy demand, i.e., $\widetilde{L}_{t}^{net} > 0$, then the buildings that need external energy need to purchase energy from the grid at the electricity retail rate. On the other hand, if the whole community has surplus energy after serving the community's load, i.e., $\widetilde{L}_{t}^{net} < 0$, the surplus energy can be fed back into the grid at the feed-in tariff.

It is considered that, at the $t$th time slot, some of the buildings in the community experience a deficit of energy while the other buildings generate renewable energy. We can group the buildings into two sets $\varnothing_{t}^{1}$ and $\varnothing_{t}^{2}$ that are defined as follows:

$$L_{b,t}^{net} > 0, \quad \forall b \in \varnothing_{t}^{1} \tag{11.3}$$

$$L_{b,t}^{net} \leq 0, \quad \forall b \in \varnothing_{t}^{2} \tag{11.4}$$

and the two sets need to satisfy the condition that $|\varnothing_{t}^{1} \cup \varnothing_{t}^{1}| = N$, where the $|\bullet|$ operation counts the number of elements in a set.

The total deficit energy in the community that needs to be served by the surplus energy or by the grid ($L_t^{def}$) can be determined based on:

$$L_t^{def} = \sum_{b \in \varnothing_t^1} L_{b,t}^{net} \tag{11.5}$$

The total surplus energy generated by the community ($P_t^{sup}$) at the $t$th time slot can be calculated as:

$$P_t^{sup} = \sum_{b \in \varnothing_t^2} \left(-1 \times L_{b,t}^{net}\right), \tag{11.6}$$

where the multiplier $-1$ is introduced for $P_t^{sup}$ to be positive.

### 11.2.2.2 Energy cost of buildings with deficit energy

The energy cost ($C_{b,t}$) of each building that experiences deficit energy during the $t$th time slot (i.e., $\forall b \in \varnothing_t^1$) can be evaluated as:

$$C_{b,t} = \begin{cases} L_{b,t}^{net} \times \rho^{es} \times \Delta t & \text{if } \widetilde{L}_t^{net} \leq 0 \\ C_{b,t}^{es} + C_{b,t}^{grid} & \text{if } \widetilde{L}_t^{net} > 0 \end{cases}, \quad \forall b \in \varnothing_t^1 \tag{11.7}$$

In this scenario, if the whole community's renewable power can cover the entire community's load ($\widetilde{L}_t^{net} \leq 0$), then the deficit energy of the building grouped in $\varnothing_t^1$ can be covered by the surplus energy in the community. Each building $b$ in the set $\varnothing_t^1$ pays for its needed energy according to the internal energy-sharing price. If the whole community's renewable power is not sufficient to cover the community's load ($\widetilde{L}_t^{net} > 0$), the buildings that have a deficit of energy need to access the surplus energy generated by other buildings and may also need to purchase energy from the grid. The building $b$'s energy cost comprises two items: payment to other building/s having surplus energy ($C_{b,t}^{es}$) and payment to the grid ($C_{b,t}^{grid}$).

To determine the payment taking place from building $b$ to the other building/s with surplus energy at the $t$th time slot (i.e., $C_{b,t}^{es}$), we first calculate the share of the building from the total surplus renewable power of the whole community at the $t$th time slot (depicted as $LS_{b,t}^1$):

$$LS_{b,t}^1 = P_t^{sup} \times \frac{L_{b,t}^{net}}{L_t^{def}}, \quad \forall b \in \varnothing_t^1 \tag{11.8}$$

which implies that, at each time slot, all the surplus renewable energy generated from the community is shared by the buildings in the community that need energy in proportion to their power demands. The value of $C_{b,t}^{es}$ then be determined as:

$$C_{b,t}^{es} = min\left(LS_{b,t}^1, L_{b,t}^{net}\right) \times \rho^{es} \times \Delta t, \quad \forall b \in \varnothing_t^1 \tag{11.9}$$

where the function $min(\cdot)$ returns the minimum one among the inputted numbers. Based on which the amount of renewable power that building $b$ can share with the community is larger than its needed energy (i.e., $LS_{b,t}^1 > L_{b,t}^{net}$). In this case, the building only pays for its energy demand at the internal energy-sharing price. Alternatively (i.e., when $LS_{b,t}^1 < L_{b,t}^{net}$), the building pays for all the shared renewable energy at the internal energy-sharing price, and the remaining deficit energy is charged by the grid:

$$C_{b,t}^{grid} = max\left(L_{b,t}^{net} - LS_{b,t}^1, 0\right) \times \rho^{grid} \times \Delta t, \quad \forall b \in \varnothing_t^1 \tag{11.10}$$

where the function max($\cdot$) returns the maximum one among the inputted numbers.

### 11.2.2.3 Revenue obtained by the buildings with surplus energy

During the $t$th time slot, buildings with surplus energy generate revenues through sharing energy with the buildings that need energy and by feeding energy back to the external grid. The revenue for each building with surplus energy $R_{b,t}$ can be estimated as:

$$R_{b,t} = \begin{cases} -1 \times L_{b,t}^{net} \times \rho^{es} \times \Delta t & \text{if } \widetilde{L}_t^{net} \geq 0 \\ R_{b,t}^{es} + R_{b,t}^{grid} & \text{if } \widetilde{L}_t^{net} < 0 \end{cases}, \quad \forall b \in \varnothing_t^2 \tag{11.11}$$

that implies that if the community's total renewable power is less than its demand ($\widetilde{L}_t^{net} \geq 0$), the renewable power generated by the buildings is used to serve the community's load. In this situation, each building that has surplus solar power receives payments based on the internal energy-sharing price. If the community's total renewable power can fully cover the community's load (i.e., $\widetilde{L}_t^{net} < 0$), then, for each building that has surplus solar power, the revenue consists of two items: the payment from other buildings that need energy ($R_{b,t}^{es}$) and the revenue from the grid ($R_{b,t}^{grid}$).

When the renewable energy generated from the whole community can fully cover the energy demand of all buildings in the community (i.e., $\widetilde{L}_t^{net} < 0$), the buildings that generate surplus renewable energy can sell a part of their surplus energy to the buildings that need energy and may also consider feeding a part of their surplus energy back to the grid that is paid by the feed-in tariff rate. To determine the payments that a building $b \in \varnothing_t^2$ receives from the other building/s that need energy at the $t$th time slot (i.e., $R_{b,t}^{es}$), we first calculate the building's share of serving the community's deficit energy (depicted as $LS_{b,t}^2$):

$$LS_{b,t}^2 = L_t^{def} \times \frac{L_{b,t}^{net}}{P_t^{sup}}, \quad \forall b \in \varnothing_t^2 \tag{11.12}$$

which implies that, at each time slot, all deficit energy in the community is mutually served by the buildings in the community that generate surplus energy in proportion to the buildings' surplus energy amounts. Based on $LS_{b,t}^1$, the value of $R_{b,t}^{es}$ can be determined as:

$$R_{b,t}^{es} = min\left(LS_{b,t}^2, \left|L_{b,t}^{net}\right|\right) \times \rho^{es} \times \Delta t, \quad \forall b \in \varnothing_t^2 \tag{11.13}$$

If the served deficit energy that building $b$ can share from the community is larger than the surplus energy that it generates (i.e., $LS_{b,t}^2 > |L_{b,t}^{net}|$), then the building only receives the revenue from serving other buildings' deficit energy at the internal energy-sharing price. The remaining surplus energy of building $b$ is then fed back to the grid and is paid at the feed-in tariff:

$$R_{b,t}^{grid} = max \left( \left| \tilde{L}_t^{net} \right| - LS_{b,t}^2, 0 \right) \times \rho^{tif} \times \Delta t, \forall b \in \emptyset_t^2 \qquad (11.14)$$

## 11.2.3  Application example

In this section, we consider an application example to highlight how buildings can benefit from implementing an internal energy-sharing pricing scheme. In this example, we consider a community consisting of five buildings, which are indexed as B1, …,B5, respectively. It is assumed that wind power sources are installed on two buildings (i.e., B1 and B2) and that the other three buildings (i.e., B3, B4, and B5) are equipped with PV solar panels. The forecasts of the 24-h wind/solar power profiles and power consumption profiles of the five buildings are shown in Fig. 11.2. It is assumed that the values of $\rho^{es}$, $\rho^{fit}$, and $\rho^{grid}$ are 30 cents/kWh, 8 cents/kWh, and 68 cents/kWh, respectively. While we consider a fixed price for purchasing energy from the grid in this case, time-varying rates could also be considered and applied.

By applying the proposed energy-sharing pricing scheme, the energy cost and revenue items of the five buildings are summarized in Table 11.2. For comparative purposes, the expected buildings' energy cost and revenue items without the implementation of the energy-sharing pricing scheme are depicted in Table 11.3. By applying the internal energy-sharing pricing scheme, the local energy economics of the community are significantly enhanced. Buildings B2 and B4 implement nearly zero-cost operations over the period. For the other three buildings, their operation costs are largely reduced when compared to the case without the internal energy-sharing pricing schemes. For example, building B3, without the implementation of the internal energy-sharing scheme, purchases its energy from the grid at a cost of $23.83. With the implementation of the internal energy-sharing scheme, building B3 can purchase 28.7 kWh from other buildings, therefore experiencing a $11.46 energy cost saving.

## 11.3  Peer-to-peer energy trading among buildings

### 11.3.1  Overview

It is possible to encourage buildings to share energy with each other by autonomously trading energy with each other in a Peer-to-Peer (P2P) approach [2–4]. The P2P energy trading paradigm emerged because of the increasing prevalence of distributed renewable energy sources because buildings are now able to trade with each other the energy generated from their on-site renewable energy sources (usually solar panels and wind turbines). Such a trading arrangement is not feasible in a traditional energy system where energy consumers simply take energy from the grid and pay for the energy at certain energy retail tariffs.

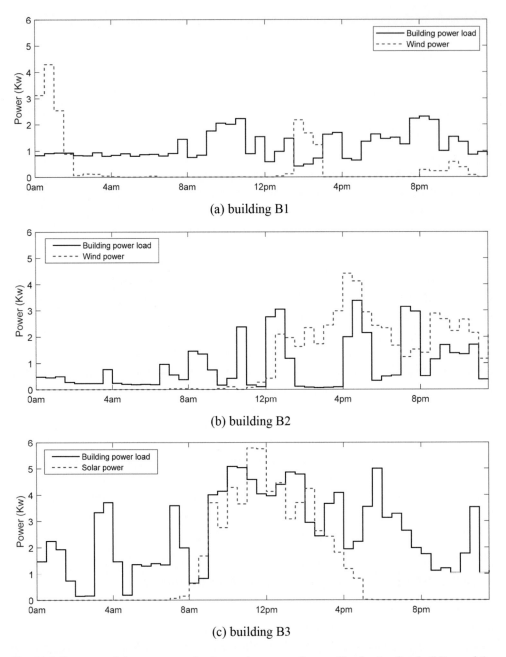

(a) building B1

(b) building B2

(c) building B3

**Fig. 11.2** Forecasts of the power production and consumption profiles for the five buildings of the applicant example.

*(Continued)*

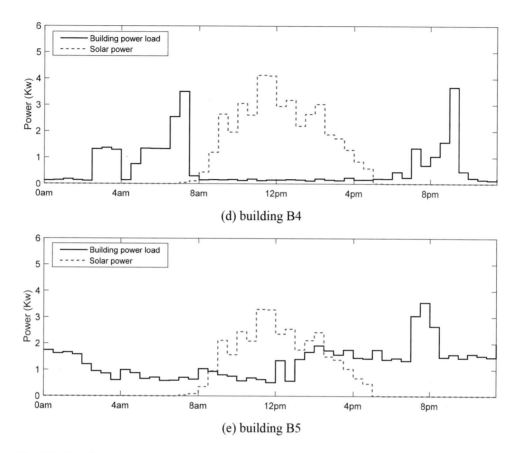

(d) building B4

(e) building B5

**Fig. 11.2, Cont'd**

**Table 11.2** Energy cost and revenue for the five buildings with the implementation of the energy-sharing pricing scheme.

|  | Payment to grid | Payment to other buildings | Revenue from grid | Revenue from other buildings | Net total value (revenue or cost) |
|---|---|---|---|---|---|
| B1 | $2.94 | $6.00 | $0.01 | $1.56 | $7.37 |
| B2 | $1.28 | $2.30 | $0.19 | $3.70 | −$0.32 |
| B3 | $4.49 | $8.63 | $0.02 | $0.73 | $12.37 |
| B4 | $0.63 | $3.69 | $0.50 | $3.80 | $0.02 |
| B5 | $2.78 | $5.64 | $0.20 | $1.53 | $6.70 |

**Table 11.3** Energy cost and revenue for the five buildings without the implementation of the energy-sharing pricing scheme.

| | Payment to grid | Payment to other buildings | Revenue from grid | Revenue from other buildings | Net total value (revenue or cost) |
|---|---|---|---|---|---|
| B1 | $16.54 | N/A | $0.43 | N/A[a] | $16.11 |
| B2 | $6.49 | N/A | $1.18 | N/A | $5.31 |
| B3 | $24.05 | N/A | $0.22 | N/A | $23.83 |
| B4 | $8.99 | N/A | $1.51 | N/A | $7.47 |
| B5 | $15.57 | N/A | $0.60 | N/A | $14.97 |

[a]N/A = "not applicable."

Obviously, the internal energy-sharing pricing mechanism presented in Section 11.2 can be also considered as an energy trading scheme for buildings. However, the internal energy-sharing pricing scheme is based on an energy trading price that is preagreed by all the involved buildings. The P2P energy trading paradigm enables buildings to negotiate and trade energy with each other as independent and autonomous agents. With the increasing popularization of dispersed renewable energy sources, P2P energy trading has been gaining popularity in recent years.

While P2P energy trading is a new energy trading paradigm with energy demand side entities (e.g., buildings, electric vehicles, and dispersed small renewable energy sources) as the participants, energy trading has been practiced in power and energy systems for decades to facilitate energy retailers and energy distribution companies to buy electricity or gas from power or gas generation companies at the wholesale price. They then sell the energy to their signed end customers at a retail price that is higher than the wholesale price. Energy trading between energy retailers and distribution companies and energy generation companies is enabled by *power markets*. The basic principle of power markets can also be used to support P2P energy trading.

### 11.3.2 Introduction to power markets

A power market [5] is a marketplace where participants trade electricity. In power markets, electricity is regarded as a tradable commodity. Power markets are practiced in many countries and are usually designed for grid-level energy trading activities. The participants are large-capacity energy entities, such as power generation companies, power distribution companies, and electricity retailers. The basic structure of the power market is depicted in Fig. 11.3. A power market is usually regulated by a market operator. For example, the Australian National Electricity Market is managed by the Australian Energy Market Operator (AEMO) [6].

There are different structures and operation rules in different power markets. Detailed information on energy markets can be found in dedicated references, e.g., [5,7,8]. We now briefly introduce a bilateral auction energy trading mechanism that is commonly

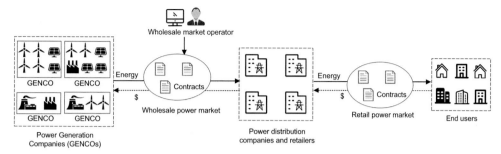

**Fig. 11.3** Basic structure of a power market.

seen in modern power markets. In such a mechanism, the market operator interacts with the electricity sellers and buyers to perform a market clearing process that relies on the four steps described in the following:

**Step 1:** The market operator collects electricity selling bids from electricity sellers (i.e., energy generation companies) and electricity purchase offers from electricity buyers (i.e., power distribution companies and retailers). Each electricity selling bid ($B$) includes a two-tuple: the seller's proposed electricity selling price $\rho^{sell}$ (in \$/kWh) and the amount of energy to be sold $E^{sell}$ (in kWh):

$$B = \left[\rho^{sell}, E^{sell}\right] \tag{11.15}$$

Similarly, each electricity purchase offer consists of a two-tuple: the electricity buyer's proposed purchase price $\rho^{buy}$ (in \$/kWh) and the amount of energy to be purchased $E^{buy}$ (kWh):

$$O = \left[\rho^{buy}, E^{buy}\right] \tag{11.16}$$

**Step 2:** The market operator stacks the received bids and offers. The electricity sellers' bids are sorted with the least cost order, i.e., ascending order of $\rho^{sell}$. The electricity buyers' bids are sorted in descending order of $\rho^{buy}$. With this approach, two bidding curves, i.e., an electricity selling curve and an electricity purchase curve, are generated as illustrated in Fig. 11.4.

**Step 3:** The market operator determines the market clearing price. It finds the point where the energy supply–demand is balanced, i.e., the sum of the energy amount in all the sellers' bids with the selling price lower than the point is equal to the energy amount in all the buyers' bids with the purchase price higher than the point. The price at that point is the energy market clearing price. In the example shown in Fig. 11.4, the market clearing price is 40 cents/kWh.

**Step 4:** The last step consists of the transaction settlement. All the electricity sellers' bids with the proposed selling price lower than the market clearing price are successfully allowed to sell electricity. They are settled with the market clearing price

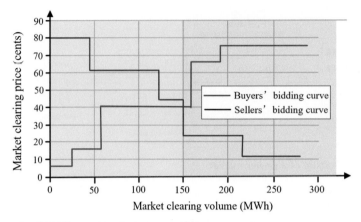

**Fig. 11.4** Example of bidding curves in power markets.

(i.e., the green part in Fig. 11.4). Similarly, all the electricity buyers' bids with the proposed purchase price higher than the market clearing price are allowed to trade energy and are settled with the market clearing price.

### 11.3.3 Local power markets in power distribution networks

The grid-level power market sets up capacity barriers. Only energy entities that have adequate energy capacity are eligible to participate in the market. For buildings and other small-capacity entities in the power distribution system, it is not possible to participate in the grid-level power market and local power markets can be established to facilitate energy trading among small-capacity energy entities in distribution networks.

The participants of local power markets include small-capacity energy producers, energy consumers, and energy prosumers as illustrated in Fig. 11.5. The engagement of prosumers leads to an important feature that distinguishes local power markets from grid-level power markets: in local power markets, a participant can have dual roles. In some time slots, it can act as an energy seller to sell energy to other buyers, while during other time slots, it can act as an energy buyer to buy energy from other sellers. Further details on the design of local power markets in distribution networks can be found in dedicated publications, e.g., [9–12].

### 11.3.4 Application example

We now consider a simple application example to illustrate how buildings can trade electricity with each other through a local power market with the bilateral auction mechanism introduced in Section 11.3.3. Let us assume that eight buildings are participating in the microenergy market. In each time slot, each building predicts its on-site electricity production from renewable energy sources and its electricity demand. The buildings then

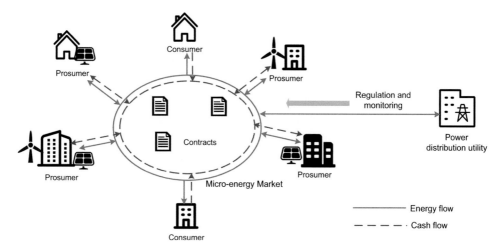

**Fig. 11.5** Schematic of a local energy market for buildings.

submit their energy-selling bids and energy purchase offers to the local power market. The market operator stacks the bids and offers and determines the market clearing price and traded energy amount in each time slot.

As an example, Table 11.4 shows the bids and offers submitted by eight buildings in two consequent time slots. The duration of each time slot can vary, depending on the specific market implementation. In this case, we consider the duration of a time slot to be 30 min. In the table, negative values of the energy volume indicate the amount of energy to be sold, while positive values indicate the energy to be purchased. It can be seen from Table 11.4 how some buildings act as sellers in one time slot and act as buyers in the other time slot. This is the characteristic of buildings acting as prosumers.

By sorting and stacking the bids and offers, the bidding curves in the two time slots can be obtained as shown in Fig. 11.6. The market clearing prices in the two time slots are

**Table 11.4** Example of bids submitted by the buildings in two time slots.

|  | Time slot 1 | | Time slot 2 | |
|---|---|---|---|---|
|  | Energy volume (kWh) | Price ($/kWh) | Energy volume (kWh) | Price ($/kWh) |
| Building 1 | −0.8 | 0.4 | 0.6 | 0.62 |
| Building 2 | 0.6 | 0.35 | 1.2 | 0.48 |
| Building 3 | 1.2 | 0.2 | −0.9 | 0.49 |
| Building 4 | −1.3 | 0.45 | −0.6 | 0.54 |
| Building 5 | −1.6 | 0.3 | 0.4 | 0.31 |
| Building 6 | 0.5 | 0.50 | −0.7 | 0.44 |
| Building 7 | 0.4 | 0.43 | 1.1 | 0.45 |
| Building 8 | 1.3 | 0.16 | −1.4 | 0.38 |

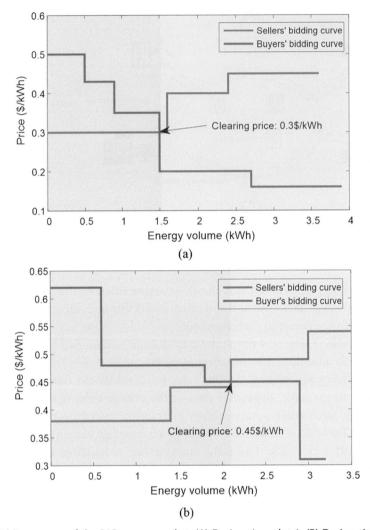

**Fig. 11.6** Bidding curves of the P2P power market: (A) During time slot 1, (B) During time slot 2.

0.45 \$/kWh and 0.3 \$/kWh, respectively. Only the sellers' bids with the proposed price lower than the market clearing price and the buyers' bids with the proposed price higher than the market clearing price are allowed to settle at the market clearing price (shown by the green part in Fig. 11.6).

The final energy trading settlement results are summarized in Table 11.5, where the "revenue/cost" column lists the revenue of the energy sellers or the cost of the energy buyers in each time slot, where negative values indicate revenues and positive values indicate costs. In time slot 1, only Building 5 successfully sells 1.5 kWh of energy to three buildings (i.e., Buildings 2, 6, and 7). In time slot 2, two buildings (i.e., Buildings 6 and 8)

**Table 11.5** Buildings' revenue/cost after the market clearing process in 2 time slots.

| | Time slot 1 | | Time slot 2 | |
|---|---|---|---|---|
| | **Traded energy (kWh)** | **Revenue/cost** | **Traded energy (kWh)** | **Revenue/cost** |
| Building 1 | 0 | 0 | 0.6 | $0.27 |
| Building 2 | 0.6 | $0.18 | 1.2 | $0.54 |
| Building 3 | 0 | 0 | 0 | 0 |
| Building 4 | 0 | 0 | 0 | 0 |
| Building 5 | −1.5 | $0.45 | 0 | 0 |
| Building 6 | 0.5 | $0.15 | −0.7 | $0.32 |
| Building 7 | 0.4 | $0.12 | 0.3 | $0.14 |
| Building 8 | 0 | 0 | −1.4 | $0.63 |

are allowed to sell energy to the other three buildings (i.e., Buildings 1, 2, and 7). For the other buildings that have still surplus energy to sell (because they did not win the bidding process), their surplus energy could be fed back to the grid and paid at feed-in tariff rates. Similarly, the buildings that need energy and that did not win the bidding have to purchase the energy from the grid.

## 11.4 Blockchain-enabled P2P energy trading

### 11.4.1 Introduction

Most of the participants in a local energy market are autonomous, small-capacity energy entities. This naturally imposes several requirements on the underlying energy trading infrastructure that can be summarized as follows:

- *Trustworthy*—The energy trading activities of the participants in local energy markets must be settled in a trusted manner, and the data security and integrity of energy trading transactions must be ensured;
- *Private*—The energy trading data is considered to be sensitive to the participants and it should not be eavesdropped on by any unauthorized third parties; and
- *Scalable*—Unlike grid-level energy markets, the number of participants in a local energy market could become very large. This requires a scalable energy trading system that can serve different numbers of participants without affecting its efficiency.

An open, distributed ledger technology called *blockchain* [13–15] has been recognized to be able to meet the earlier requirements. Its application in supporting P2P energy trading activities in power distribution networks has been actively studied and practiced over several years (e.g., [16–21]).

### 11.4.2 Introduction to blockchain

In this section, we provide a brief introduction to blockchain, and more details about blockchain can be found in dedicated references [13–15]. Blockchain was proposed

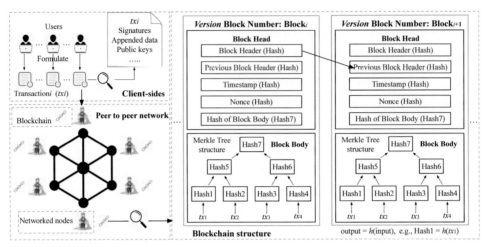

**Fig. 11.7** Schematic of a typical blockchain.

by Satoshi Nakamoto (it is believed this is a pseudonym) in 2008 [15]. In the paper, a distributed ledging technology is described, which enables the participating nodes of a P2P network to achieve secure digital coin payments without the intervention of a third party. In blockchain, each node of the P2P network maintains a copy of a ledger locally. The ledger is denoted as the "blockchain." The transactions generated in the system, which record the digital coin payment information and the account information of the traders, are packaged into blocks at regular time intervals, and the blocks are validated by the nodes and stored in the ledgers. The schematic of the blockchain is illustrated in Fig. 11.7.

Each block in a blockchain consists of two parts: a block body and a block head. The data of multiple transactions are organized following a specific structure (usually organized as a Merkle Tree structure [22]) and are stored in the block body. A Merkle Tree (also known as "hash tree") is a binary tree structure in which every leaf node is labeled with the hash value of a data block, and every node that is not a leaf is labeled with the hash value of the labels of its child nodes (as illustrated in Fig. 11.7). The block head contains the following information: (i) a block header, which is a hash value representing the digest of the block; (ii) the digest of the previous block, which is represented by a hash value; (iii) a timestamp, which is a hash value representing the generation time of the block; (iv) a random hash value called a "nonce," which is used in the consensus mechanism (introduced later); and (v) the hash of the block body. A hash value [23] is a numeric value of a fixed length that uniquely identifies data. Hash values are generated by hash functions. A hash function converts input data of arbitrary length into an encrypted output of a fixed length. A notable feature of hash values is that they cannot be used to "reversely engineer" the input data from the hash value. On the other hand, hash functions produce the same hash value when it is applied to the same data. This means it can be used to validate if the data is the same (i.e., untampered) if the hash is already known.

Transactions are broadcasted by the traders to the blockchain system together with the traders' digital signatures. The transactions are packaged into blocks and the blocks are inserted into the ledgers that are maintained by the nodes. The generation of a block and its insertion into the ledgers are achieved through a consensus mechanism that is executed by the networked nodes in a decentralized manner. The word "consensus" reflects the fact that the nodes reach an agreement on the validity of the data in the block. There are different types of consensus mechanisms, such as the Proof-of-Work (PoW) mechanism [24], the Proof-of-stake mechanism [25], and the Byzantine Fault Tolerant mechanism [26].

The blockchain proposed in Satoshi Nakamoto's paper [15] is based on the PoW mechanism. In the PoW mechanism, the nodes compete to generate blocks for packaging the transactions generated in the P2P network. Every time a node generates a candidate block to obtain such a right, it needs to find out a nonce value so that the hash of the candidate block satisfies a specific condition. This process is also called "mining" and the nodes are called "miners." The nodes set the nonce values in the head of the candidate blocks and broadcast them to the P2P network. The candidate blocks will be checked by all the nodes in the network. If more than 50% of the nodes have verified that a candidate block is valid (i.e., the nonce in the block head is the correct one), the candidate block will be regarded as a valid block. The first node that finds out the correct nonce will then receive some coins as a reward, and the candidate block generated by the node will be inserted into the blockchain and stored in the duplicated ledgers maintained by different nodes.

The transaction data stored in a blockchain is traceable and immutable. Firstly, data traceability is ensured by the blockchain's chain structure, because all the historical transaction records are stored in the blocks that are chained one-by-one. Secondly, since the blocks are chained by using their hash values, any change of the data in the body of a block will lead to the change of the block head's hash value, which will cause all the subsequent blocks to become invalid. Unless the attacker can control more than 50% of the nodes to regenerate and reinsert all the subsequent blocks using the hash of the block containing the tampered data, the data that has already been stored in the blockchain cannot be manipulated. When the number of nodes in a blockchain system is large, it is very difficult for an attacker to control more than 50% of the nodes. The mining mechanism in blockchain is very costly due to the required block generation and this tends to prevent a rational attacker from easily regenerating blocks. This structure ensures that data remains immutable when saved on a blockchain.

Many blockchain implementations are integrated with smart contracts [27]. A smart contract is a digital container that stores data and executable codes. The execution of a smart contract is event-driven: the codes in a smart contract are triggered to be executed by certain prespecified conditions. For example, in online commodity trading, a smart contract containing cash transfer codes is triggered to transfer money from the buyer's digital wallet to the seller's wallet when the buyer confirms he/she has received the commodity.

### 11.4.3 Application of blockchain in P2P energy trading

Blockchain was originally used to support peer-to-peer electronic cash trading systems that do not rely on trusted intermediaries, i.e., the *BitCoin* system [15]. In recent years, extensive academic and industrial efforts have been conducted to study the use of blockchain in supporting P2P energy trading. Such an application of blockchain in P2P energy trading is driven by the two attractive features of blockchain described in the following.

**(1)** *Decentralization*: The operation of a blockchain system is decentralized, i.e., the nodes in the P2P network collaborate to complete the transaction generation, validation, and storage without the participation of a central organization. This feature makes the blockchain technology suitable for the P2P energy trading scenario. P2P energy trading can be implemented on different scales. For example, a P2P energy market can be set up to enable a group of houses belonging to the same community to trade energy with each other, and there could be many such P2P energy markets in an urban area. As a result, the total number of participating buildings and other energy entities in P2P energy trading would be very large. In this scenario, it could be possible to have an energy market operation organization to centrally manage and verify the huge number of energy trading transactions. Taking advantage of the blockchain's decentralization nature, the participants of P2P energy trading systems can trust the market in forming and settling energy trading transactions without the supervision of a central organization.

**(2)** *Data traceability and immutability*: The mechanism of blockchain ensures that once energy trading is recorded into a blockchain, it cannot be altered by any parties. The data recorded in a blockchain is traceable, meaning that one can check and verify the historical energy trading information. These create trust among the participants and can encourage more energy entities to invest in renewable energy sources and participate in P2P energy trading.

### 11.4.4 Case studies: Deployments of blockchain-enabled P2P strategies in urban settings

#### 11.4.4.1 Overview

In the last few years, deployments of P2P solutions have occurred in different parts of the world and these have been established with the use of blockchain technology to support the P2P energy trading. Selected representative case studies are briefly introduced in the following.

### 11.4.4.2 Brooklyn microgrid project

A well-known implementation of blockchain technology for P2P energy trading is the Brooklyn Microgrid (BMG) project launched by LO3 Energy [28,29]. The microgrid consists of a decentralized group of energy resources that operate connected to and synchronous with the macro power grid (see Chapter 13 for more details on microgrids). The project established a local energy market in Brooklyn (New York, USA). The participants are residents located across three power distribution networks in the BMG's region. Within this microgrid, a participant can sell the excessive solar power produced by its rooftop solar panel to its neighbors who need energy.

The BMG system mainly includes two subsystems: a virtual community energy market platform and a physical microgrid. The virtual community energy market platform is the cyber infrastructure of the BMG. Its core component is a private blockchain system based on the Tendermint protocol [30]. The physical microgrid is an electrical power grid that is built to complement the existing grid and provide backup assistance to present power outages. In emergencies, the physical microgrid can be decoupled from the main microgrid and it can supply energy to critical facilities (e.g., hospitals and schools) at fixed electricity rates. The residences and businesses can then bid for the physical microgrid's remaining power.

The participants of BMG perform electricity trading following a bilateral auction mechanism in the virtual community energy market platform. Both electricity sellers and buyers submit bids to the market every 15 min. The market then stacks the bids based on the bidding prices and determines the clearance price. The buyers that are not settled at the clearance price are expected to receive energy from additional energy resources. All the P2P trading and transaction activities are performed at the information system level (i.e., the virtual community energy market platform) and do not affect the physical power flows of the physical microgrid. The power generation and consumption of the participants are monitored by smart meters and the payment of P2P trading is transferred to the participants' blockchain accounts.

### 11.4.4.3 PowerLedger project

PowerLedger [31] is a blockchain-enabled software platform that facilitates the trading of renewable energy and environmental commodities. The project was launched in Western Australia. PowerLedger uses blockchain-based ledgers to track energy generation and consumption of households and automatically execute settlement processes. As the first Initial Coin Offering (ICO) in Australia, the project has already raised over AU$34 million from over 15,000 buyers, while it has also gained grant support from the government.

In 2020, PowerLedger announced a 3-year deal with an Australian property developer to install the energy trading platform at 10 residential projects in Perth (Australia) to enable large-scale solar energy trading in Western Australia [32]. PowerLedger's platform is being deployed with Connected Communities Energy (a smart city initiative of Niche-living) that enables residents to manage their homes remotely.

### 11.4.4.4 Conjoule's "community network" platform

The Conjoule's "Community Network" project is another blockchain-based P2P energy trading platform [33] launched in Germany in 2018. The platform is designed to allow owners of private photovoltaic systems and neighborhood consumers to share renewable energy. It uses blockchain to ensure transparency about the source of the energy generated.

## References

[1] Energysaver, 2020. "Feed-in-tariff rates in NSW." Energysaver.nsw.gov.au. https://energysaver.nsw.gov.au/households/solar-and-battery-power/feed-tariff-rates (Accessed 08 March 2022).

[2] M.R. Alam, M. St-Hilaire, T. Kunz, Peer-to-peer energy trading among smart homes, Appl. Energy 238 (2019) 1434–1443.

[3] Y. Zhou, J. Wu, C. Long, W. Ming, State-of-the-art analysis and perspectives for peer-to-peer energy trading, Engineering 6 (7) (2020) 739–753.

[4] A.M. Worner, L. Ableitner, A. Meeuw, F. Wortmann, V. Tiefenbeck, Peer-to-peer energy trading in the real world: Market design and evaluation of the user value proposition, in: Proceedings of 40th International Conference on Information Systems (ICIS2019), Munich, Germany, Dec. 2019.

[5] B. Murray, Power Markets and Economics, John Wiley & Sons, Hoboken, NJ, 2019.

[6] Aemo, 2020. Australian energy market operator. Aemo.com.au. https://aemo.com.au/en (Accessed 02 September 2022).

[7] D.R. Biggar, M.R. Hesamzadeh, The Economics of Electricity Markets, John Wiley & Sons, Hoboken, NJ, 2014.

[8] M. Kopsakangas-Savolainen, R. Svento, Modern Energy Markets, Springer International Publishing, London, UK, 2012.

[9] C. Long, J. Wu, C. Zhang, L. Thomas, M. Cheng, N. Jenkins, Peer-to-peer energy trading in a community mirogrid, in: Proceedings of IEEE Power & Energy Society General Meeting, 2017.

[10] N. Liu, X. Yu, C. Wang, C. Li, L. Ma, J. Lei, Energy-sharing model with price-based demand response for microgrids of peer-to-peer prosumers, IEEE Trans. Power Syst. 32 (5) (2017) 3569–3583.

[11] N. Liu, X. Yu, W. Fan, C. Hu, T. Rui, Q. Chen, J. Zhang, Online energy sharing for nanogrid clusters: a Lyapunov optimization approach, IEEE Trans. Smart Grid 14 (8) (2016) 1–13.

[12] F. Luo, Z.Y. Dong, G. Liang, J. Murata, Z. Xu, A distributed energy trading system for active distribution networks based on multi-agent coalition and Blockchain, IEEE Trans. Power Syst. 34 (5) (2019) 4097–4108.

[13] D. Drescher, Blockchain Basics: A Non-technical Introduction in 25 Steps, Apress, 2017.

[14] A. Tapscott, D. Tapscott, Blockchain Revolution: How the Technology behind Bitcoin Is Changing Money, Business, and the World, Penguin Books Limited, London, UK, 2016.

[15] S. Nakamoto, Bitcoin: a peer-to-peer electronic cash system, in: Decentralized Business Review, 2008, p. 21260.

[16] F. Luo, Z.Y. Dong, G. Liang, J. Murata, Z. Xu, A distributed electricity trading system in active distribution networks based on multi-agent coalition and blockchain, IEEE Trans. Power Syst. 34 (5) (2018) 4097–4108.

[17] Z. Li, J. Kang, R. Yu, D. Ye, Q. Deng, Y. Zhang, Consortium blockchain for secure energy trading in industrial internet-of-things, IEEE Trans. Industr. Inform. 14 (8) (2018) 3690–3700.

[18] F.S. Ali, M. Aloqaily, O. Alfandi, O. Ozkasap, Cyberphysical blockchain-enabled peer-to-peer energy trading, Computer 53 (9) (2020) 56–65.

[19] A. Dorri, F. Luo, S. Kanhere, R. Jurdak, Z.Y. Dong, SPB: a secure and private Blockchain-based solution for energy trading, IEEE Commun. Mag. 50 (7) (2019) 120–126.

[20] J. Abdella, Z. Tari, A. Anwar, A. Mahmood, F. Han, An architecture and performance evaluation of blockchain-based peer-to-peer energy trading, IEEE Trans. Smart Grid 12 (4) (2021) 3364–3378.

[21] R. Khalid, N. Javaid, A. Almogren, M.U. Javed, S. Javaid, M. Zuair, A blockchain-based load balancing in decentralized hybrid P2P energy trading market in smart grid, IEEE Access 8 (2020) 47047–47062.

[22] K.S. Garewal, Markle trees, in: Practical Blockchains and Cryptocurrencies, Springer, 2020, pp. 137–148.

[23] S. Debnath, A. Chattopadhyay, S. Dutta, Brief review on journey of secured hash algorithms, in: Proceedings of 2017 4th International Conference on Opto-Electronics and Applied Optics, Kolkata, India, Nov. 2017.

[24] A. Gervais, G.O. Karame, K. Wüst, V. Glykantzis, H. Ritzdorf, S. Capkun, On the security and performance of proof of work blockchains, in: Proceedings of the 2016 ACM SIGSAC Conference on Computer and Communications Security, 2016, pp. 3–16.

[25] S. King, S. Nadal, Ppcoin: Peer-to-peer crypto-currency with proof-of-stake, Self-published paper, vol. 19, no. 1, August 2012.

[26] M. Castro, B. Liskov, Practical byzantine fault tolerance, OsDI (99) (1999) 173–186.

[27] V. Buterin, A next-generation smart contract and decentralized application platform, White Paper 3 (37) (2014). p. 2.1.

[28] E. Mengelkamp, J. Garttner, K. Rock, S. Kessler, L. Orsini, C. Weinhardt, Designing microgrid energy markets a case study: the Brooklyn microgrid, Appl. Energy 210 (2018) 870–880.

[29] Brooklyn.energy, 2019. Brooklyn microgrid. https://www.brooklyn.energy (Accessed 01 October 2020).

[30] Tendermint. 2021. Tendermint.com. https://tendermint.com (Accessed 10 August 2020).

[31] "PowerLedger." 2021. Powerledger.io. https://www.powerledger.io (Accessed 08 March 2022).

[32] Ledgerinsights, 2020. "PowerLedger in blockchain energy trading deal with Perth property developer." Ledgerinsights.com. https://www.ledgerinsights.com/power-ledger-blockchain-energy-trading-perth-property-developer-nicheliving (Accessed 22 September 2020).

[33] Utilitydrive, 2018. "Power trading among neighbors per blockchain: Innogy and Conjoule make it possible." Utilitydive.com. https://www.utilitydive.com/press-release/20180525-power-trading-among-neighbors-per-blockchain-innogy-and-conjoule-make-it-p (Accessed 22 September 2020).

# CHAPTER 12

# Building-to-grid integration

## Contents

## 12.1 Introduction

Current trends in building design are redefining the roles and functions of buildings as these are becoming complex cyber–physical. This transformation is possible due to recent advances in computing, communication, control technologies, and the widespread of Advanced Metering Infrastructure (AMI) and renewable energy sources. These developments are also supporting buildings in becoming an integrated part of power grids. Buildings' capabilities of generating and feeding energy back to the grid and performing demand response have enabled their active engagement in the grid's operations.

Previous chapters have focused on the design of Building Energy Management Systems (BEMSs) and strategies for their deployment at the building level, and we introduced simple application examples to present these techniques. In these chapters, particular attention has been given to the minimization of energy costs while trying to maintain the building occupants' comfort and satisfaction. In Chapter 11, we have introduced the possibility of energy sharing among communities of selected buildings and, in this chapter, we further extend the integration of buildings with the grid by presenting techniques in which buildings interact with the grid to provide auxiliary support to the grid's operations. This interaction and arrangement is usually referred to as *Building-to-Grid (B2G) integration*, for example [1–4]. The underlying common vision at the basis of the B2G paradigm is to create a highly renewable penetrated, digitalized, and automatic platform to maximize the harnessing of the demand response capability of buildings to support the grid's operations while maintaining adequate comfort and satisfaction for the building occupants.

This chapter provides an overview of the key features of the B2G integration. In the first part of the chapter, aspects that contributed to the development of the B2G paradigm are briefly introduced. These highlight how the B2G integration can create a highly renewable penetrated, digitalized, and automatic platform to maximize harnessing the demand response capabilities of buildings to support the grid's operation while satisfying the needs of the buildings and their occupants. In the second part of the chapter, specific scenarios in which B2G can be implemented are considered and these include possible strategies in reducing the peak-to-average ratio for both buildings and the grid as well as in responding to emergency load shedding situations. Possible strategies that deal with these types of optimization requirements are presented and complemented by means of simple application examples that illustrate the key steps of the solutions.

## 12.2 From direct load control to building-to-grid integration

As discussed in Chapter 4, conventional power Demand Side Management (DSM) techniques can be classified into two categories: price-based DSM and incentive-based DSM (with the latter also known as Direct Load Control, DLC). Price-based DSM uses time-varying or consumption-varying energy prices to encourage energy customers to shift the usage time of appliances from peak pricing hours to off-peak pricing hours. In this way, the peak power load of the grid can be reduced even if this approach heavily relies on the occupants' subjective willingness to adjust and thus cannot ensure load shifting occurs at all times. In a grid emergency condition (e.g., emergency load shedding due to grid frequency/voltage variations and emergency peak load control), DLC is usually the most widely used. With this approach, the power utility can directly control the on/off status or settings of the appliances based on the grid's operational requirements.

DLC can ensure load reduction/reshaping because the utility has full control privileges of the customer's appliances during the agreed DLC period. The main challenge of this approach is the large disturbance that it can cause to occupants because it would be difficult and impractical for the utility to precisely detect the occupants' building environments, especially when the number of controlled appliances is significantly large. The highly intrusive nature of DLC shows also less consideration for the building's autonomy. For example, a building would have its own operation policies that would need to be respected by external parties.

In this context, a BEMS can establish an adequate connection between a building and the grid. For example, a BESS can act as the delegate of building occupants because it usually manages one building or, at most, a group of buildings, and it can take the role of making energy resource operation decisions on behalf of the occupants. By relying on automatic sensors and actuators as well as wireless communication facilities, a BEMS can perform direct control operations on the building-side energy resources. The BEMS can also receive dynamic energy tariff information from the utility and, based on this information, it can control building energy resources to respond to the tariff signals. Based on its characteristics, a BEMS can support both direct load control and pricing-based demand responses.

A B2G vision where buildings can actively participate to assist the power utility in improving the grid's operations is presented in Fig. 12.1. In this arrangement, BEMSs can assist the utility by taking direct control of the building-side energy resources and can be responsible for communicating with the utility. According to the grid's operational requirements, BEMSs can schedule and control the operations of building energy resources while sufficiently considering the occupants' comfort and lifestyle requirements. BEMSs can also exchange data and information with each other to collaboratively accomplish mutual B2G objectives that could not, otherwise, be accomplished individually.

The main features of the B2G integration are depicted in Fig. 12.1 and are summarized below:

- *Autonomy preservation of buildings*: In B2G, building energy resources are managed by BEMSs rather than being remotely controlled by the utility during the agreed DLC hours. BEMSs schedule and control the operation of the building energy sources based on autonomous policies agreed upon by the relevant stakeholders, for example, occupants and building managers. BEMSs also communicate with the utility through the AMI to be aware of the grid's operational requirements. In this way, the autonomy of buildings can be sufficiently preserved, and the intrusion and disturbance to the occupants can be significantly reduced and mitigated.

- *Fine-grained load control*: Through underlying sensing and metering facilities, BEMSs are capable of monitoring and perceiving the buildings' environments and operational conditions in real-time and in a fine-grained manner. They are also able to interactively communicate with occupants to understand their preferences and lifestyles. Based on this, effective B2G operations can be achieved through fine-grained information acquisition, interactive human-machine and machine-to-machine communications, and detailed control actions. It would not be feasible to achieve such tasks with traditional DLC where the utility remotely controls several buildings.

- *Scalability*: The B2G integration vision in Fig. 12.1 is based on a decomposed structure, in which the grid's operations are supported by the control center of the grid and the collaboration of individual BEMSs. In traditional DLC, the control center of the grid makes all control decisions that require greater computational resources as the number of managed buildings increases. The B2G integration scheme allocates energy management tasks to the building-side, enabling the system to be highly flexible and scalable.

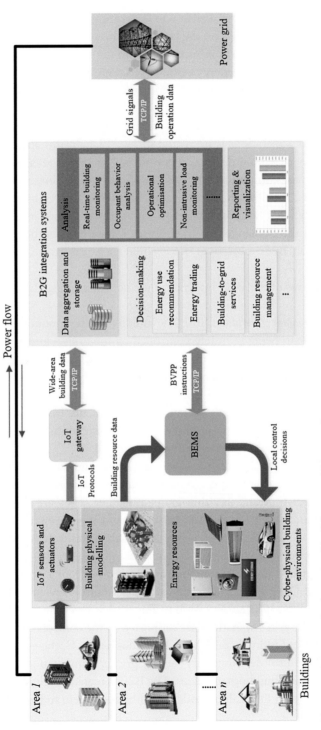

**Fig. 12.1** Conceptual schematic of building-to-grid integration.

## 12.3  Building energy management for peak-to-average ratio optimization

### 12.3.1  Overview

The grid's infrastructure (e.g., power transmission lines, power breakers, and power generators) is built to serve the grid's peak power demand. The grid's peak load has a significant impact on the grid's secure and reliable operations. For example, the occurrence of a significantly large peak power load in the grid would lead to a power network overload and the subsequent triggering of the tripping-out of the power transmission lines. In extreme scenarios, this could also result in cascading failures within the grid and it could lead to possible blackout events. Because of this, the management of the grid's peak power demand is an important task for a power system [5].

A useful variable to depict the variation in the average power load profile versus the peak power load is expressed by the Peak-to–Average Ratio (PAR) [6] which is calculated as:

$$PAR = \frac{P^{peak}}{P^{avg}} \tag{12.1}$$

where $P^{peak}$ and $P^{avg}$ are the peak and average powers, respectively, obtained from the power load profile. A large PAR value indicates a sharp load curve, while a small value for the PAR denotes a smooth load curve. Fig. 12.2 shows two load profiles that have different PAR values while they consume the same amount of energy over the period considered. From the grid's viewpoint, a smaller value of PAR is desirable, as this would be an indication of an energy demand curve with less dominant demand peaks.

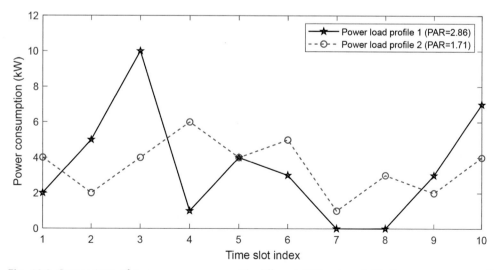

**Fig. 12.2** Comparison of two energy curves with different PAR values (with the same amount of energy consumed).

In the context of B2G, a BEMS acts as a mediator between the occupant and the grid. If the PAR of individual buildings' load profiles can be reduced, the PAR of the aggregately regional load profile can then also be decreased.

By considering an individual building, the energy cost minimization and PAR minimization could be conflicting objectives. For example, a time-varying electricity tariff would incentivize the occupant to shift many appliances from peak- to valley-price hours, leading to a high power consumption in valley-price hours. Since a time-varying energy tariff usually updates the rates on an hourly basis and an appliance's operation can run for minutes, there could exist multiple appliance operation schedules that lead to different PARs but with the same energy cost. As an example, let us consider an appliance that spends 45 min to operate and to complete its task with an energy rate of 0.5$/kWh between [5 pm, and 6 pm]. There are multiple schedules for the appliance within this time interval that leads to the same cost. The appliance could operate between [5 pm, and 5:45 pm] or between [5:15 pm, and 6 pm] (as illustrated in Fig. 12.3). Under these two schedules, the energy cost of running the appliance is identical (both are charged by 0.5$/kWh), but the PAR values of the building's power load profile could be different (because the appliance operates in different periods). This implies that it is possible to further optimize the PAR of a building's power load profile without significantly incurring additional energy costs.

Based on the above considerations, the BEMS's design introduced in the following aims to manage flexible appliances and to minimize the building's energy cost and the PAR value of the power load profile. For ease of presentation, we consider a simplified case with flexible appliances operating within a building and centrally managed by a BEMS. More complex arrangements of the appliances could be managed by extending the proposed methodology. To keep the complexity of the problem to a minimum, we consider only Time Shiftable Appliances (TSAs) (previously introduced in Chapter 9).

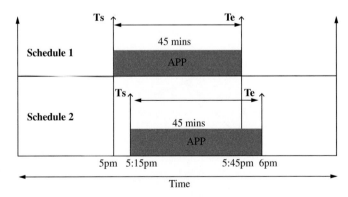

**Fig. 12.3** Diagram describing the scheduling of an appliance in different time periods and with the same energy cost ("APP" = appliances; "Ts" = starting operation time; "Te" = end operation time).

The design principle of the BEMS follows a two-stage energy management optimization approach. By considering that there are $N$ controllable appliances managed by the BEMS and that the building is charged by a time-varying electricity tariff, the first optimization stage of the BEMS' strategy is to temporarily schedule the $N$ appliances to minimize the building's energy cost. During the second optimization stage, the BEMS determines the final schedule for the appliances so that the PAR of the building's load profile is minimized while the energy cost remains equivalent to the minimum energy cost determined in the first optimization stage.

## 12.3.2  Stage 1: Minimization of the building energy cost

We can now define the energy cost minimization process for the building. For ease of reference, we reconsider the approach previously introduced in Section 9.6 in which the BEMS minimizes the building's cost charged by the time-varying electricity tariff. This enables to establish the following scheduling details at the end of the first optimization stage:

**(1)**  the temporary schedule of each controllable appliance; and
**(2)**  the minimized energy cost for the temporary appliance schedules.

## 12.3.3  Stage 2: Minimization of the PAR of a building's power load profile

### 12.3.3.1  Overview

In the second optimization stage, the BEMS re-schedules the appliances to minimize the PAR of the building's power load profile while ensuring the total energy cost remains equal to the minimum energy cost obtained from the first optimization stage. The formulation of the optimization problem for this stage is outlined in the following sections.

### 12.3.3.2  Decision variables

The second optimization stage has a total of $K$ decision variables (that correspond to the $K$ TSAs) that can be represented as a vector:

$$\boldsymbol{x} = [x_1, x_2, ..., x_K] \tag{12.2}$$

where $x_k$ ($k=1{:}K$) represents the index of the time slot when the $k$th TSA starts to operate.

### 12.3.3.3  Objective function

The objective function aims at minimizing the PAR of the building's power load over the time horizon of interest formed by $T$ time slots:

$$min\, PAR = \frac{max(\boldsymbol{P}^b)}{\bar{P}^b} \tag{12.3}$$

in which $\boldsymbol{P}^b = [P_1^b, ..., P_T^b]$ represents the total power consumption profile of the building. $P_t^b$ ($t = 1:T$) is the building's power consumption at time interval $t$ (in kW). $\overline{P}^b$ represents the building's average power consumption over the time horizon (in kW). the function $max(\cdot)$ returns the maximum value of its input vector.

The vectors $P_t^b$ and $\overline{P}^b$ are calculated as follows:

$$P_t^b = \sum_{k=1}^{K} P_{k,t} + P_t^{uc} \tag{12.4}$$

$$\overline{P}^b = \frac{\sum_{t=1}^{T} P_t^b}{T} \tag{12.5}$$

where $P_t^{uc}$ is the uncontrollable power load of the building at the $t$th time slot (in kW) that depicts the power consumed by devices and appliances not managed by the BEMS (e.g., manually controlled appliances). $P_{k,t}$ is the power consumption of the $k$th appliance at time interval $t$ (in kW) that can be determined based on:

$$P_{k,t} = \begin{cases} P_k^{op} & \text{if } x_k \leq t \leq x_k + T_k^d - 1 \\ 0 & \text{otherwise} \end{cases} \tag{12.6}$$

where $P_k^{op}$ represents the operating power of the $k$th TSA (in kW), and $T_k^d$ denotes the task execution duration of the $k$th TSA expressed as the number of time slots.

### 12.3.3.4 Constraints

The decision variables are subjected to the value boundary constraint:

$$t_k^s \leq x_k \leq t_k^e - T_k^d + 1, \ k = 1 : K \tag{12.7}$$

The following constraints are required to maintain the building's total energy cost within the desired budget (i.e., budget accounting for the budget increment $\delta$) over the whole time horizon:

$$C^{s2} \leq C^{s1} + \delta \tag{12.8}$$

$$C^{s2} = \sum_{t}^{T} \left( P_t^b \Delta t \rho_t \right) \tag{12.9}$$

where $t_k^s$ and $t_k^e$ are the lower and upper limits of the allowable operation time range of the $k$th TSA, respectively. $\Delta t$ is the duration of one time interval (in hours). $C^{s1}$ is the minimum building energy cost (in \$) generated from Stage 1 (in Section 12.3.2). $\rho_t$ is the electricity price at time interval $t$ (in \$/kWh).

We consider the building occupant can accept the energy cost is within the range of $[C^{s1}, C^{s1} + \delta]$, where $\delta$ represents the budget increment that an occupant is happy to pay on top of the minimal cost value obtained in the first stage (in \$) to contribute to the PAR minimization of the building. By taking advantage of the flexibility of the occupant-acceptable energy cost range, the BEMS can optimize the PAR of the building's power consumption profile.

## 12.3.4 Solution strategy

The first optimization stage can be solved based on Algorithm 9.4 which relies on the use of the differential evolution (DE) algorithm. The DE algorithm can also be used to solve the second stage of the optimization based on the steps presented in Algorithm 12.1.

In Algorithm 12.1, the sub-routine of *SubRoutine_GenerateTrialIndividual* produces a trial individual based on DE's heuristic rules. The function *fitness*($x, \dots$) returns the fitness value (i.e., objective function value) of an input individual $x$, and the notation "$\dots$" represents the other parameters of the function, including $C^{s1}$, $\delta$, $r$, $P^{uc}$, and the TSA models

---

**ALGORITHM 12.1 Pseudocode describing the steps for the second optimization stage of PAR-aware building energy management.**

**Start**
1. Set $T$ and $\Delta t$;
2. Input $C^{s1}$, $\delta$, and electricity rates $r = [r_t, \dots, r_T]$;
3. Input the model of the TSAs ($P_k^{op}$, $T_k^d$, $t_k^s$, $t_k^e$, $k = 1{:}K$);
4. Input the uncontrollable load profile $P^{uc} = [P_1^{uc}, \dots P_T^{uc}]$;
5. Set DE's parameters: $N$, $G$, $Cr$, and $F$;
6. **For** $i = 1{:}N$
7.    Initialize $x_i$ by randomly setting $x_{i,k}$ between $[t_k^s, t_k^e - T_k^d + 1]$;
8. **End For**
9. Set the population as $\Theta = \{x_1, \dots, x_N\}$;
10. Set the generation index $g = 1$;
11. Set the global best individual $x^* = minByFitness(\Theta, 1)$;
12. **While** ($g <= G$)
13.   **For** $i = 1{:}N$
14.     Generate the trial individual $x_i^{trial}$ by performing *SubRoutine_GenerateTrialIndividual*;
15.     Set score1 $=$ *fitness* ($x_i^{trial}, \dots$);
16.     Set score2 $=$ *fitness* ($x_i, \dots$);
17.     **If** (score1 $<$ score2)
18.       Set $x_i = x_i^{trial}$;
19.     **End If**
20.     **If** *fitness*($x_i, \dots$) $<$ *fitness*($x^*, \dots$)
21.       Set $x^* = x_i$;
22.     **End If**
23.   **End For**
24. **End While**
25. Output $x^*$.
**End**

---

$N$, $G$, $F$, and $Cr$ are control parameters of the DE algorithm: $N$ represents the population size; $G$ represents the maximum generation time; $F$ and $Cr$ represent the mutation factor and crossover factor, respectively.

---

**ALGORITHM 12.2 Pseudocode describing the function calculating the individual's fitness value.**

**Function** $v = fitness(\boldsymbol{x},...)$
1. Calculate the building's energy cost $c$ subjected to $\boldsymbol{x}$ according to Eqs. (12.4)–(12.6) and Eq. (12.9);
2. Calculate the PAR value $par$ subjected to $\boldsymbol{x}$, according to Eq. (12.1);
3. **If** $(c > C^{s1} + \delta)$
4.     Set $v$ as a very large number $v^{max}$;
5. **Else**
6.     Set $v = par$;
7. **End If**
8. **Return** $v$.
**End**

---

$(P_k^{pp}, T_k^d, t_k^s, t_k^e, k = 1{:}K)$. The pseudocode of this function is given in Algorithm 12.2. The function first checks that the constraints in Eq. (12.7) are satisfied, and then assigns the PAR value of the building's power load demand subjected to the individual $\boldsymbol{x}$ as the objective function of $\boldsymbol{x}$. In the case the constraints are not satisfied, the function assigns a very large number $v^{max}$ as the fitness value of $\boldsymbol{x}$ to ensure that this individual is not selected because of its extremely high fitness value (because this individual does not meet the problem constraints). The algorithm finally outputs the optimal schedule of the $K$ TSAs $(\boldsymbol{x}^*)$.

## 12.3.5 Application example

In this section, we consider a scenario with 6-controllable appliances located in one building to maintain the complexity of the problem to a minimum while highlighting the steps of the two-stage optimization. The configurations of the 6 TSAs are shown in Table 12.1. The time-varying electricity price is given in Table 12.2. Other parameters are provided in Table 12.3.

Figs. 12.4 and 12.5 show the temporary appliance schedules (end of the first optimization stage) and the final appliance schedules (end of the second optimization stage). It can be noted that the final appliance schedules distribute the operation time of the major appliances in the time slots more evenly while ensuring that the energy cost does not increase over the value obtained in the first optimizations stage.

Table 12.4 reports the comparison of the energy costs, PAR values, and peak power consumption of the building under two cases: (a) based on the results of the first stage optimization; and (b) obtained from the two-stage optimization process. As expected,

**Table 12.1** Configurations of the NTCAs.

| | Rated power | Task duration | Allowable operation time range |
|---|---|---|---|
| Washing machine | 0.8 kW | 50 min (10 time slots) | [1 pm, 10 pm] |
| Clothes dryer | 1.4 kW | 40 min (8 time slots) | [3 pm, 10 pm] |
| Oven | 1.2 kW | 40 min (8 time slots) | [10 am, 4 pm] |
| Pool pump | 0.9 kW | 60 min (12 time slots) | [9 am, 8 pm] |
| Rice cooker | 0.4 kW | 40 min (8 time slots) | [8 am, 12 pm] |
| Dish washer | 0.7 kW | 50 min (10 time slots) | [6 pm, 12 am] |

**Table 12.2** Time-of-use electricity tariff used in the case study.

| Time | Rate (in $/kWh) |
|---|---|
| Peak: 3–8 pm | 0.86 |
| Secondary peak: 12–3 pm | 0.52 |
| Shoulder: 7 am–12 pm, 8–10 pm | 0.27 |
| Off-peak: 10 pm–7 am | 0.08 |

**Table 12.3** Summary of other parameter settings.

| | | | DE parameters | | | |
|---|---|---|---|---|---|---|
| $T$ | $\Delta t$ | $\delta$ | $N$ | $G$ | $F$ | $Cr$ |
| 288 | 5 min | 25 cents | 300 | 100 | 0.8 | 0.2 |

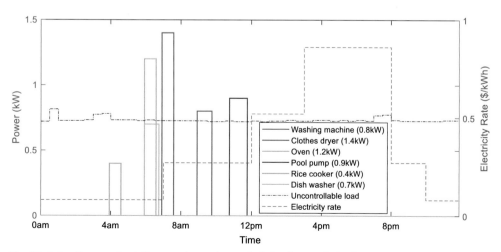

**Fig. 12.4** Appliance scheduling results at the end of the first optimization stage.

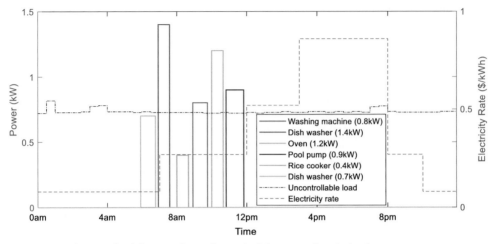

**Fig. 12.5** Appliance scheduling results at the end of the second optimization stage.

**Table 12.4** Comparison of simulation results obtained at the end of the first and second optimization stages.

|  | Total energy cost | PAR | Peak power |
|---|---|---|---|
| First optimization stage | $6.98 | 2.91 | 2.63 kW |
| Second optimization stage | $7.19 | 2.34 | 2.12 kW |

the two-stage energy optimization leads to a lower peak power consumption, and therefore, to lower PAR values. In this process, the total energy cost of the building is controlled below the occupant-specified budget limit, that is, $6.98 + $0.25 = $7.23 (where $\delta = $0.25 from Table 12.3). The two-stage BEMS facilitates the occupant to agree to spend about 20 cents to contribute to the reduction of the PAR of the building's load profile. If such a scheme could be applied to a large number of buildings, it would be beneficial to the entire energy system.

## 12.4 A building-to-grid system for emergency load shedding

### 12.4.1 Overview

The increasing energy demand of human activities poses significant stresses to modern power grids. For example, on hot summer days, people use air conditioners and this results in a very large power being transmitted through the grid, with the grid operating close to its capacity limit. In this situation, emergency load shedding [7,8] may be performed to temporarily reduce the power load of one or multiple areas and to relieve the power transmission and distribution networks' loading. Emergency load shedding can also be triggered by other events such as system faults and disturbances, for example, transmission line faults due to severe weather and drops in the grid's frequency. During these events, the grid operator could perform emergency load shedding to avoid further serious consequences, for example, system cascading and blackouts.

Traditionally, power load shedding is carried out in a simple and straightforward manner, that is, by simply interrupting the power supply to support the powering of loads in one or multiple geographical areas over a certain period (usually in the order of several hours to a day). This leads to temporary blackouts that can cause significant disturbances to building occupants and urban life in general. In fact, when performing load shedding, it is not necessary to shed all the energy loads in an area. If emergency load shedding can be performed in a more flexibility way, that is, only selected loads are interrupted to maintain the secure operation of the grid, the disturbance to the energy customers (e.g., building occupants) can be reduced. In a B2G paradigm, with the support provided by Internet-of-Things (IoT) sensors and BEMSs, emergency load shedding can be performed in a fine-grained manner to minimize its disturbance to building occupants.

## 12.4.2 Emergency load shedding approach

In this section, we present an approach that can be implemented for emergency load shedding purposes in a B2G framework. Without any loss of generality, let us consider a hot summer day when the grid operator recognizes that, by performing online load forecasting for a specific precinct consisting of a group of buildings, the precinct's power consumption is expected to be significantly high in the subsequent hour. Because of this, the grid operator launches an emergency load shedding program. Based on the system's operational condition, the BEMS assists the grid in determining the total power load shedding amount of the whole system and then allocates the load shedding amount to each utility customer entity.

Let us further assume that the load control period consists of $T$ time slots and the duration of each time slot is $\Delta t$. Based on the power consumption threshold instruction (denoted as $I = [I_1, \ldots, I_T]$, where $I_t$ ($t = 1 : T$) denotes the required power consumption threshold for the precinct (in kW)) and desired appliance use information, the BEMS performs emergency load shedding actions to assist the grid to control the total power consumption of the area. To achieve this, the BEMS collects the operation data from all controllable appliances in the buildings and uses energy management strategies to determine the operational status of the appliances at different time slots, for example, pause an appliance from the operating status or resume an appliance's operation from the pause status.

Let us consider a precinct consisting of $N$ buildings and, for each building, there are $N_n^{app}$ controllable appliances in the building ($n$ is the building index, with $n = 1:N$). We also consider that the emergency load shedding period lasts for $T$ time slots and that the duration of each time slot is denoted as $\Delta t$ (in hours). The pseudocode of the load shedding procedure is presented in Algorithm 12.3.

---

**ALGORITHM 12.3 Pseudocode describing the emergency load shedding approach.**

**Start**
1. Set up $T$ and $\Delta t$;
2. Input the power consumption threshold instruction vector $I$;
3. Perform *SubRoutine_1* to initialize the appliances' information;
4. Set the current time slot index $t=1$;
5. Perform *SubRoutine_2* to determine the appliances that are supposed to run in the current time slot;
6. Calculate the power consumption of the precinct $P_t$;
7. **If** $P_t > I_t$
8.     Perform *SubRoutine_3* to perform load shedding;
9. **End If**
10. Update the indoor temperature after the period of $\Delta t$ using the thermal model in Chapter 10 or more sophisticated ones;
11. Proceed to the next time slot $t=t+1$;
12. **If** $t > T$
13.    **Terminate**;
14. **Else**
15.    **Go to** Line 5.
16. **End If**
**End**

---

In Algorithm 12.3, load shedding is performed sequentially at each time slot of the peak power control period. At the beginning of the peak power control period, the algorithm invokes *SubRoutine_1* (specified in Algorithm 12.4) to initialize the information of the appliances in each building (Line 3), including the appliances' desired operation states ($s_{n,k}$: 0-paused; 1-operating), priorities ($\theta_{n,k}$), and operating power ($P_{n,k}^{op}$, kW). In each building, the appliances are sorted by the descending order of their priority values. To keep the complexity of the problem to a minimum, we assume that, when an appliance is in a "paused" status (i.e., $s_{n,k,t}=0$), its power consumption is taken as zero. Each appliance is also associated with a flag ($c_{n,k}$), which represents the number of time slots during which the appliance is paused by the BEMS when compared to the expected schedule that the same appliance would have operated under normal conditions, that is, without the need to implement an emergency load shedding. For example, under normal conditions, an appliance would have operated 5 time slots from a specific starting time slot to a given time slot $t$. During the emergency load shedding process, this appliance could operate for 3-time slots from the starting time to time slot $t$. This implies that this appliance has been paused for 2-time slots in this period and, therefore, the value for $c_{n,k}$ for the

---

**ALGORITHM 12.4 Pseudocode describing SubRoutine_1: initialization of the appliance information.**

**Start**
1.  **For** $n=1{:}N$
2.    Sort the $N_n^{app}$ appliances in the $n$th building with descending order;
3.    **For** $k=1{:}N_n^{app}$
4.      Set $c_{n,k}=0$;
5.      Set up $\theta_{n,k}$ and $P_{n,k}^{op}$;
6.      **For** $t=1{:}T$
7.        Set up $s_{n,k,t}\in\{0,1\}$ based on the occupant's requirements in normal operation conditions.
8.      **End For**
9.    **End For**
10. **End For**
**End**

---

appliance is $5-3=2$. This flag is initialized to zero at the beginning of the procedure and is later updated by *SubRoutine_2*.

As shown in Algorithm 12.3, at each control time slot, the BEMS first performs *SubRoutine_2* to check which appliances in each building are supposed to operate in the current time slot (Line 5). The algorithm then compares the precinct's power consumption ($P_t$) with the peak power threshold instruction $I_t$. $P_t$ is calculated as follows:

$$P_t = \sum_{n=1}^{N} \sum_{k=1}^{N_n^{app}} \left( P_{n,k}^{op} \times s_{n,k,t} \right) \tag{12.10}$$

When $P_t$ is larger than the threshold, the BEMS executes *SubRoutine_3* (described in Algorithm 12.6) to perform load shedding on the buildings to ensure that the precinct's total power consumption remains below or equal to the threshold (Lines 7–9). After load shedding, the indoor temperature is updated (Line 10) over the duration of the current time slot. In this case, we use the thermal model presented in Chapter 10, and more sophisticated ones could be used, to simulate the thermal transition process of the buildings and to calculate the indoor temperature. The above control strategy is continuously applied to each time slot until the end of the load control period is reached (Lines 11–16).

SubRoutine_2 is described in Algorithm 12.5 and it enables the BEMS to check the appliances' operating status in each building. In the algorithm, $N_n^{app}$ represents the number of controllable appliances in the $n$th building which have been sorted in descending order of the priority, and $u^{up,ls}$ represents the upper boundary of the comfort indoor temperature band in the emergency load shedding period. When dealing with a

**ALGORITHM 12.5 Pseudocode describing SubRoutine_2: determination of appliances that are supposed to operate in the current time interval.**

**Start**

1. **For** $n=1:N$
2.    **For** $k=1: N_n^{app}$
3.       **If** the $k$th appliance is an HVAC
4.          **If** $s_{n,k,t}==0$
5.             Use the thermal model presented in Chapter 10 to calculate the indoor temperature $u_{n,t}^{in}$ after $\Delta t$ by keeping it as the standby state;
6.             **If** $u_{n,t}^{in}>u^{up,ls}$
7.                Set $s_{n,z_n,t}=1$;
8.             **End If**
9.          **Else**
10.             Measure the current indoor temperature $u_{n,t}^{in}$;
11.             **If** $u_{n,t}^{in}<u^{low,ls}$
12.                Set $s_{n,z_n,t}=0$;
13.             **End If**
14.          **End If**
15.       **Else**
16.          **If** $s_{n,k,t}==0$
17.             **If** $c_{n,k}>0$
18.                Set $s_{n,k,t}=1$;
19.                Set $c_{n,k}=c_{n,k}-1$.
20.             **End If**
21.          **End If**
22.       **End IF**
23.    **End For**
24. **End For**

**End**

thermostatically controlled appliance (in this example, an HVAC system), the algorithm determines its operating status according to the indoor temperature calculated based on the thermal model presented in Chapter 10 (Lines 3–15) or with more sophisticated models. If the appliance is not an HVAC, the BEMS determines its status based on the following logic: if it is supposed to operate in the current time slot, then its status is kept as 1. Alternatively, for an appliance that is originally planned not to operate in the current time slot, the algorithm checks if the appliance has not met the occupants' desired operation time (Lines 16 and 17). If this is the case, the appliance is flagged as "supposed to operate" by setting its status variable $s_{n,k,t}-1$ and by updating its flag $c_{n,k}$ (Lines 18 and 19).

**ALGORITHM 12.6 Pseudocode describing SubRoutine_3: load shedding during one time interval.**

**Start**
1.   **For** $n=1:N$
2.      Set $z_n=0$;
3.   **End For**
4.   **While** $P_t>l_t$
5.      **For** $n=1:N$
6.         Set $z_n=z_n+1$;
7.         **While** $N_n^{app} \geq z_n$ and $s_{n,z_n,t}==0$
8.            Set $z_n=z_n+1$;
9.         **End While**
10.        **If** $z_n \leq N_n^{app}$
11.           **If** $s_{n,z_n,t}==1$
12.              **If** the $z_n$th appliance is an HVAC
13.                 Measure the current indoor temperature $u_{n,t}^{in}$;
14.                 **If** $(u_{n,t}^{in} \leq u^{up,ls})$
15.                    Set $s_{n,z_n,t}=0$;
16.                 **Else**
17.                    Set $s_{n,k,t}=0$;
18.                       Set $c_{n,k}=c_{n,k}+1$;
19.                 **End If**
20.                 Update $P_t$;
21.                 **If** $P_t \leq l_t$
22.                    **Terminate.**
23.                 **End If**
24.              **End If**
25.           **End If**
26.        **End If**
27.     **End For**
28. **End While**
**End**

At each time slot, the load shedding logics are included in *SubRoutine_3* (Algorithm 12.6). For each building, the BEMS starts by initializing the index of the appliance to be shed as zero (Lines 1–3). The BEMS then alternatively sheds one appliance at a time for each building, based on the appliances' priorities (Lines 4–27). For each building, the BEMS starts by determining the subsequent operating appliance that needs to be listed in the appliance list that is sorted in the descending order of the appliances' priorities (Lines 5–9). If the appliance is an HVAC, then it checks if it can be switched into the standby state based on the indoor temperature prediction (Lines 12–15). If the appliance is not an HVAC, it pauses the appliance and increments the pause time count flag $c_{n,k}$ by 1 (Lines 16–19). The BEMS then updates the precinct's total power consumption and checks if further load shedding is needed at the current time slot. If not, the sub-routine terminates (Lines 20–23). Alternatively, the BEMS proceeds to shed 1 appliance in the subsequent building. Such alternative load shedding for the $N$ buildings continues until the precinct's power consumption threshold is satisfied.

### 12.4.3 Application example

In this section, we present an application example to highlight the use of the algorithms introduced in the previous section. Let us assume to have a precinct formed by three buildings that act as customers and receive the load shedding instructions from the utility. The three buildings are indexed as buildings 1, 2, and 3. The BEMS in each building continuously monitors the occupants' appliance use and real-time forecasts the appliances' use in the subsequent time slots. It is assumed that the desired appliance use scenarios of the occupants in the three buildings in the subsequent period, here taken as being between 1 pm and 1:40 pm, are summarized in Tables 12.5–12.7 where each column represents a 5-min time slot. In the period of interest, there are 8 time slots, indexed from $t_1$ to $t_8$.

In Tables 12.5–12.7, the cells of the appliances that do not operate during a particular time slot are shaded in gray. The numbers in the brackets "{}" next to the appliances' names indicate the priority of the appliances set by the occupants (where 1 depicts the highest priority and, for larger numbers, the priority decreases). In the tables, the operation of the HVAC is driven by the cycling operating algorithm presented in Chapter 10, with the outdoor ambient temperature constantly set as 33 °C (just to demonstrate the

**Table 12.5** The 40-min desired appliance use scenario of Building 1 for the application example.

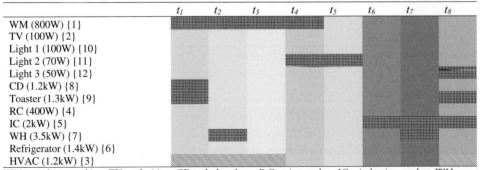

WM = washing machine, TV = television, CD = clothes dryer, RC = rice cooker, IC = induction cooker, WH = water heater, VCR = vacuum cleaner robot, PP = pool pump, AMP = amplifier, EI = electric iron.

**Table 12.6** The 40-min desired appliance use scenario of Building 2 for the application example.

| | $t_1$ | $t_2$ | $t_3$ | $t_4$ | $t_5$ | $t_6$ | $t_7$ | $t_8$ |
|---|---|---|---|---|---|---|---|---|
| WM (900W) {5} | | | | | | | | |
| TV (120W) {11} | | | | | | | | |
| Light 1 (80W) {6} | | | | | | | | |
| Light 2 (120W) {13} | | | | | | | | |
| VCR charger (700W) {7} | | | | | | | | |
| CD (1.0kW) {12} | | | | | | | | |
| RC (300W) {4} | | | | | | | | |
| IC (1.5kW) {9} | | | | | | | | |
| WH (2kW) {10} | | | | | | | | |
| Fan (90W) {14} | | | | | | | | |
| PP (750W) {8} | | | | | | | | |
| AMP (600W) {2} | | | | | | | | |
| Refrigerator (2kW) {3} | | | | | | | | |
| HVAC (1kW) {1} | | | | | | | | |

**Table 12.7** The 40-min desired appliance use scenario of Building 3 for the application example.

| | $t_1$ | $t_2$ | $t_3$ | $t_4$ | $t_5$ | $t_6$ | $t_7$ | $t_8$ |
|---|---|---|---|---|---|---|---|---|
| CM (700W) {9} | | | | | | | | |
| TV (110W) {4} | | | | | | | | |
| Light 1 (85W) {7} | | | | | | | | |
| Light 2 (100W) {8} | | | | | | | | |
| Light 3 (70W) {12} | | | | | | | | |
| CD (2.2kW) {10} | | | | | | | | |
| IC (1.7kW) {2} | | | | | | | | |
| WH (3.3kW) {5} | | | | | | | | |
| EI (2.2kW) {11} | | | | | | | | |
| AMP (900W) {6} | | | | | | | | |
| Refrigerator (1.6kW) {1} | | | | | | | | |
| AC (2kW) {3} | | | | | | | | |

procedure) and the other thermal parameters (presented in Chapter 10) are specified in Table 12.8. Fig. 12.6 shows the desired power consumption profile of the 40 min and Fig. 12.7 illustrates the precinct's desired power consumption profile by summing up the power of the appliances included in Tables 12.5–12.7.

We now apply the emergency load shedding scheme of Algorithms 12.3–12.6. Tables 12.9–12.11 show the appliance operation results of the three buildings after performing emergency load shedding. A total of 0.180 kWh, 0.47 kWh, and 1.73 kWh energy consumption values are shed for the three buildings. Fig. 12.8 shows the comparison between the profiles of the total desired power consumption and the actual power consumption of the precinct. It can be seen that, through load shedding, the precinct's power consumption is kept below the required threshold, that is, 30 kW. By determining the appliance shedding order based on the occupant-specified priorities (i.e., the appliances with low priorities are shed first, and the ones with high priorities are not necessarily shed), the scheme can significantly reduce the discomfort caused to the occupants.

**Table 12.8** Parameter settings for the HVAC's operation for the application example.

| Parameter | Building 1 | Building 2 | Building 3 |
|---|---|---|---|
| $A^{floor}$ | $80\,m^2$ | $50\,m^2$ | $100\,m^2$ |
| $A^{fen}$ | $10\,m^2$ | $10\,m^2$ | $8\,m^2$ |
| $A^{surf}$ | $400\,m^2$ | $300\,m^2$ | $500\,m^2$ |
| $V$ | $320\,m^3$ | $125\,m^3$ | $300\,m^3$ |
| $C^{air}$ | $1.005\,kJ/kg\,°C$ | $1.005\,kJ/kg\,°C$ | $1.005\,kJ/kg\,°C$ |
| $\rho^{air}$ | $1.205\,kg/m^3$ | $1.205\,kg/m^3$ | $1.205\,kg/m^3$ |
| $P^{pp}$ | $1500\,W$ | $800\,W$ | $1500\,W$ |
| $P^{std}$ | $0\,W$ | $0\,W$ | $0\,W$ |
| $\varepsilon$ | $2.8\,W/m^2$ | $3.2\,W/m^2$ | $3.6\,W/m^2$ |
| $I_t$ | $200\,W/m^2$ | $200\,W/m^2$ | $250\,W/m^2$ |
| $U$ | $2.4\,W/m^2\,°C$ | $3.8\,W/m^2\,°C$ | $4.5\,W/m^2\,°C$ |
| $ACH$ | $0.1$ | $0.1$ | $0.1$ |
| $SHGC$ | $40\%$ | $50\%$ | $38\%$ |
| $\theta$ | $0.4$ | $0.04$ | $0.04$ |
| $\delta$ | $3.91$ | $3.91$ | $3.91$ |
| $u_t^{ex}$ | $32°C$ | $32°C$ | $32°C$ |
| $u^{init}$ | $21°C$ | $20°C$ | $20.5°C$ |
| $[u^{low},u^{up}]$ | $[20°C, 22°C]$ | $[19°C, 21°C]$ | $[19°C, 22°C]$ |
| $[u^{low,ls},u^{up,ls}]$ | $[21°C, 24°C]$ | $[21°C, 24°C]$ | $[21°C, 23°C]$ |

$[u^{low},u^{up}]$ = comfort temperature band under normal conditions (i.e., the cycling operating algorithms for the HVAC, presented in Chapter 10); $[u^{low,ls},u^{up,ls}]$ = comfort indoor temperature band in the emergency load shedding period; $u^{init}$ = initial indoor temperature.

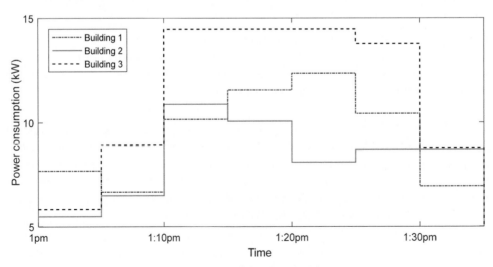

**Fig. 12.6** Desired power consumption profiles of the three buildings.

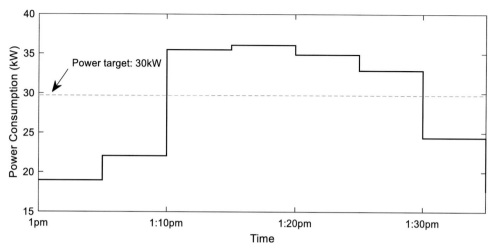

**Fig. 12.7** Desired power consumption profile of the precinct.

**Table 12.9** Appliance operation results for Building 1 after emergency load shedding.

| | $t_1$ | $t_2$ | $t_3$ | $t_4$ | $t_5$ | $t_6$ | $t_7$ | $t_8$ |
|---|---|---|---|---|---|---|---|---|
| WM (800W) {1} | | | | | | | | |
| TV (100W) {2} | | | | | | | | |
| Light 1 (100W) {10} | | | | | | | | |
| Light 2 (70W) {11} | | | | | | | | |
| Light 3 (50W) {12} | | | | | | | | |
| CD (1.2kW) {8} | | | | | | | | |
| Toaster (1.3kW) {9} | | | | | | | | |
| RC (400W) {4} | | | | | | | | |
| IC (2kW) {5} | | | | | | | | |
| WH (3.5kW) {7} | | | | | | | | |
| Refrigerator (1.4kW) {6} | | | | | | | | |
| HVAC (1.2kW) {3} | | | | | | | | |

**Table 12.10** Appliance operation results for Building 2 after emergency load shedding.

| | $t_1$ | $t_2$ | $t_3$ | $t_4$ | $t_5$ | $t_6$ | $t_7$ | $t_8$ |
|---|---|---|---|---|---|---|---|---|
| WM (900W) {5} | | | | | | | | |
| TV (120W) {11} | | | | | | | | |
| Light 1 (80W) {6} | | | | | | | | |
| Light 2 (120W) {13} | | | | | | | | |
| VCR charger (700W) {7} | | | | | | | | |
| CD (1.0kW) {12} | | | | | | | | |
| RC (300W) {4} | | | | | | | | |
| IC (1.5kW) {9} | | | | | | | | |
| WH (2kW) {10} | | | | | | | | |
| Fan (90W) {14} | | | | | | | | |
| PP (750W) {8} | | | | | | | | |
| AMP (600W) {2} | | | | | | | | |
| Refrigerator (2kW) {3} | | | | | | | | |
| HVAC (1kW) {1} | | | | | | | | |

**Table 12.11** Appliance operation results for Building 3 after emergency load shedding.

| | $t_1$ | $t_2$ | $t_3$ | $t_4$ | $t_5$ | $t_6$ | $t_7$ | $t_8$ |
|---|---|---|---|---|---|---|---|---|
| CM (700W) {9} | | | | | | | | |
| TV (110W) {4} | | | | | | | | |
| Light 1 (85W) {7} | | | | | | | | |
| Light 2 (100W) {8} | | | | | | | | |
| Light 3 (70W) {12} | | | | | | | | |
| CD (2.2kW) {10} | | | | | | | | |
| IC (1.7kW) {2} | | | | | | | | |
| WH (3.3kW) {5} | | | | | | | | |
| EI (2.2kW) {11} | | | | | | | | |
| AMP (900W) {6} | | | | | | | | |
| Refrigerator (1.6kW) {1} | | | | | | | | |
| HVAC (2kW) {3} | | | | | | | | |

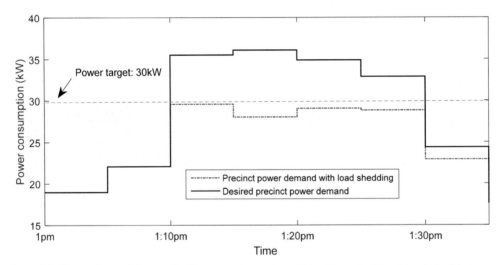

**Fig. 12.8** Comparison of the precinct's power demand profiles with and without load shedding.

# References

[1] Z. Ma, A. Clausen, Y. Lin, B.N. Jorgensen, An overview of digitalization for the building-to-grid eco-system, in: Proceedings of Energy Informatics Academy Conference Asia, 2021.

[2] A.F. Taha, N. Gatsis, B. Dong, A. Pipri, Z. Li, Building-to-grid integration framework, IEEE Trans. Smart Grid 10 (2) (2019) 1237–1249.

[3] M. Razmara, G.R. Bharati, D. Hanover, M. Shahbakhti, S. Paudyal, R.D. Robinett, Enabling demand response programs via predictive control of building-to-grid systems integrated with PV panels and energy storage systems, in: Proceedings of 2017 American Control Conference (ACC), Seattle, USA, May 2017.

[4] R. Tatro, S. Vadhva, P. Kaur, N. Shahpatel, J. Dixon, K. Alaznoon, Building to grid (B2G) at the California smart grid center, in: Proceedings of 2010 IEEE International Conference on Information Reuse & Integration, Las Vegas, NV, USA, Aug. 2010.

[5] M. Uddin, M.F. Romlie, M.F. Abdullah, S.A. Halim, A.H.A. Bakar, T.C. Kwang, A review on peak load shaving strategies, Renew. Sustain. Energy Rev. 82 (2018) 3323–3332.

[6] Peak-to-Average Ratio, 2019. Encyclopedia.com. https://www.encyclopedia.com/computing/dictionaries-thesauruses-pictures-and-press-releases/peak-average-ratio (Accessed 08 March 2022).

[7] P. Pinceti, Emergency load-shedding algorithm for large industrial plants, Control. Eng. Pract. 10 (2) (2002) 175–181.

[8] Y. Dong, X. Xie, K. Wang, B. Zhou, Q. Jiang, An emergency-demand-response based under speed load shedding scheme to improve short-term voltage stability, IEEE Trans. Power Syst. 32 (5) (2017) 3726–3735.

# CHAPTER 13

# Buildings and microgrids

## Contents

## 13.1 Introduction

Over the past decades, power grids have been operating following the structure depicted in Fig. 4.3: the electricity is generated from the central power plants and is transmitted over long distances to load areas where the local substation distributes the electricity to buildings and other energy loads. Such a vertical operation structure is currently being modified with the wider acceptance and penetration of distributed renewable energy sources where the electricity does not need to be generated and transmitted from the large power plants and it can be produced from distributed renewable energy sources placed in the load area. As a result of this trend, the concept of *microgrids* [1,2] has gained extensive popularity. For illustrative purposes, a schematic describing the main aspects of a microgrid is presented in Fig. 13.1. The use of the term "micro" is to distinguish a microgrid from the broad grid that is also denoted as the macro grid in this book. A microgrid aggregates and manages a variety of energy resources in a certain geographical area. The energy resources include both energy generation sources and energy loads.

This chapter provides an overview of the relationship between modern buildings and microgrids. In particular, in the first part of the chapter, we introduce a couple of available definitions of smart grids and present the concept of microgrids together with their benefits and possible operation modes. We then discuss how buildings can operate and integrate within a microgrid system and highlight the benefits of such arrangements. In the final part of the chapter, we consider one of the main advantages associated with

*Building Energy Management Systems and Techniques*
https://doi.org/10.1016/B978-0-323-96107-3.00011-4

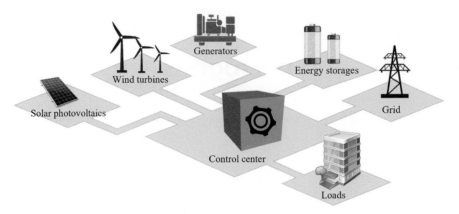

**Fig. 13.1** Conceptual schematic of a microgrid.

microgrids which relies on their ability to deal with power outages. A possible strategy that can be implemented in such scenarios is presented to highlight the possible roles played by the buildings and the microgrid. A simple application example is included to highlight some of the key features of the solution. The proposed approach could also be further expanded by adopting more sophisticated algorithms and strategies.

## 13.2 Introduction to microgrids

### 13.2.1 Overview

There are several definitions available for the concept of microgrid and we briefly introduce a couple of selected ones in the following. The US Department of Energy Microgrid Exchange Group defines a microgrid as "a group of interconnected loads and Distributed Energy Resources (DERs) within clearly defined electrical boundaries that acts as a single controllable entity with respect to the grid. A microgrid can connect and disconnect from the grid to enable it to operate in both connected or island-mode." [3] A definition produced within an EU Research Project depicts a microgrid as "comprising Low-Voltage (LV) distribution systems with DERs (microturbines, fuel cells, Photovoltaics (PV), etc.), storage devices (batteries, flywheels), energy storage system, and flexible loads. Such systems can operate either connected or disconnected from the main grid. The operation of microsources in the network can provide benefits to the overall system performance, if managed and coordinated efficiently." [4].

These definitions highlight several common features of microgrids that can be summarized as follows:

- *Distributed*: Energy resources in a microgrid are usually located in different locations rather than placed in a central plant. The distributed energy resources are physically connected by power connection lines and other power electronic devices.

— *Autonomous and self-organized*: Microgrids are essentially autonomous energy systems. Each microgrid can have its local operation policies and can be equipped with energy management systems that can monitor and control its on-site energy resources.
— *Self-energy supply capability*: A microgrid is capable of self-producing energy through its local energy generation resources. These energy generation resources usually include distributed renewable energy sources, small-capacity fossil-fueled power generators, and energy storage devices. For example, the lighting systems in streets in a microgrid could be powered by the wind power generated by the nearby urban wind turbines.

The concept of microgrid emerged because of the widespread deployment of distributed renewable energy sources (typically distributed wind turbines and solar panels). Traditionally, buildings produce an energy load that is covered by external energy systems (i.e., macro power grids or gas networks) and do not possess on-site energy generation sources. With such an arrangement, energy is generated from the energy plants that are usually distant from the energy load centers (e.g., urban areas) and it is delivered to the loads (e.g., buildings) by the energy transmission infrastructure. With the availability of distributed renewable energy sources, energy loads can be directly served by the renewable energy sources deployed on-site or nearby. Such a self-energy supply pattern can reduce the amount of energy that is transmitted from remote energy plants and, therefore, reduce the energy losses due to long-distance transmission. Based on these considerations, it can be noted how the characteristics of a microgrid can be regarded as equivalent to those of a power grid (that comprises energy generation sources and energy loads and is equipped with electrical and mechanical devices, e.g., transformers, power feeders, and power breakers, to aggregate them together) that operates at a smaller scale, therefore the inclusion of the term "micro" to distinguish microgrids from power grids.

## 13.2.2 Operation modes and benefits

Depending on whether a microgrid is connected to the external macro grid or not, a microgrid can operate in two main modes:
— *Grid-connected operation mode*: In this mode, the microgrid is connected to the macro grid and can have bi-directional power exchanges with it. The microgrid can import energy from the macro grid when necessary as well as export energy to the macro grid; and
— *Off-grid operation mode* (also known as *islanded operation mode*): In this mode, the microgrid is decoupled from the macro grid and relies only on its internal energy resources to feed the load.

A microgrid can switch between these two operation modes over time.

Microgrids can be beneficial to an energy system (e.g., from the macro power grid to end users). Microgrids can improve the reliability and resilience of the grid. For example, when system faults are detected in an area, the connection line between the area and the macro grid can be tripped by protection schemes. As a result, the area is decoupled from the macro grid. This can isolate the fault and prevent the occurrence of cascading system failures. The isolated area can continue to operate as an islanded microgrid by using local energy production sources to serve the local demand. Microgrids can also support managing electricity supply and demand. By feeding locally generated energy to cover internal energy loads within a microgrid, the amount of energy imported from the grid can be reduced. This can avoid losses due to long-term power transmission and reduce the power reserve capacity and cost of the macro grid. Microgrids help to integrate renewable energy, creating real chances to cut greenhouse gas emissions and to reduce the dependency of the whole energy system on fossil fuels. End users can benefit from the local energy supply provided by microgrids in terms of reduced energy bills.

The operation of a microgrid is usually automated by a centralized Energy Management System (EMS) [5]. The microgrid EMS acts as the "brain" of the microgrid, e.g., it monitors the microgrid's operational condition and coordinates the internal energy generation sources to serve the buildings and other energy loads. In some instances, a building itself can be considered to act as a small microgrid. In such a situation, the microgrid EMS coincides with the Building Energy Management System (BEMS).

## 13.3 Buildings in a microgrid

### 13.3.1 Overview

Buildings can be regarded as important components of a microgrid. In most situations, buildings act as the main energy consumer in a microgrid system and the internal energy generation sources within the microgrid generate energy to serve the buildings' energy demand.

Rather than passively taking energy from the generation sources, buildings in modern microgrid systems can also actively communicate with other entities in the microgrid through advanced information and communication facilities and can participate in the microgrid's operation [6]. To improve the overall efficiency of a microgrid, it is important to understand the factors that affect the buildings' energy consumption so that the microgrid can make economic generation scheduling (i.e., to minimize the energy generation cost while meeting the energy demand) on the energy generation sources and the building managers can make energy saving plans for the buildings.

To facilitate the efficient operation of microgrids, accurate energy metering in buildings at multi-scales (i.e., from microlevels to macro levels) need to be implemented, so that the microgrid can precisely understand the buildings' energy demand and optimally allocate energy generation among different energy generation sources.

At the macro level, the energy consumption of the entire building needs to be accurately monitored. This can be achieved by deploying the Advanced Metering Infrastructure (AMI) system backboned by smart meters. An understanding of the entire building's energy consumption can assist the microgrid designer in performing capacity planning for the energy generation sources. For the microgrid, it can help the microgrid operation to generate scheduling plans for the microgrid's energy generation sources.

At the microlevel, finer-grained energy measurements are desired at individual plug-in appliances and devices in buildings. These measurements can help the building manager and microgrid operator to understand the occupant's energy use behaviors and lifestyles. Based on this, they can prepare internal energy management plans to achieve energy saving/shifting for the building. Currently, there are two approaches for individual appliance-level energy measuring. One approach is to install submeters or IoT sensors at the individual appliances. This approach can achieve accurate energy measuring, but it is costly as the meters/sensors and the associated infrastructure are usually not cheap. Many meters/sensors would be required to achieve a significant coverage of an entire building. Another approach is to rely on Non-Intrusive Appliance Load Monitoring (NILM) technologies [7,8] that use specific techniques (e.g., machine learning techniques) to disaggregate individual appliances' power consumption profiles from the entire building's power consumption profile. Using NILM, the individual appliances' usage footprints can be identified with low cost because it does not require installing submeters or sensors on individual appliances. The accuracy of NILM in identifying appliance usage profiles needs to be carefully evaluated.

Sophisticated energy management schemes would be developed to coordinately manage the buildings' energy resources with other energy resources in the microgrid, to maximize the overall efficiency of the microgrid [9–14]. To achieve this, the data recorded by the meters/sensors installed on the buildings and other resources can be sent to the microgrid EMS which analyzes the data to determine the microgrid's operational condition. The microgrid EMS can then formulate the global energy management operations to manage the demand response of the buildings and to schedule the plans for the other energy resources in the microgrid.

## 13.3.2 Case study: The UCSD campus microgrid

In the last few years, microgrid projects have been set up around the world. An example of this is the microgrid in the campus of The University of California (UCSD) [15]. The microgrid system has been implemented to support the carbon-neutral campus initiative. The microgrid uses various on-site energy generation sources to supply electricity, heating, and cooling for the buildings in the campus. The UCSD microgrid provides flexible and reliable power distribution service to the campus. It generates more than 85% of the electricity used on campus annually. The microgrid generates power from multiple sources, including a 30-MW natural-gas-fired combined heat and power generation plant, a 2.8-MW renewable fuel cell that generates combustion-free electricity from waste methane gas, and 2.4-MW solar arrays.

The UCSD microgrid is equipped with a computerized EMS. It communicates with all major buildings in the campus and centrally monitors and controls the heating, ventilation, and air conditioning (HVAC) systems in the buildings. The EMS is programmed to achieve energy use reduction in evenings, weekends, and holidays. The EMS also assists the buildings in reducing power consumption in peak hours, maximizes their energy conservation, and performs efficient room temperature management.

### 13.3.3 Energy management of microgrids with power flow analysis

Energy management strategies can be designed for managing the operations of energy resources within a microgrid. These manageable energy resources could include distributed power generators, energy storage systems, electric vehicles, and controllable appliances and devices that are placed in the buildings within the microgrid. Peer-to-peer (P2P) energy trading principles can also be applied to enable energy trading among multiple autonomous microgrid systems. For example, Ref. [9] designs a P2P energy trading system for microgrids. In the system, the EMS of each microgrid schedules its internal energy resources to determine the amount of energy it needs to buy or it intends to sell. Based on this, the EMSs of the microgrids then communicate with each other to negotiate the energy trading amount and price.

When performing energy management for a single building (e.g., the methodologies in Chapters 8–10), the BEMS manages the operation of the energy resources in the same geographical place (i.e., within the building), and there are no power exchanges between the building and other energy entities (except for the grid). In a microgrid, the energy resources and energy loads could be located in different places and are connected by electrical devices (e.g., electrical buses and connection lines), forming a small power network. In this situation, the power flow in the network needs to be considered in the microgrid energy management process. Power flow refers to the distribution of electric power in an interconnected system—this includes the direction and amount of power transmitted in each power transmission line and the magnitude of the voltage in the buses. Power flow represents the operating state of the electric system at a specific time point. Given the power output of each power generation source and the power demand of each energy load in the system, numerical analysis is usually performed to determine how the power flow will be distributed in the system. This is called the "power flow analysis" (also known as the "load flow analysis"). The commonly used numerical methods for power flow analysis include the Newton-Rapson method and the Gauss–Seidel method. More details about power flow analysis can be found in dedicated Refs. [16, 17].

When scheduling the energy resources in a microgrid, power flow analysis usually needs to be included. This is because the microgrid's operation must satisfy the security requirements of the underlying electrical network of the microgrid. For example, when scheduling the power output of the energy generation sources and power consumption of the loads, it must be ensured that the transmitted power on each line in the microgrid remains below the line's maximum transmitted power capacity and the voltage magnitude is kept within a secure value range. If the power flow analysis shows that any of these security criteria are violated, the energy resources need to be rescheduled. Energy management strategies based on power flow analysis and applicable to microgrids can be found in Refs. [18–20].

## 13.4 Operating a building as an off-grid microgrid during power outages

### 13.4.1 Overview

Buildings are capable of generating and storing energy for themselves when equipped with renewable energy sources and energy storage devices. In such a scenario, a building could be managed with the same strategies considered for microgrids. In some Refs. [21,22], such a scenario is referred to as "nano grid." While a microgrid could consist of multiple buildings and other energy resources located in a certain geographical area, the term "nano grid" here is used to depict a very small microgrid formed by a single building. This representation describes the fact that a building acts as a tiny power grid system that consists of a variety of power generation and consumption resources, and other relevant electrical components.

As a microgrid, a building can operate in the grid-connected mode previously introduced in Section 13.2.2. A building equipped with energy generation/storage devices can also operate as an off-grid microgrid. In this situation, the building's energy demand is served by its on-site energy sources and energy storage devices. This scenario is commonly seen in grid power outage events. Traditionally, power outage events lead to blackouts, causing significant disturbances to people due to interruption of power supply. By integrating power generation and storage devices, the building can be capable of performing self-energy supply to minimize the impact of power outage events on the building occupants.

Power outage events can be classified into two types: (i) unplanned outages and (ii) planned outages. Unplanned outages are usually caused by unexpected accidents and disturbances. For example, a falling branch would trip down a power transmission line and cause blackouts in a certain area. Climate change has led to more and more extreme weather days that might lead, for example, to significant cooling or heating demands. In parallel, the growing energy demand of urban areas is increasing the stress induced on energy grids. These factors tend to cause more frequent power outage events. For example, statistical data show that the average number of blackout events in the USA doubles every five years [23]. Blackouts significantly affect people's production and living activities. A planned outage is scheduled by the power utility due to certain reasons, such as device maintenance and network load management. In planned outage events, end users are usually notified of the starting time and duration information of the electricity interruption from the utility in advance. Planned outages are not uncommon. For example, between 2015 and 2016, there have been roughly 1700 planned power outages across Australia due to equipment maintenance [24].

Energy storage systems play an important role in operating a building as a microgrid. They can be used to absorb the excessive renewable energy generated from the building's on-site renewable energy sources and discharge it for the building's later use. In this way, the dependency of the building on the external macro grid can be reduced. In particular, energy storage systems can provide energy backup support to operate a building as an off-grid microgrid and to reduce the disturbance of power outage events to the building occupants.

In addition to conventional Battery Energy Storage Systems (BESSs) that are being widely used in buildings, Electric Vehicles (EVs) have been gaining a growing popularity and can also act as BESSs in the periods when they are plugged into the building. Such an arrangement is usually referred to as *Vehicle-to-Home* (*V2H*) or *Vehicle-to-Building* (*V2B*) integration [25]. With V2H/V2B integration, EVs can also provide energy backup services to buildings. This is particularly useful in power outage events when EVs can facilitate buildings to operate as off-grid microgrids. Extensive research and development efforts [26–28] have been devoted to demonstrating the effectiveness of V2H/V2B integration in operating residential buildings as off-grid microgrids. In this way, the operations of critical appliances can be given priority in an outage event. If the EV's battery has sufficient capacity or there are multiple EVs available, all energy demands of the building could be served without interruptions to occupants.

## 13.4.2 Application example

In this section, we present an application example to demonstrate the scenario of using a BESS (it can also be a plugged-in EV) to accommodate the rooftop PV solar power and to serve a building's energy demand over a planned outage. To minimize the complexity of the problem, we consider the building as a residential house. In a planned outage, the utility notifies the user of the start and end time of the outage event. Based on the known information, the user can define energy usage plans to be implemented during the outage period.

We consider that the building occupant intends to run $N$ household appliances during the outage period. The appliances are assigned with different priorities by the occupant, here represented by positive integers, i.e., 1 represents the highest priority (the larger the integer is, the lower the priority). We use a binary variable $s_{a,t}$ to represent the state of an arbitrary appliance $a$ at the $t$th time slot: 1-on; 0-off. We also use $P_a^{app}$ to denote the rated power of appliance $a$ (in kW). The properties defining the model representation of a

**Table 13.1** Properties defining the model representation of a BESS and variables representing its operational conditions.

| Notation | Definition |
|---|---|
| $P^{bess,rc}$ | Charging power capacity (in kW) |
| $P^{bess,rd}$ | Discharging power capacity (in kW) |
| $\eta^c$ | Charging efficiency (in %) |
| $\eta^d$ | Discharging efficiency (in %) |
| $E^{bess,r}$ | Energy capacity of the BESS (in kWh) |
| $SOC^{lower}$ | Lower allowable State-of-Charge (SOC) limit (in %) |
| $SOC^{upper}$ | Upper allowable SOC limit (in %) |
| $P_t^{avlb}$ | Available power (i.e., maximum dischargeable power) of the BESS in time slot $t$ (in kW) |
| $SOC_t$ | SOC of the BESS in time slot $t$ (in %) |
| $P_t^{bess}$ | Charging/discharging power of the BESS in the time slot (in kW) |

BESS and the variables representing the BESS's operational condition are shown in Table 13.1. Following the sign convention used in Chapter 8, we assume that the positive and negative values of $P_t^{bess}$ indicate the charging and discharging power of the BESS, respectively.

Based on the appliances' power consumptions and priorities and the rooftop solar panel's power output, the BEMS schedules the BESS' charging/discharging to serve the household power demand at each time slot. The energy management logic is included in Algorithm 13.1.

In Algorithm 13.1, the function min(·) returns the minimum value of the inputted numerical values. The BESS is controlled sequentially in each time slot. In each time interval, the BEMS monitors the current solar power output (Line 4). It then retrieves the set of appliances the occupant intends to run at the current time slot and sorts them in the order of highest to lowest priorities (Lines 5 and 6). The BEMS tries to use the BESS to serve the appliances according to their priorities. For each of the sorted appliances, the BEMS checks if the current available power (solar power output plus the BESS's maximum dischargeable power) is adequate to serve the appliance. If this is the case, then the appliance is served or, otherwise, the appliance is not served (Lines 10–16). If the current solar power output is adequate for serving all the required appliances, then the surplus solar power is charged into the BESS for storage (Lines 17 and 18) or, otherwise, the BESS is discharged (Lines 19 and 20). After processing the appliances, the BESS' state-of-charge is updated (Line 22). Finally, the BEMS checks if the outage event is over. If this is the case, it outputs the scheduling decisions or, otherwise, it moves to the next time interval and it repeats the previous energy management logic (Lines 24–28).

---

**ALGORITHM 13.1 Pseudocode for the operations of a BEMS with a household BESS and controllable appliances during a power outage event.**

**Start**
1.  Input the rated power and priority of each appliance;
2.  Input the duration of each time slot $\Delta t$ (hour);
3.  Set time index $t=1$;
4.  Measure the solar power output $P_t^{pv}$;
5.  Get the set of the appliances that need to be served at the $t$th time slot;
6.  Sort the appliances according to ascending order of the priority number.
7.  Set $P_t^{avlb} = P_t^{pv} + min\left(\dfrac{E^{bess,r} \times \left(SOC_t - SOC^{lower}\right) \times \eta^d}{\Delta t}, P^{bess,rd}\right)$;
8.  Set the power consumed by the appliances as $P_t^{com} = 0$;
9.  **For** each sorted appliance $a$
10.    **If** $P_t^{avlb} \geq P_a^{app}$
11.       Set $P_t^{avlb} = P_t^{avlb} - P_a^{app}$;
12.       Set $P_t^{com} = P_t^{com} + P_a^{app}$;
13.       Set $s_{a,t} = 1$;
14.    **Else**
15.       Set $s_{a,t} = 0$;
16.    **End If**
17.    **If** $P_t^{pv} \geq P_t^{com}$
18.       Set $P_t^{bess} = min\left(\left(P_t^{pv} - P_t^{com}\right), P^{bess,rc}, \dfrac{E^{bess,r} \times \left(SOC_t - SOC^{lower}\right)}{\Delta t \times \eta^c}\right)$;
19.    **Else**
20.       Set $P_t^{bess} = -1 \times min\left(\left(P_t^{com} - P_t^{pv}\right), P^{bess,rd}, \dfrac{E^{bess,r} \times \left(SOC^{upper} - SOC_t\right) \times \eta^d}{\Delta t}\right)$;
21.    **End If**
22.    Update the SOC of the BESS using Eqs. (8.2) and (8.3) in Chapter 8;
23. **End For**
24. **If** the outage finishes?
25.    **Terminate**;
26. **Else**
27.    Proceed to the next time slot and set $t=t+1$;
28.    **Go to** Line 4;
29. **End**
**End**

---

We now consider simulation scenarios to implement the energy management system design of Algorithm 13.1. It is assumed that the outage event starts at 3 pm and ends at 6 pm, lasting for 3 h. The control time interval is set to be 10 min, i.e., 1/6 h. Therefore, there are a total of 18 control time slots, indexed as $t_1$, $t_2$, ..., $t_{18}$. The model parameters of the BESS in the building are shown in Table 13.2. The 3-h PV solar power output data is given in Table 13.3.

**Table 13.2** Parameter settings of the BESS.

| Parameter | Value |
|---|---|
| $P^{bess,rc}$ | 4 kW |
| $P^{bess,rd}$ | 4 kW |
| $\eta^c$ | 100% |
| $\eta^d$ | 100% |
| $E^{bess,r}$ | 12 kWh |
| $SOC^{lower}$ | 0.1 |
| $SOC^{upper}$ | 0.9 |
| $SOC^{init}$ | 0.4 |

**Table 13.3** 3-Hour PV solar power output data.

| | $t_1$ | $t_2$ | $t_3$ | $t_4$ | $t_5$ | $t_6$ | $t_7$ | $t_8$ | $t_9$ |
|---|---|---|---|---|---|---|---|---|---|
| Power output (kW) | 1.9 | 2.4 | 3.0 | 1.6 | 1.8 | 2.8 | 2.4 | 1.1 | 0.6 |

| | $t_{10}$ | $t_{11}$ | $t_{12}$ | $t_{13}$ | $t_{14}$ | $t_{15}$ | $t_{16}$ | $t_{17}$ | $t_{18}$ |
|---|---|---|---|---|---|---|---|---|---|
| Power output (kW) | 1.4 | 1.2 | 0.2 | 0 | 0 | 0 | 0 | 0 | 0 |

**Table 13.4** Use time and priorities of the controllable appliances.

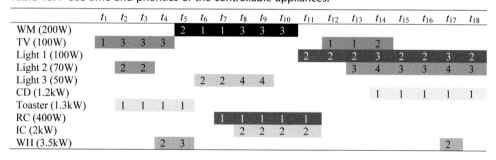

Note: "WM" is washing machine; "TV" is television; "CD" is clothes dryer; "RC" is rice cooker; "IC" is induction cooker; "WH" is water heater.

The controllable appliances that are intended to be used by the occupant over the outage period are shown in Table 13.4. The usage time of each appliance is highlighted by a color, and the number in the square indicates the priority number of the appliance in the time slot.

By running Algorithm 13.1, the operation decisions for the appliances are determined as shown in Table 13.5. Each gray, grid-styled square means that the appliance is not served at that time slot. Fig. 13.2 shows the charging/discharging power profile and the corresponding SOC profile of the BESS. It can be seen that almost all the appliances are well-served by the PV solar power and the BESS. Only the water heater, which is a

**Table 13.5** Operation decisions for the appliances.

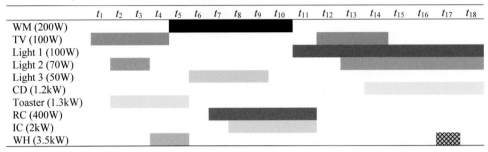

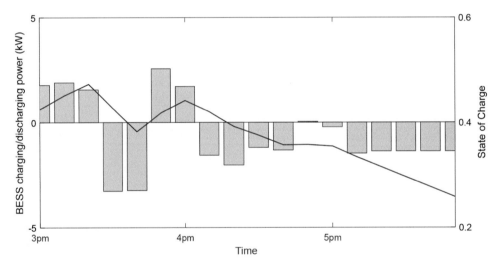

**Fig. 13.2** Charging/discharging profile and SOC profile of the BESS (positive—charging; negative—discharging).

high power consumption appliance (3.5 kW) is not served in the time period between 5:40 and 5:50 pm (the 17th time slot). In this case, the occupant's energy demand is 5.86 kWh, with only 5.28 kWh being served.

In addition to the amount of power generated by the building's on-site renewable energy sources, the building microgrid's operation significantly depends on the size of the BESS and the BESS' initial SOC at the beginning of the outage period. Table 13.6 shows the served energy over the outage period under different size settings of the BESS and the PV solar power output data given in Table 13.3. The initial SOC is set to be 40% for all settings. It can be seen that when the power capacity and energy capacity of the BESS reach 5 kW and 20 kWh, respectively, the energy demand is nearly fully served. The use of a larger-sized BESS would lead to a higher purchase cost. Table 13.7 shows the operation decisions for the appliances when $P^{bess,\,rc} = P^{bess,\,rd} = 2\,\text{kW}$

**Table 13.6** Served energy over the power outage period under different BESS sizes.

| | $P^{bess,rc}=P^{bess,\ rc}=2\,kW;$ $E^{bess,r}=6\,kWh$ | $P^{bess,rc}=P^{bess,\ rc}=4\,kW;$ $E^{bess,r}=12\,kWh$ | $P^{bess,rc}=P^{bess,\ rc}=5\,kW;$ $E^{bess,r}=20\,kWh$ |
|---|---|---|---|
| Total energy demand | 5.86 kWh | 5.86 kWh | 5.86 kWh |
| Served energy | 4.10 kWh | 5.28 kWh | 5.83 kWh |

**Table 13.7** Operation decisions of the appliances with $P^{bess,rc}=P^{bess,rc}=2\,kW$ and $E^{bess,r}=6\,kWh$.

**Table 13.8** Served energy over the power outage period under different initial SOC values.

| | $SOC^{init}=0.2$ | $SOC^{init}=0.4$ | $SOC^{init}=0.6$ |
|---|---|---|---|
| Total energy demand | 5.86 kWh | 5.86 kWh | 5.86 kWh |
| Served energy | 4.11 kWh | 5.83 kWh | 5.86 kWh |

and $E^{bess,re}=6\,kWh$. It can be seen that when the BESS' capacity is relatively small, the water heater is not served across the outage period and Light 3 is not served between 4:20 and 4:30 pm (the 9th time slot).

Table 13.8 shows the served energy over the outage period under different initial SOC values of a BESS with the size of $P^{bess,rc}=P^{bess,rd}=5\,kW$ and $E^{bess,re}=20\,kWh$. It can be seen that with larger initial SOC values, the BESS has more energy to serve the building. Under our scenario settings, when the initial SOC value reaches 0.6, the building microgrid can satisfy its own energy demand.

## References

[1] R.H. Lasseter, P. Paigi, Microgrid: a conceptual solution, in: Proceedings of 2004 IEEE 35th Annual Power Electronics, Aachen, Germany, June 2004.
[2] B.S. Hartono, Y. Budiyanto, R. Setiabudy, Review of microgrid technology, in: Proceedings of 2013 International Conference on QiR, Yogyakarta, Indonesia, June 2013.

[3] DOE Microgrid Worship Report, U.S. Department of Energy Office of Electricity Delivery and Energy Reliability, Chicago Illinois, USA, Jun. 30–31, 2012. [Online]. Available: https://www.energy.gov/sites/prod/files/2012%20Microgrid%20Workshop%20Report%2009102012.pdf (Accessed 11 July 2023).

[4] N.D. Hatziargyriou, G. Strbac, J.A.P. Lopes, Microgrids—Large scale integration of micro-generation to low voltage grids, in: Proceedings of CIGRE 2006, Paris, France, Aug. 2006.

[5] J. Jimeno, J. Anduaga, J. Oyarzabal, A.G. de Muro, A architecture of a microgrid energy management system, Eur. T. Electr. Power 21 (2) (2011) 1142–1158.

[6] Y. Agarwal, T. Weng, and R.K. Gupta, Understanding the role of buildings in a smart microgrid, in: Proceedings of 2011 Design, Automation & Test in Europe, 2011.

[7] W. Kong, Z.Y. Dong, D. Hill, J. Zhao, F. Luo, An extensible approach for non-intrusive load disaggregation with smart meter data, IEEE Trans. Smart Grid 9 (4) (2018) 3362–3372.

[8] W. Kong, Z.Y. Dong, D. Hill, F. Luo, Y. Xu, Improving nonintrusive load monitoring efficiency via a hybrid programming method, IEEE Trans. Industr. Inform. 12 (6) (2016) 2148–2157.

[9] F. Luo, Y. Chen, Z. Xu, G. Liang, Y. Zheng, J. Qiu, Multiagent-based cooperative control framework for microgrids' energy imbalance, IEEE Trans. Industr. Inform. 13 (3) (2017) 1046–1056.

[10] F. Luo, Z. Xu, K. Meng, Z.Y. Dong, Optimal operation scheduling for microgrid with high penetrations of solar power and thermostatically controlled loads, Sci. Technol. Built Environ. 22 (6) (2016) 666–673.

[11] I. Prodan, E. Zio, A model predictive control framework for reliable microgrid energy management, Int. J. Electr. Power Energy Syst. 61 (399–409) (2014).

[12] E. Kuznetsova, Y.F. Li, C. Ruiz, E. Zio, G. Ault, K. Bell, Reinforcement learning for microgrid energy management, Energy 59 (133–146) (2013).

[13] E. Kuznetsova, Y.F. Li, C. Ruiz, E. Zio, An integrated framework of agent-based modelling and modelling and robust optimization for microgrid energy management, Appl. Energy 129 (2014) 70–88.

[14] S. Leonori, A. Martino, F.M.F. Mascioli, A. Rizzi, Microgrid energy management systems design by computational intelligence techniques, Appl. Energy 277 (2020).

[15] B. Washom, J. Dilliot, D. Weil, J. Kleissl, N. Balac, W. Torre, C. Richter, Ivory tower of power: microgrid implementation at the university of California, San Diego, IEEE Power Energy Mag. 11 (4) (2013) 28–32.

[16] J.D. Glover, T.J. Overbye, M.S. Sarma, Power System Analysis and Design, sixth ed., Cengage Learning, Inc., USA, 2016.

[17] J.C. Das, Load Flow Optimization and Optimal Power Flow, CRC Press, 2022.

[18] Y. Zhang, H.J. Jia, L. Guo, Energy management strategy of islanded microgrid based on power flow control, in: Proceedings of 2012 IEEE PES Innovative Smart Grid Technologies, 2012.

[19] S. Chopra, G.M. Vanaprasad, G.D.A. Tinajero, N. Bazmohammadi, J.C. Vasquez, J.M. Guerrero, Power-flow-based energy management of hierarchically controlled islanded AC microgrids, Int. J. Electr. Power Energy Syst. 141 (2022) 108140.

[20] M. Sechilariu, F. Locment, Urban DC Microgrid: Intelligent Control and Power Flow Optimization, Butterworth-Heinemann, 2016.

[21] Y. Yerasimou, M. Kynigos, V. Efthymiou and G.E. Georghiou, Design of a smart nanogrid for increasing energy efficiency of buildings, Energies, vol. 14, no. 2, 3683, 2921.

[22] L.R. Jie, R.T. Naayagi, Nanogrid for energy aware buildings, in: Proceedings of 2019 IEEE PES GTD Grand International Conference and Exposition Asia, 2019.

[23] U.S. Energy Information Administration, n.d. Eia.gov. https://www.eia.gov/electricity, accessed on 28/07/2018 (Accessed 08 March 2022).

[24] Ausgrid Reported Power Outages. n.d. Ausgrid.com.au. https://www.ausgrid.com.au/poweroutages (Accessed 28 July 2018).

[25] D. Borge-Diez, D. Icaza, E. Acikkalp, H. Amaris, Combined vehicle to building (V2B) and vehicle to home (V2H) strategy to increase electric vehicle market share, Energy 237 (2021).

[26] D.P. Tttle, R.L. Fares, R. Baldick, M.E. Webber, Plug-in vehicle to home (V2H) duration and power output capability, in: Proceedings of IEEE Transportation Electrification Conference and Expo (ITEC), USA, Jun. 2013, pp. 1–7.

[27] R. Roche, F. Berthold, F. Gao, F. Wang, A model and strategy to improve smart home energy resilience during outages using vehicle-to-home, in: IEEE International Electric Vehicle Conference (IEVC), Florence, 2014, pp. 1–6.

[28] H.Y. Shin, B. Ross, Plug-in electric vehicle to home (V2H) operation under a grid outage, IEEE Trans. Smart Grid 8 (4) (2016) 2032–2041.

# CHAPTER 14

# Occupant-to-grid integration

## Contents

## 14.1 Introduction of occupant-to-grid integration

In previous chapters, we have discussed a wide range of strategies in which modern buildings can take advantage of the latest technologies to optimize their energy consumption and to exploit renewable energy sources. In this chapter, we consider how building occupants can be integrated into the process and how they can be engaged in supporting energy-efficient building management strategies. We refer to this as the Occupant-to-Grid (O2G) integration.

For an effective O2G implementation, it is important to establish suitable interaction methodologies that foster the engagement of building occupants to adjust their behavior for the promotion of energy-efficient buildings (Fig. 14.1). The concept of O2G represents a subgroup of the earlier concept of *consumer-to-grid* integration [1].

The energy management techniques presented in the previous chapters are mainly based on machine-to-machine communication. In these instances, the Building Energy Management System (BEMS) communicates with the building side energy resources and performs control actions on them, to optimize buildings' energy performance while

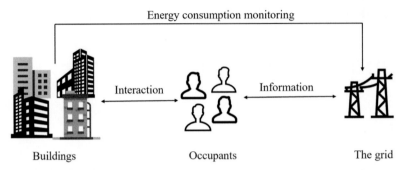

**Fig. 14.1** Illustration of occupant-to-grid integration.

satisfying the building occupants' requirements. The O2G integration involves machine-to-human communication and, in such scenarios, the BEMS communicates with the occupants to influence their behaviors that would directly or indirectly affect the energy efficiency of buildings. Such an influence can be achieved using different strategies, for example, by analyzing the occupants' energy-related behaviors and preferences, and by guiding them through personalized recommendations. Other strategies could rely on reaching occupants through different communication channels, including personal conversations, letters, smartphone interactions, and information made available through displays installed in the building, e.g., [1–3].

In the first part of the chapter, we consider methodologies that are used to support the implementation of O2G integration by placing particular attention on Personalized Recommendation (PR) technologies, e.g., [4]. The fundamental principles of PR are introduced to enable its implementation in the building energy domain. By applying PR in a building environment, it is possible to develop energy-oriented personalized recommendation systems for building occupants. These approaches analyze the patterns of occupants' energy use and behaviors as well as their preferences for energy products and services. Based on this analysis, the recommendation systems can then generate energy-efficient suggestions and recommendations to the occupants and assist their energy-related choices. In the second part of the chapter, application examples are provided to highlight the ease of use of PR technologies implemented in the context of O2G integration.

## 14.2 Introduction to personalized recommendation technology

### 14.2.1 Background

It is common in a decision-making scenario to seek recommendations from experts or trusted parties to inform the decision process. Historically, such recommendations significantly rely on human interactions. With the wider use of the Internet, recommendation technologies have been developed to support e-commerce activities in assisting potential customers in identifying the most suitable products. For example, let us consider the simple scenario in which we have to select a restaurant for dining and, to do so, we rely on the recommendations of friends or people in that geographical area. The recommendation to us would then depend on their experience of the potential restaurants that we could visit and on their understanding of our dining preferences. Such a process relies on a filtering process applied to the available information about the restaurants and their dining experiences, as well as the evaluation of our dining preferences. With current recommendation technologies, this process can be easily automated and this has been the case for nearly two decades for e-commerce activities.

At present, recommendation strategies are required to process a large amount of information that includes details on the wide range of possible items that could be recommended, the experiences of previous customers concerning these items, and the estimated preferences of the potential customer to whom the recommendation is prepared for.

The initial work on PR is dated back to the mid-1990s [5]. Since then, it has significantly grown and has been successfully applied to a wide range of domains (e.g., [6–9]), and this chapter considers its use in the domain of O2G integration.

### 14.2.2 Formulation of a personalized recommendation problem

The core task of a personalized recommendation problem is to filter out one or more items from a set of candidate items subjected to a target user's preference. In this section, we present a basic formulation that can be used to describe a personalized recommendation problem. The key components defining such a formulation are defined as follows:

- the set of users, identified in the following as $C$;
- the set of candidate items that can be recommended, referred to as $S$; and
- the utility function, denoted as $u$, that measures the usefulness of item $s$ to user $c$, i.e., $u: C \times S \rightarrow R$, where $R$ is a totally ordered set of non-negative integers or real numbers.

With the adopted notation, a personalized recommendation problem can then be expressed formally as follows [4,10]:

$$s_c^{\prime} = \underset{s \in S}{argmax}[u(c, s)], \tag{14.1}$$

where the function $argmax(\cdot)$ finds the variable maximizing an objective function (in this case, the variable is $s$ and the objective function is $u(c,s)$). The recommendation problem formulated in Eq. (14.1) aims at finding the items included in the set $S$ that maximize the utility $u(c, s)$ of the target user $c$ ($c \in C$).

In PR, the utility of an item is usually represented by a rating given by a user. It can be expressed as a binary value that indicates a condition of "like" or "dislike," or it can be a numerical value to indicate the preference degree of a user on a particular item.

In most PR problems, the utility $u$ is often not defined on the whole $C \times S$ space, but only over a subset of it. This is because the user usually only rates a subset of items. Because of this, the working mechanism of a recommendation system involves the process of predicting the rating or preference that a user would give to an item. Based on these predictions, it recommends the items with the highest predicted rating/preference values to the user.

Different personalized recommendation techniques have been developed to date. These methods can be generally categorized into two main classes: (i) item-based recommendation and (ii) peer-based recommendation, and these are briefly introduced in the following.

### 14.2.3 Profiling users and items

In the implementation of a PR, a profile needs to be generated for each item and user to be considered. For example, an item profile is a set of attributes that characterizes the item and this could consist of the following four properties when considering a basketball: (1) brand: Wilson; (2) size: 7; (3) color: orange, represented as 3 RGB numbers [255, 165, 0]; and (4) weight: 620 g. The brand "Wilson" is a categorical data rather than a numerical value and we can use the "one-hot" encoding scheme to encode it as a binary string [11]. A one-hot is a group of bits among which the combinations of values allow only one use of the "1" bit with all other bits being assigned the value of "0." By assuming the whole set of the possible basketball brands is {"Wilson," "Spalding," "Nike"}, then the brand property of a basketball with the "Wilson" brand is expressed as "100." With the same approach, the brand property of a basketball with the "Spalding" brand would be expressed as "010." For illustrative purposes, Fig. 14.2 illustrates the item profile of the basketball.

Similarly, a user profile is usually expressed as a term vector, representing the features of a user. For example, Fig. 14.3 illustrates the profile of a user that, to keep the example as simple as possible, is based on the definition of three properties: age, height, and weight.

| Features |
| --- |
| Brand: Wilson |
| Size: 7 |
| Color: Orange |
| Wight: 620 |

Item profile $x = [1,0,0, 7, 255,165,0, 620]$

**Fig. 14.2** Illustration of an item profile.

User profile $u = [34, 175, 67]$

**Fig. 14.3** Illustration of a user profile.

In the examples of Figs. 14.2 and 14.3, we have considered numerical numbers to describe each item. It is possible, when implementing PR methodologies, to consider other variable types, such as text and words, e.g., [12,13].

## 14.2.4 Item-based recommendation

*The item-based recommendation* is also known as *content-based recommendation*. It recommends items to a target user based on the similar items the user rated previously. In particular, the target user's historical ratings and preferences are analyzed, and then the items from the candidate set that are highly similar to the user's historically preferred ones are recommended to the user.

Let us assume that we like to listen to music and that one day we go to a music shop. The shop owner is a friend of ours who knows that we have previously purchased several albums of the "Metallica" and "Arch Enemy" bands. In our current visit to the shop, the owner recommended two albums from the "Children of Bodom" band to us. This is because the "Children of Bodom" band is highly similar to the "Metallica" and "Arch Enemy" (that we like) in terms of music style (all are heavy metal). This represents a simple example from our ordinary lives that shows the basic process involved in an item-based recommendation.

The item-based recommendation makes a rating prediction on a specific item (denoted as the "target item") based on measuring the similarity between the target item and a set of items with known ratings given by the target user. The pseudocode describing the item-based recommendation process is presented in Algorithm 14.1.

## 14.2.5 Peer-based recommendation

The item-based recommendation relies on the target user's historical preference knowledge to define the recommendation. With the *peer-based recommendation*, it is possible to establish recommendations based on the collective knowledge of a group of "peers" (i.e., other users) of the target user. The basic principle of a peer-based recommendation is to infer the target user's preference for the candidate items from the known preferences of the users who have similar profiles (i.e., the "peers") with the target user. To illustrate this, let us imagine going to a restaurant for dinner where

---

**ALGORITHM 14.1 Pseudocode describing the item-based recommendation process.**

**Start**
1. Generate the set of items with known ratings $\Phi$;
2. Generate the set of items with unknown ratings $\Theta$;
3. **For** each item $x$ in $\Theta$
4.    Generate the profile of $x$;
5.    **For** each item $y$ in $\Phi$
6.       Generate the profile of $y$;
7.       Calculate the similarity degree between the profiles of $x$ and $y$, denoted as $s_{x,y}$;
8.    **End For**
9.    Generate the predicted rating of $x$ (denoted as $r_x$) based on $s_{x,y}$, $y \in \Phi$;
10. **End For**
11. Sort all the candidate items (with known or predicted ratings) with descending order of their ratings;
12. Recommend the top $N$ items with highest ratings to the target user.
**End**

---

the waiter wants to recommend some dishes to us. Although he does not know our taste, the shop's sales records show that many other customers of the same gender and similar age as us have usually ordered one of the following dishes: fish and chips, seafood pasta, and Beijing duck. Based on this knowledge, the waiter has reasons to believe that we could also like to eat these dishes and recommends them to us. This example represents a simple scenario in which a peer-based recommendation is established. The basic concept underlying this approach relies on the consideration that if the target user's peers like one item, then there are reasons to believe that the target user could like it as well. Peer-based recommendation uses collaborative filtering techniques to predict the unknown rating that the target user gives to an item based on known ratings given by the peers. Algorithm 14.2 presents the pseudocode to perform a peer-based recommendation.

## 14.2.6 Similarity measure and rating prediction

Both item- and peer-based recommendations involve measuring the similarity between the profiles of two items or two users (Line 5 in Algorithm 14.1 and Line 5 in Algorithm 14.2, respectively). Several methods can accomplish such a task. One of the most commonly used approaches is based on the *cosine similarity measure* that can be expressed as follows:

> **ALGORITHM 14.2 Pseudocode describing a peer-based recommendation process.**
>
> **Start**
> 1. Generate the set of peers of the target user, denoted as $\boldsymbol{\Phi}$;
> 2. Generate the profile of the target user $u$;
> 3. Generate the set of the items with unknown ratings from $u$, denoted as $\boldsymbol{\Theta}$;
> 4. **For** each peer $z$ in $\boldsymbol{\Phi}$
> 5.     Generate the profile of $z$;
> 6.     Calculate the similarity degree between the profiles of $u$ and $z$, denoted as $s_{u,z}$;
> 7.     Obtain the rating the peer $z$ gives to the item $y$, denoted as $r_{z,y}$;
> 8. **End For**
> 9. **For** each item $y$ in $\boldsymbol{\Theta}$
> 10.    Generate the predicted rating for $u$ on $y$ based on $s_{u,z}$ and $r_{z,y}$, $z \in \boldsymbol{\Phi}$, denoted as $r_{u,y}$;
> 11. **End For**
> 12. Sort all the candidate items with descending order of $r_{u,y}$;
> 13. Recommend the top $N$ items with highest ratings to the target user $u$.
> **End**

$$sim(x, y) = cos\,(x, y) = \frac{x \cdot y}{\|x\|^2 \times \|y\|^2} = \frac{\sum_{k=1}^{K} x_k y_k}{\sqrt{\sum_{k=1}^{K} x_k^2} \times \sqrt{\sum_{k=1}^{K} y_k^2}}, \qquad (14.2)$$

where $x$ and $y$ are two item/user profiles described by $K$ properties.

By denoting $c$ as the target user and $s$ as the target item with an unknown rating, our objective is to predict the rating the target user would like to give to the target item (denoted as $r_{c,s}$). In item-based recommendation, the prediction can be made by aggregating the known ratings given by the target user on other items:

$$r_{c,s} = \underset{s' \in \boldsymbol{\Theta}}{aggr}(r_{c,s'}), \qquad (14.3)$$

where $\boldsymbol{\Theta}$ represents the set of items with ratings given by the target user $c$; $s'$ is an item in $\boldsymbol{\Theta}$; $r_{c,s'}$ depicts the rating already given by the target user $c$ on the item $s'$; and $aggr(\cdot)$ is the aggregation function.

For peer-based recommendation, the prediction for $r_{c,s}$ is made by aggregating the known ratings given by the other users on the target item:

$$r_{c,s} = \underset{c' \in \boldsymbol{\Phi}}{aggr}(r_{c',s}), \qquad (14.4)$$

where $\boldsymbol{\Phi}$ represents the set of users who have given ratings on the target item $s$; $c'$ depicts a user in $\boldsymbol{\Phi}$; and $r_{c',s}$ is the rating that has already been given by another user $c'$ on the item $s$.

Several functions can be used as the aggregation function. For example, the following expressions represent simple aggregation functions that can be used for item-based recommendations:

$$r_{c,s} = \frac{1}{|\Theta|} \sum_{s' \in \Theta} r_{c,s'}, \tag{14.5}$$

$$r_{c,s} = k \sum_{s' \in \Theta} (sim(ss') \times r_{c,s'}), \tag{14.6}$$

where $sim(s, s')$ is the similarity calculation function defined in Eq. (14.2) and it returns the similarity value between the items $s$ and $s'$; and $k$ is the weighting factor, which usually takes the value of $k = 1/\sum_{s' \in \Theta} |sim(s, s')|$. Eq. (14.5) takes the average of all the known ratings, while Eq. (14.6), widely used in these applications, takes the weighted sum of all the known ratings by considering the similarity values between items/users.

For peer-based recommendations, Eqs. (14.5) and (14.6) become:

$$r_{c,s} = \frac{1}{|\Phi|} \sum_{c' \in \Phi} r_{c',s}, \tag{14.7}$$

$$r_{c,s} = k \sum_{c' \in \Phi} (sim(cc') \times r_{c',s}), \tag{14.8}$$

where $sim(c, c')$ measures the similarity degree between the users $c$ and $c'$; and $k$ usually takes the value of $k = 1/\sum_{c' \in \Phi} |sim(c, c')|$.

It is worth noting that the earlier item- and peer-based recommendation algorithms are basic methodologies for the implementation of PR strategies and more sophisticated strategies can be found in dedicated publications, e.g., [14,15].

## 14.3 Application example: Personalized building-side renewable investment recommendation system

In this section, we present a simple application example to demonstrate how PR technology can be utilized to assist an occupant in performing information filtering on energy-oriented products and services. For this purpose, we design a personalized renewable investment recommendation system that recommends the most suitable rooftop Photovoltaic (PV) solar panels to a target user by using the peer-based recommendation method.

From the occupant's viewpoint, if a user intends to purchase a rooftop PV solar panel, the following factors should be considered: the panel's price, expected product lifetime, capacity, and brand. Of course, other factors might also influence this decision. Since many PV solar panel products are available on the market, it is challenging for the user to compare available products and to identify the most suitable one. The system developed in this section aims to support this process by recommending solar panel products to a target user (by extracting preference knowledge from a group of users). The schematic of the proposed system is illustrated in Fig. 14.4.

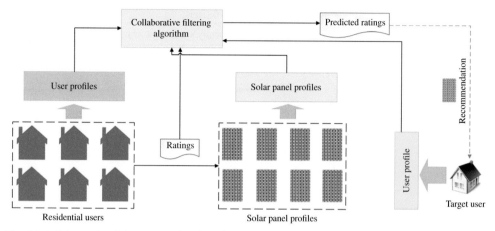

**Fig. 14.4** Schematic of the personalized rooftop solar panel investment recommendation system.

In this application example, we try to keep the complexity of the model and calculations to a minimum. For this purpose, we consider only two factors that affect a user's choice in the purchase of a PV solar panel and these consist of the number of family members and the family income. Intuitively, a family with more members often has a larger energy demand, while a family with a higher income could afford PV solar panels at higher prices. Based on these assumptions, we can establish the *user profile* for an arbitrary user $u$ ($\mathbf{PF}_u$) as a two-dimensional vector:

$$\mathbf{PF}_u = [n_u m_u], \tag{14.9}$$

where $n_u$ and $m_u$ are the user's number of family members and averaged annual family income (in \$).

Since there is more than one feature in the user's profile, we use the cosine-based similarity calculation formula to measure the similarity between two arbitrary users $u$ and $v$:

$$
\begin{aligned}
sim(uv) &= \cos(\mathbf{PF}_u \mathbf{PF}_v) = \frac{\mathbf{PF}_u \cdot \mathbf{PF}_v}{||\mathbf{PF}_u||_2 \times ||\mathbf{PF}_v||_2} \\
&= \frac{\sum_{k=1}^{K}(PF_{u,k} \cdot PF_{v,k})}{\sqrt{\sum_{k=1}^{K}(PF_{u,k})^2} \times \sqrt{\sum_{k=1}^{K}(PF_{v,k})^2}},
\end{aligned} \tag{14.10}
$$

where $sim(u, v)$ represents the similarity degree between users $u$ and $v$; $K$ is the total number of elements in the user profile (in this case, $K = 2$); $PF_{u,k}$ represents the $k$th element in $\mathbf{PF}_u$, and $PF_{u,1} = n_u$ and $PF_{u,2} = m_u$.

We consider a total of 10 users (including the target user), whose information is summarized in Table 14.1. The target user is assumed to have a moderate number of family members (i.e., 3) and a moderate annual income (i.e., \$100,000). Since the two

**Table 14.1** Description of the profiles of the 10 users.

|  | No. of family members | Family's average annual income ($) |
|---|---|---|
| Target user | 3 (0.5) | 100,000 (0.5) |
| User 1 | 1 (0.17) | 120,000 (0.6) |
| User 2 | 3 (0.5) | 150,000 (0.75) |
| User 3 | 4 (0.67) | 70,000 (0.35) |
| User 4 | 3 (0.5) | 50,000 (0.25) |
| User 5 | 6 (1) | 200,000 (1) |
| User 6 | 5 (0.83) | 170,000 (0.85) |
| User 7 | 3 (0.5) | 130,000 (0.15) |
| User 8 | 2 (0.3) | 80,000 (0.4) |
| User 9 | 2 (0.3) | 90,000 (0.45) |

Note: Normalized values are shown in brackets.

properties of $PF_u$ are expressed in different scales (the number of family members is usually less than 10, and the family's annual income is in the order of thousands of dollars), we normalize these values when substituting them in Eq. (14.10) to perform the similarity calculations. We choose 6 and 200,000 as the normalization base for $n_u$ and $m_u$, respectively, and the normalized values are shown in brackets in Table 14.1.

Let us assume that there are eight rooftop PV solar panel products available on the market and the relevant product information is presented in Table 14.2. We also assume that the user only considers the two key properties of a solar panel that are: the power capacity and the price. These properties then form the PV solar panel's profile. As shown in Table 14.2, solar panels with the same capacity could have different prices because, for example, the manufacturer could provide different warranty years.

It Is also considered that the market has collected the ratings given by the other nine users (i.e., the "peers" of the target user) on the eight rooftop solar panel products of Table 14.3. The ratings are expressed as an integer in the range of [1, 5] with the larger value indicating a higher preference.

**Table 14.2** Description of the profiles of the eight PV solar panels assumed to be available on the market.

|  | Capacity (kW) | Price ($) |
|---|---|---|
| Panel 1 | 5 | 6500 |
| Panel 2 | 5 | 8500 |
| Panel 3 | 4 | 5700 |
| Panel 4 | 4 | 6800 |
| Panel 5 | 3.5 | 4000 |
| Panel 6 | 3.5 | 5000 |
| Panel 7 | 3.5 | 5800 |
| Panel 8 | 3 | 3400 |

**Table 14.3** Summary of the ratings given by the nine users on the eight PV solar panel products.

|        | Panel 1 | Panel 2 | Panel 3 | Panel 4 | Panel 5 | Panel 6 | Panel 7 | Panel 8 |
|--------|---------|---------|---------|---------|---------|---------|---------|---------|
| User 1 | 1 | 1 | 3 | 2 | 4 | 3 | 3 | 5 |
| User 2 | 1 | 2 | 3 | 3 | 4 | 5 | 3 | 3 |
| User 3 | 2 | 1 | 5 | 3 | 4 | 3 | 1 | 2 |
| User 4 | 1 | 1 | 2 | 3 | 5 | 5 | 3 | 4 |
| User 5 | 4 | 5 | 3 | 4 | 1 | 1 | 1 | 1 |
| User 6 | 5 | 4 | 4 | 5 | 2 | 3 | 1 | 1 |
| User 7 | 1 | 1 | 2 | 1 | 5 | 4 | 3 | 3 |
| User 8 | 1 | 1 | 1 | 1 | 4 | 3 | 2 | 5 |
| User 9 | 1 | 1 | 1 | 1 | 4 | 4 | 3 | 5 |

Table 14.3 generally reflects the preference trend that we mentioned earlier as, for example, demonstrated by the fact that users with higher income and/or large numbers of family members would prefer solar panels with larger capacities and higher prices. The purpose of collaborative filtering is then to discover this trend from the collective knowledge of a group of users and to use it to infer the target's user preference.

Since we do not know the target user's preference (in terms of rating values) for the solar panels, we use the user-based recommendation technique to aggregate the preference knowledge in Table 14.3 and predict the target user's preference. By taking solar panel 1 as an example, the following steps need to be performed (based on Algorithm 14.2) to predict the target user's rating on this specific panel.

- Step 1: Input the ratings given in Table 14.3.
- Step 2: Generate the profile of the target user: [0.5, 0.5].
- Step 3: Generate the profile of the peers according to Table 14.1: User 1: [0.17, 0.6]; User 2: [0.5, 0.75]; User 3: [0.67, 0.35]; User 4: [0.5, 0.25]; User 5: [1, 1]; User 6: [0.83, 0.85]; User 7: [0.5, 0.15]; User 8: [0.3, 0.4]; User 9: [0.3, 0.45].
- Step 4: Use Eq. (14.10) to calculate the similarity degree between each peer and the target user. The results are summarized in Table 14.4.
- Step 5: Calculate the weighting factor $k$: $k = 1/(0.87 + 0.98 + 0.95 + 0.95 + 1 + 1 + 0.88 + 0.99 + 0.98) = 1/8.6 = 0.12$.
- Step 6: Use Eq. (14.8) to calculate the predicted rating $r$ for the target user on the PV solar panel 1: $r = 0.12 * (0.87*1 + 0.98*1 + 0.95*2 + 0.95*1 + 1*4 + 1*5 + 0.88*1 + 0.99*1 + 0.98*1) = 0.87$.

**Table 14.4** Calculated similarity degrees between the target user and the nine users.

|             | User 1 | User 2 | User 3 | User 4 | User 5 | User 6 | User 7 | User 8 | User 9 |
|-------------|--------|--------|--------|--------|--------|--------|--------|--------|--------|
| Target user | 0.87 | 0.98 | 0.95 | 0.95 | 1.0 | 1.0 | 0.88 | 0.99 | 0.98 |

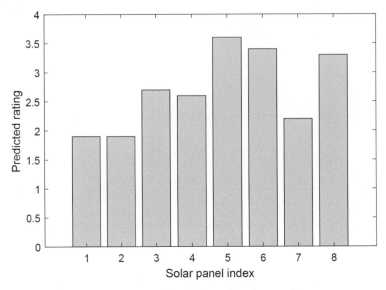

**Fig. 14.5** Predicted ratings for the target user on the eight solar panels.

We can then apply the previous steps to predict the target user's ratings on the other solar panels. Fig. 14.5 summarizes the predicted ratings for all solar panel products. Since the target user has a moderate income level and several family members, it is predicted that the user would prefer the solar panels with medium levels of price and capacity better represented by panels 5, 6, and 8.

## 14.4 Other applications of personalized recommendation in occupant-side energy systems

### 14.4.1 Overview

In the previous section, we have designed a simple personalized recommendation system to demonstrate how PR technology can be used to provide decision-making assistance to the occupant on investing in rooftop PV solar panels.

Recent research has produced more sophisticated applications of PR in occupant-side energy systems. Although these contributions do not explicitly express their relevance in the O2G integration domain, they can be regarded as implementations of the O2G paradigm. Some of these contributions are now considered in the following because of their impact on the deployment of O2G integration.

## 14.4.2 Smart grid recommender system

The concept of a *smart grid recommender system* has shown that PR can assist the stake-holders of a smart grid in filtering information to improve the energy efficiency of the whole energy system [16]. Details on the technical implementations of different energy-oriented personalized recommendation systems can be found in dedicated publications, e.g., [17–21].

## 14.4.3 Energy saving appliance recommendation system

A personalized recommendation system based on content-based recommendation and load monitoring techniques has been proposed in [17]. The system recommends the advertisements of energy-saving appliances to a building occupant. The system is designed to identify the appliances that the occupant needs or uses the most and that are not energy-efficient. The recommendation system then encourages the user to replace them with more energy-efficient appliances available on the market. The system uses load monitoring technology to monitor individual appliances' usage profiles and it then relies on a rule-based classifier to identify the appliances that are frequently used by the user and that require high power consumption. Based on the background data of the occupant (including family structure, and income level), the system automatically generates a user profile, which includes textual descriptions of the appliances that the user will be encouraged to replace. The system performs text mining from a collection of energy-saving styled appliance mode descriptions to filter out the energy-saving appliances that mostly match the user profile, and recommends them to the target user.

## 14.4.4 Electricity retail plan recommendation systems

A peer-based recommendation technique has been proposed for the establishment of electricity tariff plan recommendation systems for end energy users, e.g., [18–20]. In a deregulated electricity retail market, retailers publish many electricity retail plans for end users to sign contracts. The plans usually are from different retailers and can involve, among others, different pricing rates and different renewable penetration rates. A typical example is the "Power-to-Choose, Texas" platform [22]%%%, where end users can view the electricity retail plans published by several energy retailers in Texas (USA) and can select their preferred ones to sign the contracts, in a similar manner as they would be doing e-shopping on eBay. Since the electricity plans are expected to differ from each other and since it is most likely that end users would not have the expertise to distinguish the fine differences among the plan configurations, the use of a personalized recommendation system becomes a useful tool to help an end user to select a suitable plan from many candidate plans.

Example implementations are now briefly described. An electricity retail plan recommendation based on a simulated energy consumption dataset and a group of retail plans from the "Power-to-Choose Texas" platform is presented in [18]. It assumes that there are certain relationships between the users' energy consumption patterns and their preferences on the electricity plans. Based on this, the system uses peer-based recommendations to find out about these relationships from the known preferences and energy consumption data of a group of users and predicts the target user's ratings on the electricity plans. A recommendation for Time-of-Use (TOU) electricity tariffs is proposed in [19]. With this approach, it is assumed that many TOU tariffs on the market possess different segment structures. It then assumes that a user signs a TOU tariff and consumes daily energy under the tariff's structure, but the user is not familiar with the other TOU tariffs on the market that could be more suitable. Based on this, the system designs a matrix factorization-based collaborative filtering algorithm to predict how the target user could shift the energy consumption profile if choosing to sign a TOU tariff $p$ (where $p$ depicts a specific TOU tariff) based on the power consumption profiles of the target user's peers who have already signed $p$ and the similarities between the target user and the peers. The system then predicts the target user's potential interest in $p$ based on the cost savings that can be achieved with $p$ and the shifted energy consumption. The system then sorts the TOU tariffs based on the predicted interests and makes recommendations to the target user. A machine learning-based collaborative filtering method for electricity tariff plan recommendation is illustrated in [20]. This contribution considers a set of candidate electricity tariff plans and a group of residential users. For each user, the system acquires the user's appliance use profiles and interprets the user's "rating" on a candidate plan based on the energy cost associated with the plan. The system also measures the similarity between two users by comparing their appliance usage patterns. Based on this, the system recommends electricity tariff plans to the target user from the rating knowledge obtained from similar users.

### 14.4.5 Household appliance usage behavior recommendation system

In this section, we describe a personalized recommendation system that generates energy-efficient household appliance use time plans for a target user who is charged by a TOU electricity tariff [21]. This approach recognizes that, although TOU tariffs have been widely used, many occupants are still not familiar with demand response techniques. For example, a user could be using the washing machine at noon without realizing that this time would incur the highest electricity rate. To address such a situation, the system is designed for recommending cost-efficient appliance use time plans that fit the target user's lifestyle to the target user, while aiming at increasing the demand response awareness. The system starts by collecting a group of end users' appliance usage data through IoT sensors or non-intrusive appliance load monitoring technology. It then analyzes the appliance use data by categorizing it into two categories: (1) critical non-shiftable appliances, such as televisions, lights, and computers; and (2) time-shiftable appliances, such as washing machines, dishwashers, and clothes dryers. It is recognized that the critical appliances' operation time is fully determined by the occupant and reflects the occupant's lifestyle, while the time-shiftable appliances' operation time reflects the occupant's demand response. The system then uses a collaborative filtering algorithm to measure the users' lifestyle similarity based on their usage profiles of critical non-shiftable appliances, and it generates recommendations of the shiftable appliances' operation time for the target user.

## 14.4.6 Home battery energy storage recommendation system

As introduced in Chapter 8, Battery Energy Storage Systems (BESSs) have been increasingly deployed in buildings to provide emergency energy backup service to the building occupant and to improve the building's on-site renewable energy sources by absorbing the surplus renewable energy into the BESS and by discharging it for the building's later use. With technological and manufacturing advancements, BESS products have been becoming more and more affordable. The increasing applications of BESSs available for the residential sector are fostering the home BESS market. More and more commercial home BESSs have been produced by different manufacturers (e.g., the direct current coupled home BESS products by LG Chem [23] and the Power Wall home battery systems by Tesla [24]).

The more and more home BESS products provide more options for occupants to choose from. At the same time, this also imposes a challenge to occupants to identify the most suitable ones from different candidate home BESS products whose number could be large and could possess different capacities and other technical configurations. To provide information filtering support to occupants in such an environment, Guo et al. [25] describes a personalized recommendation system prototype for recommending suitable home BESSs to a target occupant by considering his/her energy demand characteristics and the technical configurations of different home BESS products available on the market. The recommendation system is designed based on the collaborative filtering principle. It firstly collects the following data of a group of occupants: (i) energy consumption; (ii) the occupants' profiles (e.g., the number of family members and family income level); and (iii) the occupants' preferences on different home BESS products in the market—this is represented as the ratings the occupants have given to the home BESS products (e.g., via online platforms). For a specific target occupant, the recommendation system uses a neural collaborative filtering technique [26] to infer the target occupant's preference for the home BESS products he/she has not ever rated by considering the preferences of other occupants who have similar profiles and energy consumption characteristics with the target occupant. Based on this, the system filters out a list of home BESS products with the highest inferred rates and recommends them to the target occupant.

# References

[1] P. Palensky, D. Dietrich, Demand side management: demand response, intelligent energy systems, and smart loads, IEEE Trans. Industr. Inform. 7 (3) (2011) 381–388.

[2] Smart grids model region Salzburg, n.d., Cleanenergysolutions.org (web archive link, 03 December 2020). https://cleanenergysolutions.org/training/smart-grids-model-region-salzburg (Accessed 03 December 2020).

[3] S. Darby, The Effectiveness of Feedback on Energy Consumption: A Review for Defra of the Literature on Metering, Billing and Direct Displays, Environmental Change Institute, University of Oxford, Oxford, UK, April 2006. [Online]. Available: https://www.eci.ox.ac.uk/research/energy/downloads/smart-metering-report.pdf. (Accessed 11 July 2023).

[4] J. Bobadilla, F. Ortega, A. Hernando, A. Gutierrez, Recommender systems survey, Knowl.-Based Syst. 46 (2013) 109–132.

[5] W. Hill, L. Stead, M. Rosenstein, G. Furnas, Recommending and evaluating choices in a virtual community of use, in: Proceedings of Conference on Human Factors in Computing Systems, 1995.

[6] X. Wang, F. Luo, Y. Qian, G. Ranzi, A personalized electronic movie recommendation system based on support vector machine and improved particle swarm optimization, PloS One 11 (2016) 1–17.

[7] J. Liao, W. Zhao, F. Luo, J. Wen, M. Gao, X. Li, J. Zeng, SocdialLGN: light graph convolution network for social recommendation, Inform. Sci. 589 (2022) 595–607.

[8] C.A. Gomez-Uribe, N. Hunt, The netflix recommender system: algorithms, business value, and innovation, ACM Trans. Manag. Inf. Syst. 6 (4) (2016) 1–19.

[9] J. Beel, S. Langer, M. Genzmehr, B. Gipp, C. Breitinger, A. Nurnberger, Research paper recommender system evaluation: a quantitative literature survey, in: Proceedings of the International Workshop on Reproducibility and Replication in Recommender Systems Evaluation, Oct. 2013.

[10] G. Adomavicius, A. Tuzhilin, Toward the next generation of recommender systems: a survey of the state-of-the-art and possible extensions, IEEE Trans. Knowl. Data Eng. (2005).

[11] K. Ramasubramanian, J. Moolayil, Applied Supervised Learning with R. Packt, 2019.

[12] C.D. Manning, P. Raghavan, H. Schutze, An Introduction to Information Retrieval, Cambridge University Press, Cambridge, UK, 2008.

[13] G. Ignatow, R. Mihalcea, An Introduction to Text Mining: Research Design, Data Collection, and Analysis, SAGE Publications, Thousand Oaks, CA, 2017.

[14] C.C. Aggarwal, Recommender Systems: The Textbook, Springer International Publishing, New York, USA, 2016.

[15] D. Tikk, I. Cantador, S. Berkovsky, Collaborative Recommendations: Algorithms, Practical Challenges and Applications, World Scientific Publishing, Singapore, 2018.

[16] F. Luo, G. Ranzi, X. Wang, Z.Y. Dong, Service recommendation in smart grid: Vision, technologies, and applications, in: Proceedings of 9th International Conference on Service Science, Chongqing, Oct 2016.

[17] F. Luo, G. Ranzi, W. Kong, Z.Y. Dong, S. Wang, J. Zhao, Non-intrusive energy saving appliance recommender system for smart grid residential users, IET Gener. Transm. Distrib. 11 (7) (2017) 1786–1793.

[18] F. Luo, G. Ranzi, X. Wang, and Z.Y. Dong, Social information filtering based electricity retail plan recommender system for smart grid end users, IEEE Trans. Smart Grid, vol. PP. no. 99, 2017.

[19] Y. Zhang, K. Meng, W. Kong, Z.Y. Dong, Collaborative filtering-based electricity plan recommender system, IEEE Trans. Industr. Inform. 15 (3) (2019) 1393–1404.

[20] S. Li, F. Luo, J. Yang, G. Ranzi, J. Wen, A personalized electricity tariff recommender system based on advanced metering infrastructure and collaborative filtering, Int. J. Electr. Energy Syst. 113 (2019) 403–410.

[21] F. Luo, G. Ranzi, W. Kong, G. Liang, Z.Y. Dong, A personalized residential energy use recommendation system based on load monitoring and collaborative filtering, IEEE Trans. Industr. Inform. 17 (2) (2021) 1253. 1262.

[22] Power-to-choose platform, n.d., Powertochoose.org. https://www.powertochoose.org (Accessed 02 December 2022).

[23] LG Energy Solution, n.d. [online]. Available: https://www.lgessbattery.com/m/us/main/main.lg (Accessed 29 June 2023).

[24] Tesla Power Wall, n.d. [Online]. Available: https://www.tesla.com/powerwall (Accessed 29 June 2023).

[25] X. Guo, F. Luo, Z. Zhao, Z.Y. Dong, Personalized home BESS recommender system based on neural collaborative filtering, in: Proceedings of IEEE 2nd International Conference on Power, Electronics and Computer Application, Shenyang, China, Jan. 2022.

[26] X. He, L. Liao, H. Zhang, L. Nie, X. Hu, T.S. Chua, Neural collaborative filtering, in: Proceedings of 26th International Conference on World Wide Web, 2017, pp. 173–182.

# Index

Note: Page numbers followed by *f* indicate figures, *t* indicate tables, and *b* indicate boxes.